D0141532

An Introduction to Arthropod Pest Control

Arthropod pests are responsible for huge annual losses in global crop production and for transmitting a number of infectious diseases. The control of such pests is therefore of the utmost importance. *An Introduction to Arthropod Pest Control* provides an up-to-date, detailed overview of current approaches to pest control, including chemical pest control, the use of biological and biorational control agents, as well as the latest developments in biotechnology. The book specifically emphasises the techniques available for controlling pests using examples of crop pests, animal pests and pests that transmit disease, from a wide range of countries. The book is intended as a standard introductory text for undergraduate and postgraduate students in the fields of pest control, entomology, crop protection and agricultural and environmental sciences. It is also aimed at professional pest control practitioners and government employees working in extension services.

J.R.M. THACKER is Senior Lecturer in the Department of Biological Sciences, University of Paisley in Scotland, UK. He has researched in the field of pest control for the past 15 years and has carried out fieldwork in Brazil, Morocco, Oman, Saudi Arabia and the USA, in addition to his work in the UK. His primary research areas comprise pesticide application, pesticide formulations and pesticide side-effects. He has published widely in peer-reviewed journals and has also written articles for the popular press. This is his first book.

An Introduction to Arthropod Pest Control

J.R.M. Thacker
Dept of Biological Sciences
University of Paisley, UK

CAMBRIDGE
UNIVERSITY PRESS

University Printing House, Cambridge CB2 8BS, United Kingdom

Published in the United States of America by Cambridge University Press, New York

Cambridge University Press is part of the University of Cambridge.

It furthers the University's mission by disseminating knowledge in the pursuit of education, learning and research at the highest international levels of excellence.

www.cambridge.org
Information on this title: www.cambridge.org/9780521567879

© J.R.M. Thacker 2002

This publication is in copyright. Subject to statutory exception
and to the provisions of relevant collective licensing agreements,
no reproduction of any part may take place without the written
permission of Cambridge University Press.

First published 2002

A catalogue record for this publication is available from the British Library

Library of Congress Cataloguing in Publication data

Thacker, J. R. M. (Jonathan Richard MacDougall), 1962–
An introduction to arthropod pest control / J.R.M. Thacker.
p. cm.
Includes bibliographical references and index.
ISBN 0 521 56106 X (hbk.) – ISBN 0 521 56787 4 (pbk.)
1. Arthropod pests – Control. I. Title.
SB931.T52 2002
632.7–dc21 2002023436

ISBN 978-0-521-56106-8 Hardback
ISBN 978-0-521-56787-9 Paperback

Cambridge University Press has no responsibility for the persistence or accuracy of URLs for external or third-party internet websites referred to in this publication, and does not guarantee that any content on such websites is, or will remain, accurate or appropriate.

This book is dedicated to my family

Contents

Plate sections between pages 144 and 145, and pages 240 to 241.

Preface

The aim of this book is to provide a basic introduction to the techniques that can be used to control arthropod pests. The book was written because I wanted to try and compress information from a diversity of sources into a single volume. Having designed and taught introductory pest control courses for the past 10 years, both in the UK and abroad, I found that the literature from which I obtained source material was often unmanageable from the perspective of the student. I found myself recommending one text for pesticide chemistry, one for pesticide application, one for biological control, one for pest management, etc., in addition to supplementary reading from journal articles. My goal therefore was to produce a single text that students could use as a basic source of introductory material.

The book is structured so that each of the techniques that comprise arthropod pest control is considered separately. As such the book takes a techniques- rather than a systems-based approach. In the 'real world' of course it is systems that are dealt with. However, from a pest control perspective I believe that it is fundamentally important to have a detailed grasp of the tools that you have available before you begin to use them. In essence, the book provides background preparation for advanced study of the subject. The techniques that are considered comprise botanical pesticides, synthetic pesticides, biocontrol agents, microorganisms, pheromones, growth regulators, sterile insect release, host-plant resistance and cultural methods. The book begins with a brief historical account of pest control and ends with discussion of the very latest developments in pest control biotechnology.

Each of the chapters is based on material from more than 50 publications. As a result of this, I made a decision early on not to cite every single article or book that I had used. Rather, I provide a list of further reading for those who want to explore each of the techniques discussed in more detail. This decision was made not only for the sake of space but also because I am yet to meet anyone who truly uses the extensive referencing that is provided in many of the books that are published today.

The intended target audience for the text comprises students taking introductory courses in pest control, pest control professionals and amateurs with an interest in this subject. I have tried to make the book as practical as possible by providing tables that list the products that are available for pest control in the appropriate chapters. In addition to a comprehensive subject index I have also provided indices for the pest and beneficial species covered. The book is intended to be global in nature and the specific examples that are used include pests of medical and agricultural importance that are taken from African, American, Asian and European environments. Enjoy.

J.R.M. Thacker
January 2002

Acknowledgements

Many thanks go to Dr Des Nicholl, Head of Biological Sciences, University of Paisley, who was the first person to convince me that I should embark on the task of writing a textbook. Many thanks also to the five reviewers who first considered the book proposal and to the two sets of reviewers who read the draft manuscript. The book benefited greatly from their input. The photographs used in the book were supplied, free of charge, by Micron Sprayers Ltd, by CibaGeigy, by Brian Federici at the University of California Riverside, by Jim Kalisch at the University of Nebraska, by Jean Adams and Vic D'Amico at Cornell University and by the Agricultural Research Service of the US Department of Agriculture. Financial support for the production of colour plates was provided by the Carnegie Foundation. Many thanks also to the following publishing companies for the use of copyright material: Annual Reviews, Oxford University Press, John Wiley & Sons, Elsevier, Prentice Hall, Pearson Education, British Crop Protection Council Publications. Finally, many thanks also to the team at Cambridge University Press for their support during the preparation of this book.

1

A brief history of arthropod pest control

Introduction

Two of the most important challenges facing humanity in the twenty-first century comprise food production and disease control. These are challenges that are associated with the increasing global human population (Figure 1.1) and with the control of arthropod pests, the subject of this book. The importance of these challenges cannot be overemphasised.

Arthropod pests are responsible for global pre- and postharvest crop losses of approximately 20–50% of potential production and for transmitting a number of the world's most important diseases (Table 1.1). For example, it has been estimated that the protozoan organism causing malaria infects approximately 500 million people worldwide – almost 10% of the people on earth.

It is undoubtedly the case that humanity's problems with arthropod pests are not new and they certainly predate the development of agriculture approximately 10 000–16 000 years ago. Arthropods first appear in the fossil record over 500 million years ago during the Cambrian[1] period at the start of the Palaeozoic. The oldest insect fossils to have been found so far are dated to the Devonian, a period that began 400 million years ago. Modern insects begin to appear in the fossil record about 280 million years ago. However, the time scale over which different insect orders are detected in the fossil record

[1] Geological time is split into four major phases that are known as the Precambrian, Palaeozoic, Mesozoic and the Cenozoic. Each of these major phases is then further subdivided, e.g. the Cambrian is the first phase or subdivision of the Palaeozoic. See glossary for further details.

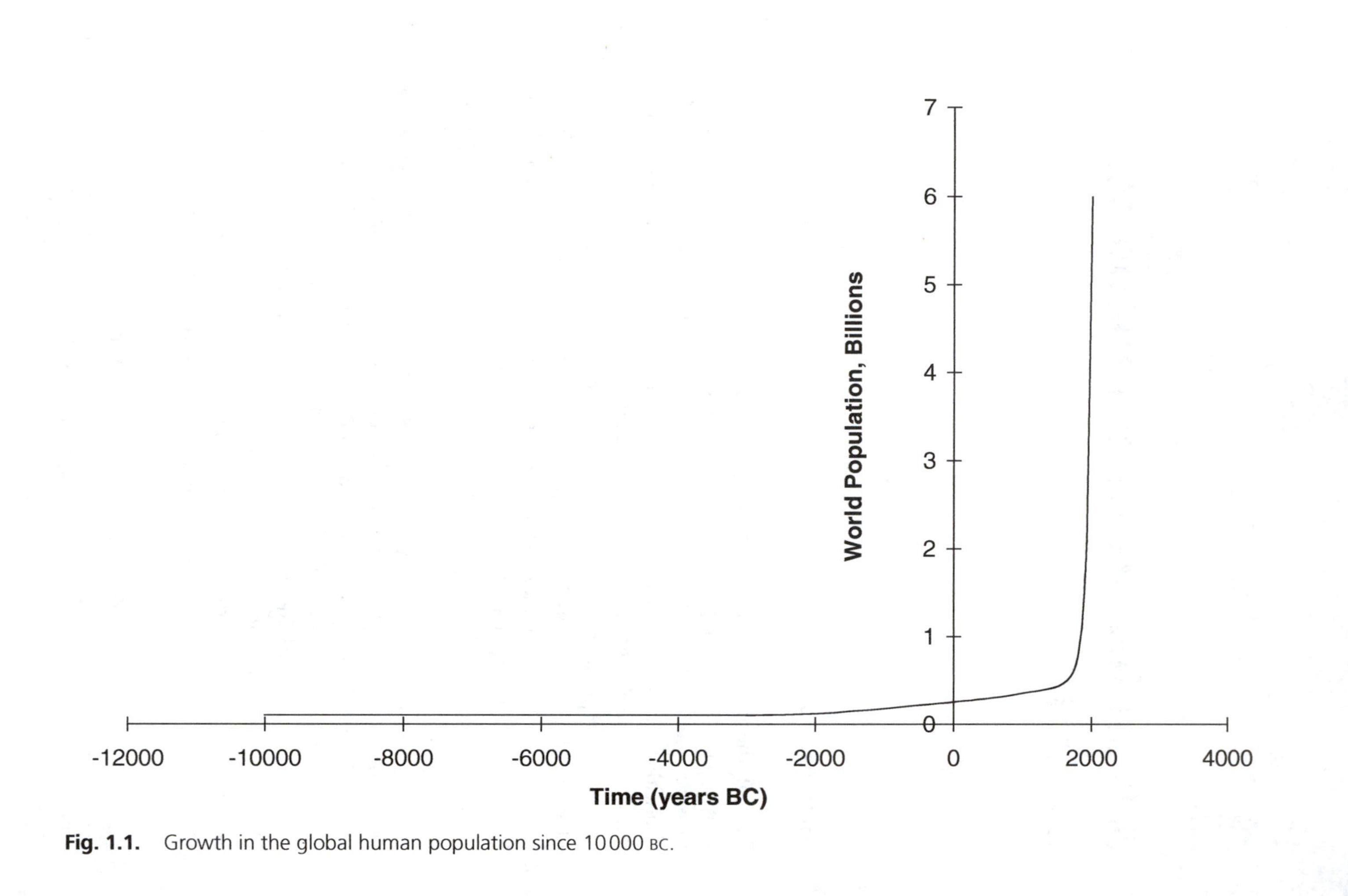

Fig. 1.1. Growth in the global human population since 10 000 BC.

Table 1.1. *The world's most important arthropod-borne diseases*

Disease	Vector	Agent	Cases	Deaths
Malaria	Mosquito	Protozoa (*Plasmodium*)	300–500 m	1.5–2.7 m
Filariasis	Mosquito	Filarial worms	120 m	NA[a]
Dengue	Mosquito	Virus	50 m	0.5 m
Onchocerciasis	Blackfly	Filarial worms	18 m	NA[b]
Chagas	Triatomine bugs	Protozoa (*Trypanosoma*)	16–18 m	20000–50000
Leishmaniasis	Sand flies	Protozoa (*Leishmania*)	12 m	NA[a]
Trypanosomiasis	Tsetse flies	Protozoa (*Trypanosoma*)	300000–500000	NA[a]
Yellow fever	Mosquito	Virus	200000	30000

Notes:
[a] Accurate estimates not available.
[b] Onchocerciasis or 'river blindness' causes chronic rather than acute illness. The WHO estimates that, of the 18 m people infected, 6.5 m suffer from severe dermatitis and 270000 are blind.

Source: Data collected from the World Health Organization (WHO) *World Health Report 1998.*

stretches up until 75 million years ago, during the Cretaceous period at the end of the Mesozoic. For example, silverfish (order Thysanura) are detected in the fossil record from the Devonian while caterpillars (order Lepidoptera) are not detected until the Cretaceous.

By contrast, *Homo sapiens* has been around for about a 100000 years. It should therefore come as no great surprise to know that so far we have not been able to conquer the problems that arthropod pests pose. These animals have had far more time than humans to evolve and adapt to life on earth.

This first chapter gives a brief historical account of arthropod pest control. The chapter begins by considering the importance of the development of agriculture as a method for food acquisition. The cultivation of plants and the domestication of animals would have vastly increased the opportunities for humans to associate with arthropod pests. This is followed by a description of some of the earliest recorded examples of attempts to control arthropod pests. A brief review of pest control from 1600 to 2000 AD is then given. The chapter ends with a description of some of the modern high-tech methods of pest control that have been developed for use in the twenty-first century. In short, this introduction serves as a foreword to the book as a whole.

Neolithic agricultural development and pest control

The earliest evidence of domesticated plants and animals, the basis of agricultural development, dates from between 16 000 and 10 000 years ago. The earliest evidence comes from Mesopotamia[2] and from the Nile delta in Egypt. The later evidence comes at the start of the Neolithic period (8000 BC) from Europe, a period that is dated from the end of the European ice age. Wherever agricultural development began, it was certainly invented more than once because there are no links that can be accurately made between the farmers of such far-removed areas as the Americas and the Middle East.

As a method for food acquisition, agriculture would have had profound and long-lasting effects upon the development of human populations and consequently upon the development of problems associated with arthropod pests. The change from a hunter–gatherer lifestyle to that of a farmer would have had many implications. First, an increased food supply would mean that more individuals could be supported within the family unit. One result of this would be that families would tend to increase in size as it would be advantageous to have more hands to work the land. Second, growing crops would have required the development of a more permanent home base from which to run a farm. This would have led to the development of early fixed settlements, a prelude to villages, towns and cities. Jericho in Palestine is the oldest documented permanent human settlement and was founded around 9000 BC. Third, a more reliable and guaranteed food source would have left more time for other pursuits that are often characteristic of the development of civilisations. These would have included artistic pursuits such as writing and music. Fourth, lessons would have begun to be learnt about soils, the climate and successful agronomic practices. This would have been especially the case as populations expanded and moved to occupy new geographical areas. In Europe,

[2] Mesopotamia – the region between the rivers Tigris and Euphrates, including parts of what is now eastern Syria, south-eastern Turkey and almost all of Iraq – is also thought to be the traditional site of the garden of Eden.

where people practised slash-and-burn agriculture it would have been necessary to move on a regular basis anyway. Fifth and last, increased crowding (of people, crops and animals) would have exacerbated problems associated with arthropod pests and so would have stimulated the development of the first selective breeding and domestication programmes and the first attempts to try to control pest species. In short, the development of agriculture represented a major step forward in human cultural evolution. It was also a development that would, for the first time, have brought humans into mass contact with the arthropod pests that would have used their crops and their animals for food and reproduction. Such contact would undoubtedly have led to the development of attempts to control these noxious organisms.

Early attempts at pest control

The earliest attempts to boost (or sustain) agricultural production concentrated upon agronomic practices that ensured an adequate water supply, the use of fertile soil and the choice of the most well-adapted cultivars. Progress with the control of pest species would therefore have been slow, although practices such as rotations and cultivar selection would undoubtedly have helped. The usual response by people to large-scale attack by pests was to suffer or move. For example, the exodus of the Israelites (*c.* 1300 BC) described in the Old Testament has been attributed to plagues of locusts, flies and lice that consumed crops and spread disease among the inhabitants of the Nile.

The use of chemicals to control pests can be traced back at least 4000 years. For example, the Hindu book, the *Rig Veda*, written in India in 2000 BC, makes reference to the use of poisonous plants for pest control. It is also known that plants were used as sources of insecticidal compounds by the Egyptians during the time of the Pharaohs. Ancient Romans are credited with having used false or white hellebore as a rodenticide. Homer, in 1000 BC, mentions the use of sulphur as a fumigant while Pliny, in 77 AD, makes reference to the use of arsenic, soda and olive oil. Lastly, in 970 AD, the Arab scholar Abu Mansur described over 450 plant products with toxicological and/or pharmacological properties.[3] Despite such knowledge, though, progress in pest control until at least the 1500s was agonisingly slow. Agricultural development had been critical in the development of early civilisations and empires in Asia, the Middle East and South America. However, such development had largely

[3] For example, see review by Yang, R.Z. & Tang, C.S. (1988). Plants used for pest control in China: a literature review. *Economic Botany*, **42**, 376–406. See also glossary for more detail on Abu Mansur.

Table 1.2. *Exchange of agricultural products – selected examples from the Columbian exchanges (1492–1503)*

New World → Old World	Old World → New World
Avocado	Barley
Chocolate	Banana
Corn	Cattle
Peanut	Chicken
Peppers	Citrus
Potato	Lettuce
Sunflower	Onion
Tobacco	Pear
Tomato	Wheat

Source: Modified from Tribe, D. (1994). *Feeding and Greening the World.* Wallingford: CAB International.

occurred because of improvements in agronomic practices, most notably in the provision of good nutrient supplies for crops. This is not surprising since it was agronomic practices that were the greatest constraint to increased yields. Pest control, using cultural techniques such as crop rotations, would have happened. However, this control occurred more because of a desire to improve yields through better agronomy than as a result of any preplanned pest control strategy.

The development of agriculture and pest control in western Europe during this time period was a completely different matter. This region was something of a rural backwater that was years behind the empires that had already developed elsewhere in the world. Technical developments such as horseshoes, fixed-mouldboard ploughs and watermills were gradually introduced to farmers in Europe. However, early essays on agriculture, written by the classical scholars Cato, Varro and Columella, were still being used in Europe right up until the sixteenth century.[4]

The events of greatest significance to pest control at this time came at the

[4] In fact, many of these essays were still in use through the seventeenth and eighteenth centuries, particularly after English translations of the original Latin texts became more widely available. See glossary for further details.

Table 1.3. *Examples of major pests that have invaded North America from Europe, Asia and South America from the eighteenth century to the present*

Common name	Latin name	Date of arrival
Housefly	*Musca domestica*	1769
Codling moth	*Cydia pomonella*	c. 1800
Cabbageworm	*Pieris rapae*	c. 1860
Cottony cushion scale	*Icerya purchasi*	1868
Gypsy moth	*Lymantria dispar*	1869
Boll weevil	*Anthonomus grandis*	1892
European corn borer	*Ostrinia nubilalis*	1908
Pink bollworm	*Pectinophora gossypiella*	1915
Cereal leaf beetle	*Oulema melanoplus*	1940
Mediterranean fruit fly	*Cerratitis capitata*	1975
Russian wheat aphid	*Diuraphis noxia*	1986

end of the fifteenth century with the four voyages of Christopher Columbus to the New World (1492–1503). These voyages led to the exchange of plants and animals between the Old and the New World and, consequently, to the exchange of insect pests. Tables 1.2 and 1.3 list of some of the plants and insect pests that were exchanged between continents as a result of these and other voyages. It was these and later exchanges that eventually led to the development of modern systems of plant quarantine. The movement of pest species also led to the development of pest control based upon the use of predatory species imported from a pest's country of origin (classical biological control – see Chapter 5).

Pest control after the sixteenth century

The explorations of the New World and the opening-up of trade routes with Asia not only led to the movement of pests but also to the discovery of new means for controlling pests. Many native cultures were already using extracts from plants for the control of arthropod pests and early explorers rapidly exploited such technology. For example, the first explorers of the Americas observed that native Indians in Venezuela were using the powdered seeds of a lily, *Sabadilla officinarum*, to protect crops from insect attack. This observation led to the export of this crop to Europe, and to the use of the extract for pest control right up until the middle of the twentieth century. Similar events happened with the discovery that nicotine was widely used in North America for pest control, that quassia (from *Quassia amara*) was widely used in Central America and that sweet flag (*Acorus calamus*) was widely used in China and India. One result of these explorations was that, during this period, plant-based chemical control of arthropod pests began to increase. A more complete list of the plants in use that were discovered during these European explorations is given in Table 1.4.

These plant-based extracts really dominated the pest control market in Europe and the colonies up until the end of the nineteenth century. However, from the sixteenth century onwards, inorganic compounds (some of which had been mentioned 2000 years earlier) began to become more widely available and hence, more widely used. For example, arsenic mixed with honey was used as an ant bait from the mid-1600s onwards. Copper arsenite, lead arsenate and calcium arsenate all became widely available from the end of the nineteenth century onwards. In the early 1900s sodium fluoride and cryolite (an aluminium salt of fluorine) were marketed for pest control. Finally, a number of other formulations based on sodium, mercury, copper and tin were also developed.

Many of these inorganic chemicals for pest control had two features in common: they had high mammalian toxicities and they acted as stomach poisons (they needed to be consumed by pest species to be effective). Both of these characteristics led to a decline in their use and they were eventually replaced by more selective and effective synthetic organic compounds. Most of the inorganic pesticides that remain in use today (2002) are fungicides.

Developments in the products available for controlling arthropod pests were matched in the eighteenth and nineteenth centuries by technological developments in agricultural equipment and by the development of scientific approaches towards farming. For example, the French inventor Victor Vermorel designed and marketed one of the first commercial crop sprayers in 1880. One result of these developments was that farmers began to experiment with farming. As a result, food output (in Europe and North America) and the global human population continued to increase while pest control on farms began to become more effective.

The really dramatic changes in agricultural output (in Europe and North America) and in arthropod pest control (globally) however were to have their origins in events that happened before and during World War II. These changes were to herald what was perhaps to be the zenith of chemical pest control during the years 1945–70. During this time period, four major groups

Table 1.4. *Insecticidal plants discovered by Europeans after the sixteenth century*

Plant[a]	Active compound	Date of discovery	Native use[b]	European use[c]
Sabadilla officinarum	Sabadilla	c. 1500s	Crop protection (powdered seeds, South America)	Crop protection
Nicotiana tabacum	Nicotine	Late 1500s	Crop protection (Crude liquid extracts, North America)	Crop protection
Quassia amara	Quassin	Late 1700s	Aphid control (Extracts from wood chips used in Central America)	Aphid control
Heliopsis longpipes	Heliopsin	Early 1800s	Leaves burnt, used as a fumigant (Mexico, for fly control)	Not widely used
Ryania speciosa	Ryanodine	1940s	Stem used to make poison for arrows (Amazon basin)	Used against Lepidoptera
Calceolaria andina	Napthoquinones	1990s	Unknown native use (plants from Chile)	None so far
Derris chinensis	Derris	Mid-1900s	Fish poison (East Asia)	Crop protection
Acorus calamus	Not yet determined	Early 1600s	Insect repellent and crop protection	Insect repellent

Table 1.4. (*cont.*)

Plant[a]	Active compound	Date of discovery	Native use[b]	European use[c]
Tagetes minuta	Thiophenes	1600s	Fly control	Fly control and intercropping
Chrysanthemum cinerariaefolium	Pyrethrum	c. 1800	Fly control and crop protection	Public health and crop protection
Azadiractica indica	Neem	1970s	Public health and crop protection	Public health and crop protection

Notes:

[a] Plants referred to are the first plants identified to contain the active chemical. It is now known that many plant species can produce the same active ingredient.

[b] Native uses are given in broad terms, i.e. crop protection would refer to a general use against a range of crop pests.

[c] European use refers to whether the compound became widely used in Europe, at least once. There are many other plant species that are known to produce chemicals that are toxic to pests but these have not yet been widely used in Europe.

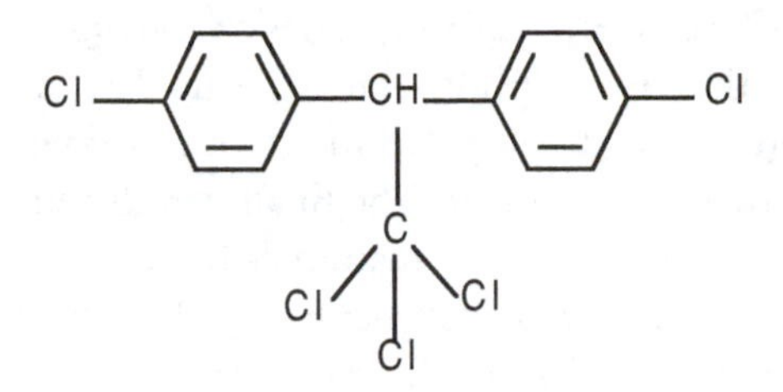

Fig. 1.2. Dichlorodiphenyltrichloroethane (DDT). The chemical structure is shown because this is probably the best-known pesticide worldwide. The structure exerts its toxic effects by disrupting the passage of nerve impulses along axons.

of synthetic organic insecticides – organochlorides, organophosphates, carbamates and pyrethroids – were all first discovered and developed for widespread use. All of these major groups of insecticide are still in widespread use around the world today.

The development of modern chemical pesticides

Modern twentieth century arthropod pest control began with the discovery in 1939 that dicholorodiphenyltrichloroethane (DDT) was toxic to insects (Figure 1.2). The insecticidal properties of this chemical compound were recorded by Paul Müller, a research scientist working for the Geigy Chemical Company. He was awarded the Nobel Prize for Medicine in 1948 for his research. This chemical, which is perhaps the best-known insecticide worldwide, revolutionised pest control. It is relatively cheap to produce, has a broad spectrum of activity and is selective in its toxicity. DDT was first used during the war to protect troops from diseases such as yellow fever, typhus, elephantiasis and malaria. The success of the pesticide was better than anyone had thought possible and after the war DDT was used widely to combat these diseases in the civilian population. For example, in India alone cases of malaria declined from 75 million to fewer than 5 million *per annum* in a decade. It is because of this and other successes that DDT continues to be used today for vector control in a number of developing countries. After the war, DDT was also widely used in agriculture to protect crops from pests. The discovery of the toxic properties of DDT spawned research into the toxic properties of related organic molecules. The result of this research was that the organochloride insecticides became a major force in pest control throughout the late 1940s and 1950s.

At the same time as the discovery of the insecticidal properties of the organochlorides came the discovery of the organophosphate insecticides. A German chemist G. Schrader had been looking for a replacement for nicotine. This chemical was in short supply at the time and in 1941 he finally produced his first compound, schradan. Schradan is very toxic to mammals but its discovery led to the development of a group of insecticides that has finally numbered more than 100 active ingredients in over 10000 different formulations. However, since this research was linked with wartime German studies on the organophosphorus nerve gases (sarin[5] and tabun), this was hardly an auspicious start for these chemicals (see Chapters 3 and 4 for more detail). The organophosphate insecticides are still in wide use all over the world today.

The third major group of synthetic organic insecticides to be developed were the carbamates. These chemicals were originally developed by the Geigy Corporation in 1951 but it was not until 1956 that the first commercial product, carbaryl, was marketed by Union Carbide. The development of synthetic carbamates took place because it was already known that the chemical physostigmine (a naturally occurring carbamate found in *Physostigma venenosum*, the calabar bean) had powerful anticholinesterase activity. This naturally occurring carbamate could not be used for pest control because it is unable to penetrate pest species' nervous systems although it was (successfully) used in witchcraft trials by ordeal in West Africa. Research into synthetic carbamates eventually produced over 20 active ingredients, many of which are still in use today.

The fourth major group of chemical insecticides to be developed during the time period following the end of World War II were the synthetic pyrethroids. These insecticides were developed as a result of attempts to improve the chemical stability of naturally occurring pyrethrum, a compound that was extracted from the flower heads of chrysanthemums. The first synthetic pyrethroid, allethrin, was introduced for pest control in 1949. This early research led to further improvements in the chemical stability of this group throughout the next 30 years. Today, the pyrethroids comprise almost 40 different active ingredients and in many developed countries they are the most widely used insecticides for pest control.

In addition to these four insecticidal groups (organochlorides, organophosphates, carbamates and pyrethroids) we can now also add the neonicotinoids. These compounds were successfully developed and commercialised

[5] On 20 March 1995 a Japanese cult called the Aum Shinrikyo released sarin in the Tokyo subway system in an attempt to kill members of the police force who were planning several raids on cult facilities. The release of the gas injured 3800 people and killed 12.

during the 1990s and, in some markets at least, are now becoming the insecticides of choice (see Chapter 3 for more detail).

These five major groups of insecticide continue today to dominate the chemicals used for arthropod pest control, although regional variations exist in the group of choice. In terms of their development we can characterise at least three major changes within the synthetic chemical pest control market. These changes are as follows:

- A decrease in the rate of application of insecticides (Figure 1.3). Lead arsenate, which was widely used in Europe at the beginning of the twentieth century, was typically applied at a rate of 10000 g/ha. By contrast, alphacypermethrin, an insecticide in wide use in Europe at the end of the twentieth century can be applied at a rate of 10g/ha.
- An increase in the research and development costs associated with bringing a new active ingredient to the market (Table 1.5). The reasons for this increase include: (1) a decline in the rate of discovery of novel molecules for pest control; and (2) an increase in the number and complexity of tests required prior to product approval (Table 1.6).
- A decline in the number, but increase in size, of the companies involved in pesticide research and development work. At the end of 1998 Hoechst and Rhône-Poulenc joined forces to create Aventis and in November 2000 Zeneca and Novartis merged to create Syngenta. With sales in 2000 of *c.* \$4 billion and \$6 billion, respectively, these are now two of the world's largest agrochemical companies (see glossary for more details).

Within agriculture, there is absolutely no doubt that the discovery and development of these and other pesticides have made, and will continue to make, an enormous contribution towards massive increases in crop yields that have taken place in countries all over the world. There is also no doubt that these chemicals have made, and will continue to make, a substantial contribution towards the control of vectors of disease. Crop breeding, government support and legislation, increased fertiliser use, mechanisation and improved agronomic practices have all also contributed to agricultural production and to pest control. However, the contribution that chemical control of arthropod pests has made to boosting agricultural productivity worldwide can still not be overstated. By 1998 the global chemical crop protection market was valued at *c.* £25 billion.

Despite the many successes that have occurred with chemicals that are used for pest control, such technology did not prove to be the panacea that most people thought it might be. Within 20 years of the end of World War II alternatives were being sought in Europe and North America to the use of

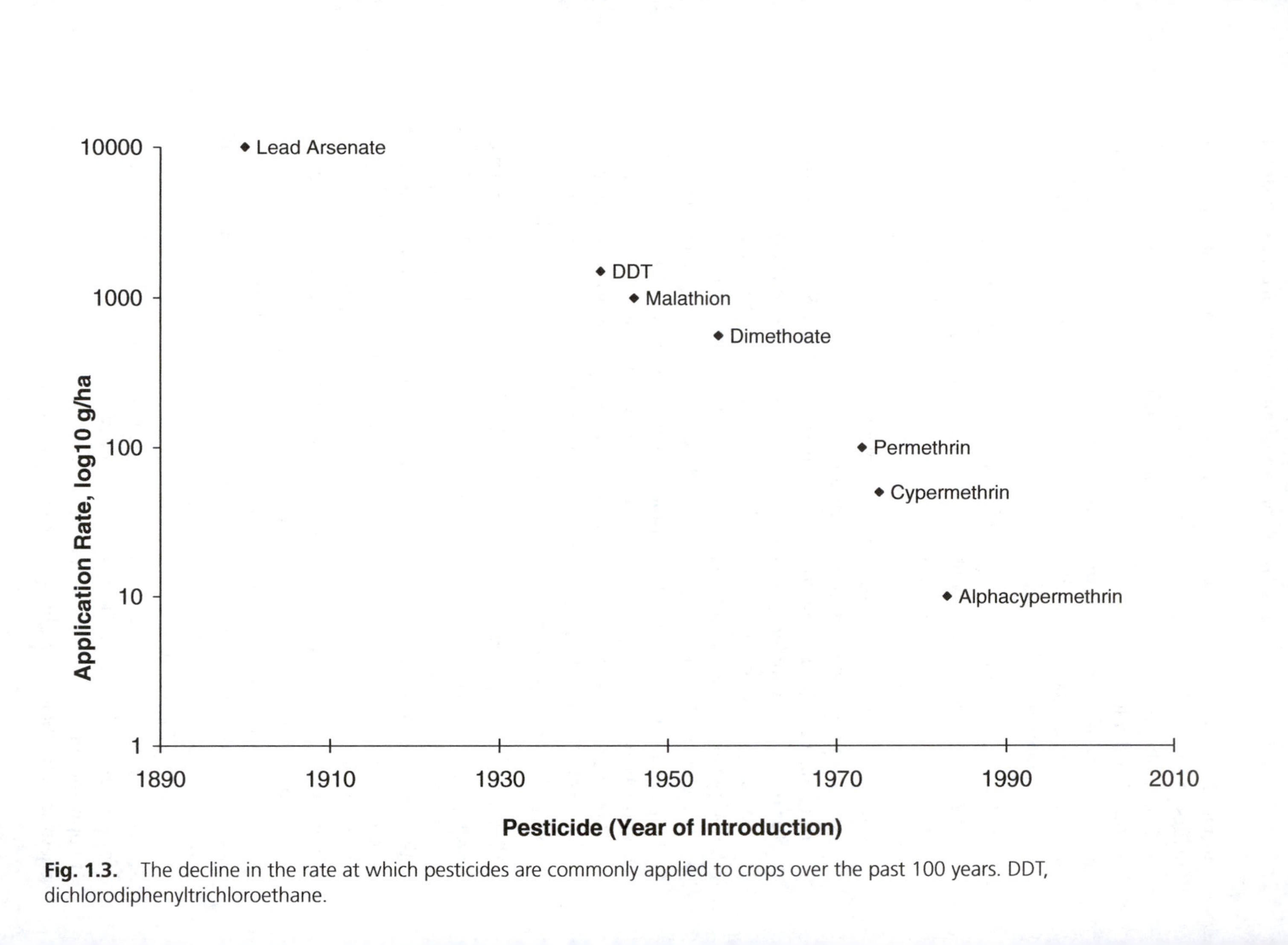

Fig. 1.3. The decline in the rate at which pesticides are commonly applied to crops over the past 100 years. DDT, dichlorodiphenyltrichloroethane.

Table 1.5. *Rate and cost of discovery of new insecticides 1950s–1990s*

Year	Rate of discovery[a]	Approximate cost
1956	1 in 1800	£0.5 million
1964	1 in 3600	£1.5 million
1970	1 in 7400	NA[b]
1972	1 in 10 000	NA[b]
1977	1 in 12 000	£10 million
1987	1 in 16 000	£10–15 million
1989	1 in 20 000	£20 million
1996	1 in 30 000	£30–45 million
1998	1 in 50 000	£50–60 million

Notes:
[a] Rate comprises the number of chemical compounds that need to be screened in bioassays in order to identify one that is useful for further development.
[b] Accurate data not available.

Table 1.6. *Increase in the number of pesticide toxicity tests required by registration authorities in Europe*

1950s	1980s	1990s
Rat feeding test	Rat feeding test	Rat and dog acute and chronic tests
Rat acute toxicity	Rat acute toxicity	Bird acute oral toxicity
	Dog feeding test	Bird 5-day dietary toxicity
	Dog acute toxicity	Bird subchronic and reproductive toxicity
	Teratogenic effects	Fish acute toxicity test
	Metabolic studies	Fish life cycle toxicity test
		Fish early-life stage toxicity test
		Fish 28-day chronic toxicity (juveniles)
		Fish bioconcentration toxicity tests
		Aquatic invertebrates acute toxicity and 21-day chronic toxicity test
		Algal growth rate toxicity test
		Midge larvae acute or chronic toxicity
		Bees acute oral and contact toxicity
		Bee brood feeding tests
		Arthropods residual exposure tests
		Earthworm acute toxicity test

chemicals for the control of arthropod pests. This search led not only to the development of new approaches towards pest control but also to the development of a new philosophical paradigm within which to apply control tactics. This philosophical paradigm is known as integrated pest management (IPM).

The development of integrated pest management

The development of alternatives to the use of chemicals for the control of arthropod pests began in earnest in the developed world following the publication in 1962 of Rachel Carson's book *Silent Spring*. This book articulated, for the first time, the increasingly widespread belief that all was not well with what was often indiscriminate and prophylactic chemical-based insect control. Resistance to insecticides had already been documented in 1914 and was becoming increasingly common (Figure 1.4), resurgent pests were becoming more common, the public was becoming increasingly concerned about residues in food and there was a general concern about the environmental impacts associated with widespread pesticide use. In short, questions began to be asked about a pest control tactic that relied on one technique (a chemical pesticide application).

The result of this questioning was the development of a new philosophical approach towards pest control. This approach was called IPM. Aided by the development of new economic and ecological models and by the development of new techniques for pest control, IPM began to develop over the period 1960–90. The essence of IPM in agricultural situations is that techniques for pest control are integrated or blended. This integration should result in: (1) pest damage that is below levels that would be economically damaging; (2) a minimum adverse impact of a chemical upon the environment (including impacts on nontarget species); and (3) a food production system that is sustainable in the medium to long term. Given such a definition, IPM can either be exceedingly complex or exceedingly simple.

Inherent in the definition of IPM is the concept of economic damage and models were developed in the 1960s that could be used to incorporate this concept. Two of the most important parameters within these models are the economic injury level (EIL) and the economic threshold (ET). These terms were first formally proposed by Stern and colleagues in 1959. The former refers to the point (in pest density) at which the cost of damage caused equals the cost of the control measure to be applied. The latter parameter recognises that control tactics take time to work and the ET is therefore the point at

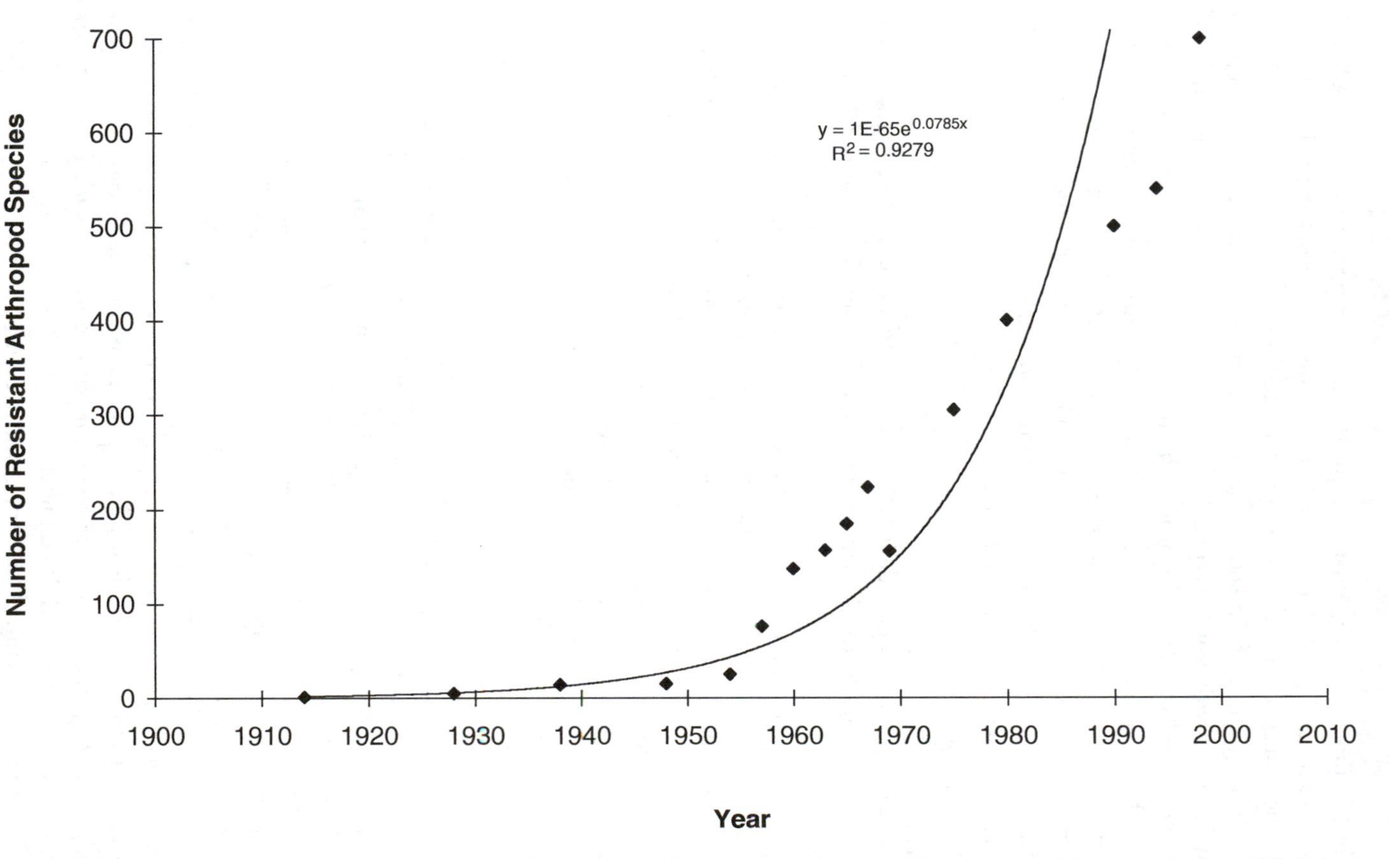

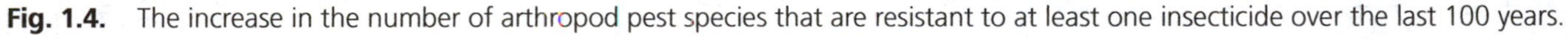

Fig. 1.4. The increase in the number of arthropod pest species that are resistant to at least one insecticide over the last 100 years.

which control measures must be applied in order to prevent a pest population from reaching the EIL. The ET is therefore always at or below the EIL. Figure 1.5 shows a schematic of these respective levels.

The magnitude of the separation between the EIL and the ET will depend on the speed with which a control tactic will work, with the current density of the pest population, with its propensity to increase before it is controlled and with the economic value of the damage being caused. The simplest models that have been produced incorporate data on pest density, product price, a damage function (a measure of damage per pest or pests), the efficacy of the control measure and the cost of the control measure. These data are incorporated as shown in the equation below:

$$\phi = C/PDK$$

where ϕ = the ET (a pest density), C = cost of control, P = the market value of the product, D = the damage function and K = the efficacy of the treatment applied.

The damage function is a measure of the yield loss that will occur in relation to injury caused by a particular pest density. This function is usually highly dependent upon environmental variables and it is typical for these to be incorporated into predictive models of crop losses and economic thresholds.

While economic models may have been useful in getting growers to think about the economics of control, they are very difficult to construct and implement accurately in a field-based situation, particularly with an individual who is known to be risk-averse (i.e. a farmer). The problem of course is that all of the parameters that go into construction of the economic model can vary. Market conditions (the determinant of product price) frequently fluctuate, with the result that the price on the day on which decisions have to be made with regard to a control measure may be vastly different from the price that the farmer gets for the harvested crop. Both agronomic practices (good or bad husbandry, etc.) and geographical location (climate and market) will affect economic models of pest damage. Costs of control may vary widely between countries depending upon whether or not government subsidies are available for crop protection. Costs will also vary depending upon whether an allowance is made for environmental impact or not, i.e. what is the cash value attached to the environmental damage caused by the use of a control measure? In most cases this will be zero but this does not mean that this will always be so. Finally, consumer tastes are known to vary widely in terms of both product quality and desirability.

Despite the difficulties inherent in constructing good economic models for pest control, farmers in many countries are beginning to think in terms of ETs

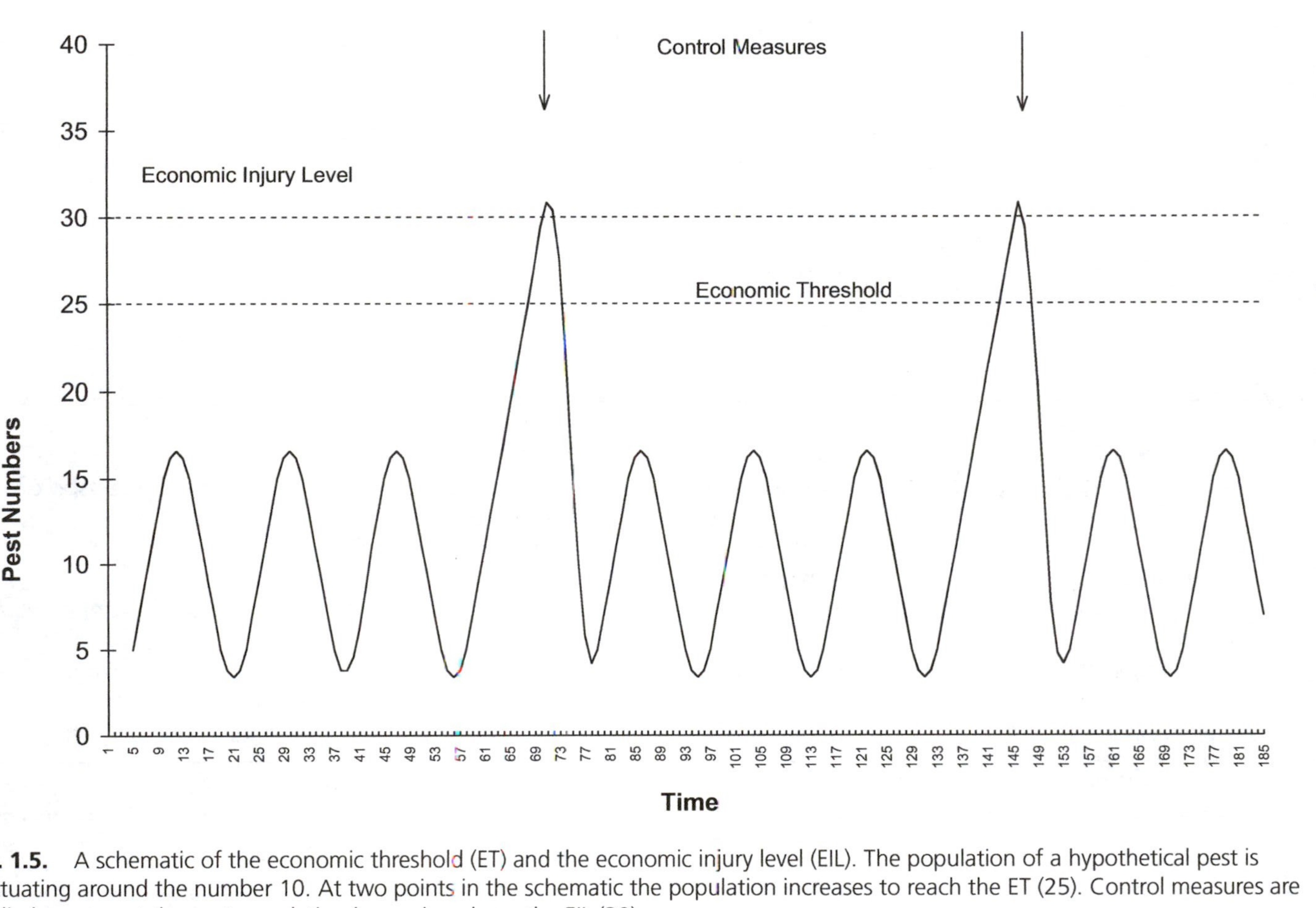

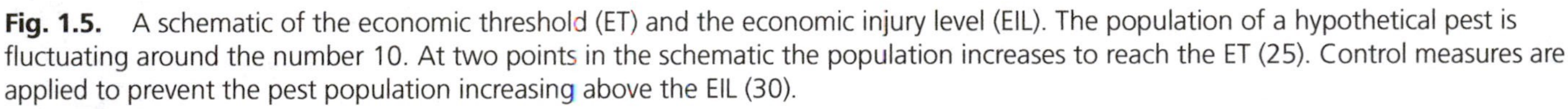

Fig. 1.5. A schematic of the economic threshold (ET) and the economic injury level (EIL). The population of a hypothetical pest is fluctuating around the number 10. At two points in the schematic the population increases to reach the ET (25). Control measures are applied to prevent the pest population increasing above the EIL (30).

Table 1.7. *Example Economic Thresholds*

Species	Common name	Economic threshold	Source
Ostrinia nubilalis	European corn borer	10–20 egg masses per 100 corn plants	Texas Agricultural Extension Service
Nezara viridula	Stink bug	1 stink bug per foot (30 cm) of a row in soybeans	Texas Agricultural Extension Service
Empoasca fabae	Alfalfa leafhopper	Spray when the number of hoppers in 10 sweep samples is equal to or above the height of the crop in inches (or centimetres). Multiply above by 4 if resistant Alfalfa is planted	Pioneer Hi-Bred International
Hypera postica	Alfalfa weevil	Use a 15-in. (38-cm) diameter sweep net and make 10 sweeps in Alfalfa when crop is 25 cm high. Spray if more than 20 larvae caught	Utah State University

Table 1.8. *Economic injury level for corn earworm in sorghum based on number of larvae per head*

Crop value ($) per acre	Cost of control ($) per acre								
	100	120	140	160	180	200	220	240	260
2	5	4	3	3	3	2	2	2	2
3	8	5	5	5	4	4	3	3	3
4	10	8	7	6	6	5	4	4	4
5	12	10	9	8	7	6	6	5	5
6	15	12	11	9	8	8	7	6	6
7	17	14	12	11	10	9	8	7	7
8	20	17	14	12	11	10	9	8	8
9	22	19	16	14	12	11	10	9	9
10	25	21	18	16	14	12	11	10	10

Source: Data collected from the web page of North Dakota State University. See http://www.ndsu.nodak.edu.

for crop protection. Many government organisations are also taking a similar line through their extension services. For example, the US Department of Agriculture had the aim of getting 75% of its acreage for crops in grass seed production systems under IPM by the year 2000.

Although in many cases IPM has really only meant judicious pesticide use, the change in approach to pest control has brought many benefits for farmers, consumers and the environment. Tables 1.7 and 1.8 give examples of ETs that have been developed for some of the world's most important pest species. The tables highlight the variability inherent in determining ETs.

The second important parameter within the definition of IPM concerns the use of ecological information within the decision-making process. Such information had already been used in the construction of economic models for pest control, particularly in relation to species' fecundity and rates of population increase. However, in the 1970s further attempts to bolster IPM programmes were given when it was argued that pests could be classified according to the *r–K* continuum. This continuum had originally been described in the late 1960s as a tool that could be used to understand why species had different life-history strategies. The result for pest control was that control measures could be selected depending upon where species fell within the continuum (Table 1.9). An essential factor in such a selection process therefore was that there was a range of control measures

Table 1.9. *Control strategies for pests based on the* r–K *continuum*

r-selected ⟵	Intermediate ⟶	*K*-selected
Species with a high fecundity. Control is likely to involve pesticides. Attempts to target these pests during ecological bottlenecks may be worthwhile	Species whose numbers are kept low through the action of natural enemies. Outbreaks occur when natural control is disrupted. Control should be based upon restoring natural enemies (biological)	Species well adapted to their environments. Control should be based on cultural, genetic techniques and/or the use of pesticides

Source: Based on Conway, G. R. (1977) Mathematical models in applied ecology. *Nature* **269**; 291–7. This continuum represents a theoretical model within which species can be characterised. For example, houseflies, locusts and armyworms are considered to be *r*-selected pests. Many aphids are considered to be intermediate pests because of the diversity of predatory species that feed on these species. Finally, tsetse flies and codling moths are considered to be examples of *K*-selected pests. For more information consult Southwood, T.R.E. (1977). The relevance of population dynamic theory to pest status. In *Origins of Pest, Parasite, Disease and Weed Problems,* ed. J.M. Cherrett & C.R. Sagar, pp. 35–54, Oxford: Blackwell Publications. Also, see glossary for more details.

to select from in the first place. An expansion in the range of control measures available to growers in the developed world had begun at the start of the 1960s.

Alternatives to chemical pest control

The range of other techniques that can be used either with, or instead of, chemical pesticides is wide. They include the following: cultural and biological techniques, the use of pheromones, the use of genetic manipulation, the use of microbes and insect growth regulators, planting resistant cultivars and the of use of various legislative techniques to prevent problems happening on a regional or country-wide basis. Some of these techniques, of course, are not new. However, with the end of the Second World War many farmers in

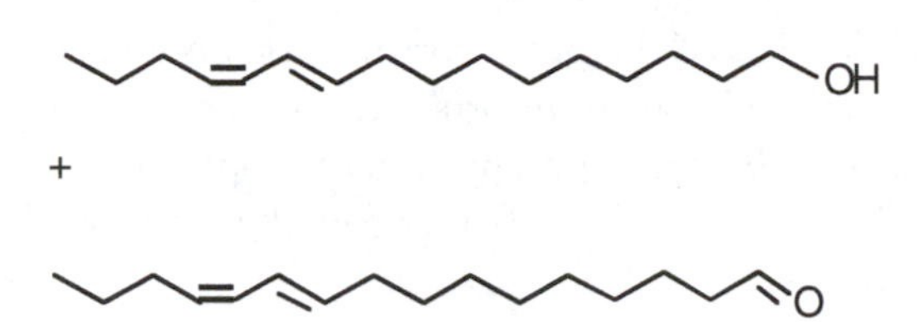

Fig. 1.6. Sex attractant pheromone released by the female *Bombyx mori* to attract males. The pheromone is a blend of two compounds and is shown because it was the first to be chemically characterised. Pheromones are discussed in detail in Chapter 7.

developed countries had switched to an almost total reliance on chemical pesticides. In many cases traditional agronomic practices that were of utility for pest control (such as crop rotations) had been almost completely abandoned. This meant that under the new philosophy of IPM many of the more traditional practices would have to be revisited. A good example of this concerns biological control.

The idea of using predatory species to control pests can be dated to at least the twelfth century when farmers in the Orient used ants to protect fruit trees from pests. This practice continues today. More recently, the cottony cushion scale, *Icerya purchasi*, has been controlled in California since the end of the nineteenth century by a predatory beetle imported from Australia. Ideas concerning the use of biological control agents were certainly not new in the mid-1960s. However, for a large number of farmers embracing biological control meant that their whole approach to pest control would have to change.

Some of the techniques that were developed during this time period, however, were new. For example, the first chemical identification of an insect pheromone had taken place in 1959 (Figure 1.6; sex attractant of silkworm *Bombyx mori*) and by the early 1980s many more pheromones had been developed for commercial use, particularly to monitor insect numbers. In the mid-1960s the first trials with genetically manipulated individuals of the American screwworm *Cochliomyia hominivorax* had been undertaken. This work was to lead directly to attempts in the 1990s to initiate a programme of tsetse fly control using similar methods in Africa. In the mid-1970s some of the first microbial agents were registered for use against crop pests and insect growth regulators had also become available.

By the beginning of the 1990s therefore most farmers had an array of tactics at their disposal that could be used in the fight against arthropod pests. They also had a framework and philosophy (IPM) within which to use these tactics. Chemical pesticides still dominated the pest control market but users were beginning to apply these chemicals far more prudently – an approach that has been helped, at least partially, by developments in more sophisticated application equipment. This is the situation which, from a global perspective, is still pretty much the case today.

However, in the mid-1990s pest control took its next major leap forward with the development of the first genetically engineered crop plants with inbuilt resistance to insect pests. It is highly likely that these plants and other high-tech developments based on recombinant DNA technology will take pest control into the future.

High-tech pest control for the twenty-first century

Recombinant DNA technology can assist pest control in a number of ways. Plants can be genetically engineered with resistance to crop pests, plants can be engineered to be resistant to pesticides that would otherwise be phytotoxic and beneficial predatory insects can be engineered to be resistant to pesticides. Of these techniques, the first two have reached commercialisation in at least some countries. For example, by 1998 the US Department of Agriculture had approved almost 30 genetically modified crop lines for use in agriculture. Almost a third of these crops have been modified to be resistant to attack from crop pests. Table 1.10 shows the crops that were deregulated between 1996 and 2002.

In a large number of cases the genetic modification was brought about by transferring the gene coding for the delta-endotoxin, produced by the bacteria *Bacillus thuringiensis*, into the crop plant (see Chapters 6 and 13 for more details). These first transgenic insect-resistant plants are claimed to be resistant to attack from various lepidopterous or coleopterous pest species. The majority of the transgenic crop plants that have been approved though are resistant to herbicides. The remainder are either resistant to attack by viruses or have some form of modification that alters the quality of the harvested product. It is highly likely that such developments will continue and will improve the control of crop pests well into the future. By 1998, 19% of the North American maize crop (*c.* 33 million ha) comprised genetically modified plants that expressed the *B. thuringiensis* endotoxin. Whether these developments are of help to those who most need to develop their agricultural output, however, remains to be seen. The cost of engineered seed may be just one of many factors that prevents resource-poor farmers in developing countries from exploiting this technology.

Table 1.10. *Genetically modified crops deregulated by the US Department of Agriculture since 1996 as of January 2002*[a]

Crop	Genetic modification	Date of Approval
Corn	Phosphinothricin herbicide-tolerant	April 1999
Rice	Phosphinothricin herbicide-tolerant	April 1999
Oilseed rape	Glyphosate herbicide-tolerant	January 1999
Beet	Glyphosate herbicide-tolerant	December 1998
Oilseed rape	Phosphinothricin herbicide-tolerant	January 1998
Corn	Glyphosate herbicide-tolerant	November 1997
Chicory	Male sterility	October 1997
Soybean	Oil profile altered	May 1997
Cotton	Bromoxynil herbicide-tolerant/ lepidopteran insect-resistant	March 1997
Soybean	Phosphinothricin herbicide-tolerant	July 1996
Papaya	Papaya ringspot virus-resistant	September 1996
Squash (marrow)	Cucumber mosaic virus-resistant/ watermelon mosaic virus-resistant/ zucchini yellow mosaic virus-resistant	June 1996
Potato	Colorado potato beetle-resistant	May 1996
Tomato	Altered fruit-ripening	March 1996
Corn	Male sterility	February 1996
Cotton	Sulphonylurea herbicide-tolerant	January 1996
Corn	Lepidopteran insect-resistant	January 1996

Note:
[a] The very first crop to be deregulated in the USA was the Flavr Savr tomato in October 1992.

Source: Data from USDA website http://www.aphis.usda.gov/bbep/bp/not_reg.html. A number of the above crops have had more than one approval from application by different companies. The data therefore indicate the first approval for the crop and the modification concerned.

Summary

In summary, we can identify at least four major phases in the development of techniques for arthropod pest control. First, early attempts at pest control were based on various cultural and agronomic practices that may, in some instances, have been helped by the use of chemicals, particularly from the sixteenth century onwards. The chemicals used during this first phase were either plant-based or inorganic in composition. This phase lasted from the beginning of recorded history of pest control to the beginning of the middle of the twentieth century.

Second, from 1940 to 1965, synthetic chemical pesticides dominated approaches towards arthropod pest control. These chemicals were seen by many as a cure-all and, at that time at least, completely revolutionised farming by boosting productivity to previously unthought-of levels. These same chemicals also made substantial advances in helping to suppress various arthropod vectors of disease.

The third phase is the era of IPM. The emphasis on prophylactic use of chemicals declined while a vast array of other methods of pest control began to receive increased attention. Chemicals still continued to dominate within arthropod pest control but their use was dependent upon a far more rational decision-making process. This is the situation that prevails today.

Fourth, and last, genetically modified crops have been used to protect against insect attack. This high-tech science is still in its seminal stages and whether this technology truly revolutionises pest control or whether it becomes another technique that is part of the IPM paradigm remains to be seen. This book will end by making some predictions about where this technology may take us in this future.

Further reading

Brooks, G.T. & Roberts, T.R. (eds) (1999). *Pesticide Chemistry and Bioscience: The Food–Environment Challenge.* London: Royal Society of Chemistry.

Carson, R. (1962). *Silent Spring.* London: Hamish Hamilton.

Chapman, R.F. (2000). Entomology in the twentieth century. *Annual Review of Entomology*, **45**, 261–85.

Conway, G.R. (1977). Mathematical models in applied ecology. *Nature*, **269**, 291–7.

Dent, D. (2000). *Insect Pest Management*, 2nd edn. Wallingford: CAB International.

Ganlzelmeier, H. (1999). Plant protection – current state of techniques and innovations. In *Pesticide Chemistry and Bioscience: The Food–Environment Challenge*, ed. G.T. Brooks & T.R. Roberts, pp. 100–19. London: Royal Society of Chemistry.

Horn, D.J. (1988). *Ecological Approach to Pest Management.* New York: Guilford Press.

Metcalf, R.L. & Luckman, W.H. (1994). *Introduction to Insect Pest Management.* New York: John Wiley.

Pedigo, L.P. (1999). *Entomology and Pest Management*, 3rd edn. Upper Saddle River, New Jersey: Prentice-Hall.

Rechcigl, J.E. & Rechcigl, N.A. (eds) (2000). *Biological and Biotechnological Control of Insect Pests*. Boca Raton, Florida: Lewis.

Rechcigl, J.E. & Rechcigl, N.A. (eds) (2000). *Insect Pest Management, Techniques for Environmental Protection*. Boca Raton, Florida: Lewis.

Southwood, T.R.E. (1977). The relevance of population dynamic theory to pest status. In *Origins of Pest, Parasite, Disease and Weed Problems*, ed. J.M. Cherrett & C.R. Sagar, pp. 35–54. Oxford: Blackwell.

Stern, V.M., Ray, F.S., Van Den Bosch, R. & Hagan, K.S. (1959). The integrated control concept. *Hilgardia*, **29**, 81–101.

Van Driesche, R.G. & Bellows, T.S. (1996). *Biological Control*. New York: Chapman & Hall.

Van Emden, H.F. & Peakall, D.B. (1996). *Beyond Silent Spring – Integrated Pest Management and Chemical Safety*. London: Chapman & Hall.

Yang, R.Z. & Tang, C.S. (1988). Plants used for pest control in China: a literature review. *Economic Botany*, **42**, 376–406.

Yudelman, M., Ratta, A. & Nygaard, D. (1998). *Pest Management and Food Production*. Washington, DC: International Food Policy Research Institute.

2

Botanical insecticides

Introduction

The chemicals that plants produce to protect themselves against insect attack belong to a group that includes compounds known as secondary plant substances. These chemicals are a subset of what are known as phytochemicals (plant-based chemicals). Within the context of arthropod pest control they have also been referred to as botanical insecticides.

Secondary plant substances are produced as byproducts of major biochemical pathways (see Chapter 10). Chemically, they include alkaloids, terpenoids and phenolics as well as a number of other compounds. Many of these chemicals have been successfully exploited by humanity for the control of arthropod pests. The most widely used and studied botanical insecticides are discussed in this chapter.

The use of plant extracts as insecticides can be dated back at least 4000 years (see Chapter 1). It is highly probable, however, that the exploitation of the toxicological properties of plants has an even older history. Prior to the onset of agriculture as a means of food production native peoples[1] had already deified trees and many plant-based extracts were believed to possess special powers, particularly for healing. With the development of agriculture links would have rapidly been made between food production and pest control. These links would have provided an opportunity for this specialised knowledge to be used directly in protecting crops from pests.

[1] Throughout this chapter, reference to native peoples or to native cultures refers to local preparation and use of plant extracts for pest control. See also glossary for comments on deification.

The earliest documented examples of plants being used as pesticides occur in China, Egypt, Asia and Europe. None of these uses involves large-scale commercial production of botanical insecticides. This did not take place until the sixteenth century when European explorers began to discover the plants that native cultures were using for pest control (see Chapter 1). Widespread commercial use of botanical insecticides in western Europe and North America continued from the sixteenth century up until the Second World War. But, with the advent of modern synthetic insecticides in the 1940s, the use of botanical insecticides has declined.

Some large-scale plant screening programmes did take place throughout the 1940s, 1950s and 1960s, particularly in the USA and in China. However, although many of these programmes identified a number of potentially useful extracts, most lacked sufficient funding to be sustained. The one notable exception to this was the discovery and successful commercialisation of ryanodine by Merck. In contrast to large-scale commercial production of botanical insecticides, native peoples have sustained their use of plant extracts since records began. In almost all cases these extracts are processed and formulated at the local level. That native usage of botanical insecticides continues is at least one reason for the current upsurge of interest in these chemicals.

Other reasons for the current revival of interest in botanical insecticides include: the success story that neem extract (from the neem tree *Azadirachta indica)* has become; the high costs associated with developing new synthetic insecticides; the small proportion of plants that have been assayed for toxicological activity; and the fact that carbamate and pyrethroid insecticides were developed from studies on plant-based chemicals. For example, conservative estimates put the number of plant species on our planet between 250 000 and 500 000. Approximately 2400 plants have been recorded as being useful for pest control. Of these, only a tiny proportion (less than 10) have been developed commercially. It is highly likely therefore that novel and potent molecules that can be used for pest control remain to be discovered.

The revival of interest in botanical insecticides can be split into two approaches. The first comprises conventional western agrochemical companies looking for new leads for insecticides. The second comprises organisations within developing countries that are hoping to promote and exploit native biotechnologies. Although native peoples continue to use botanical insecticides, it has been argued that their use has not developed commercially as much as it might have. Neem-based extracts are the current exception. Box 2.1 gives some of the reasons for the lack of commercial development of many botanical insecticides.

This chapter begins by considering the plants (Table 2.1) that became

Box 2.1

Reasons for the lack of commercial development of some botanical insecticides

1. Many farmers regard botanical pesticides as old-fashioned. Modern pest control should involve using modern synthetic chemicals.
2. Most botanical insecticides are not on official lists for crop protection products. Developing countries often use official lists produced in western European countries. Botanical insecticides are not included in these lists because they have variable and unknown chemical constituents.
3. Many botanical insecticides do not appear to give the dramatic kills that synthetic insecticides do.
4. Many botanical insecticides are rapidly inactivated by exposure to light and air.
5. There is sometimes a problem with seasonal availability of botanical insecticides.
6. Most of the research on botanical insecticides has been qualitative rather than quantitative. Farmers need to know how to formulate products, the doses that should be used and formulation information. As yet, much of these data is still to be collected.
7. Accurate toxicity data for botanical insecticides are often lacking.

widely used in North America and Europe following discoveries made by European explorers. Other plants that are still used by native peoples today as sources of botanical insecticides are then described. The use of essential oils for pest control is briefly discussed. Finally, the chapter concludes with a brief consideration of some of the issues associated with using botanical insecticides at the start of the twenty-first century.

Sabadilla

Sabadilla was used in Europe from the sixteenth century through to the middle of the twentieth century. The decline in its use was caused by the discovery of modern synthetic insecticides, particularly dichlorodiphenyltrichloroethane (DDT).

The first European explorers of the Americas observed that native South American Indians used powdered seeds to protect their crops from insect attack. Insecticidal formulations were originally obtained from *Sabadilla officinarum*, a lily that grows wild in Central and South America. The commercial source of sabadilla however was a related plant, *Schoenocaulon officinale,* which

Table 2.1. *Insecticidal plants used in native cultures*[a]

Family	Plant	Common name
Annonaceae	*Annona muricata*	Soursop
	Annona reticulata	Custard apple
	Annona squamosa	Sweetsop
Araceae	*Acorus calamus*	Sweet flag
Compositae	*Chrysanthemum cinerariaefolium*	Chrysanthemum
Flacourtiaceae	*Ryania speciosa*	Ryania
Guttiferaceae	*Mammea americana*	Mammey
Labiatae	*Minthostachys glabrescens*	Muna
	Minthostachys mollis	–
	Mentha spp.	Spearmint
Leguminoseae	*Derris eliptica*	Derris
	Derris malaccensis	–
	Derris uliginosa	–
Liliaceae	*Allium sativum*	Garlic
	Schoenocaulon officinale	Sabadilla
Meliaceae	*Azadirachta indica*	Neem
	Melia azederach	Persian lilac
Simarubaceae	*Quassia amara*	Quassia
	Aeschrion excelsa	'Not known'
	Picrasma excelsa	'Not known'
Solanaceae	*Capsicum frutescens*	Chilli pepper
Zingiberaceae	*Curcuma domestica*	Turmeric

Note:
[a] Plants listed as active against a wide variety of pests in Stoll, G. (1992). *Natural Plant Protection*. Weikersheim, Germany: Verlag Josef Margraf.

subsequently became widely cultivated in Venezuela. At the peak of its use in the early 1940s over 120 000 tons of sabadilla were exported to the USA alone. As well as being used for crop protection, powdered seeds were used to treat headlice throughout the eighteenth and nineteenth centuries.

The active ingredients of sabadilla are ceveratrum alkaloids, mainly cevadine and veratridine (Figure 2.1). These alkaloids are rapidly destroyed by sunlight, although the potency of crude extracts can be enhanced by heating (75–80 °C for 4 h) and by pretreating the formulation with lime. A highly toxic extract can be made by heating 500 g of pulverised seed in 4 l of kerosene for 1 h at a temperature of 150 °C. If the pulverised seed is pretreated with lime a temperature of 60 °C can be used.

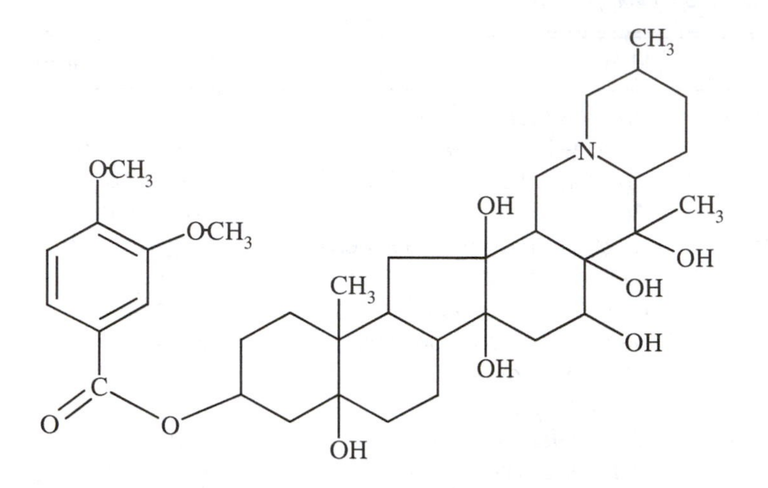

Fig. 2.1. Chemical structure of the alkaloid veratridine.

Sabadilla acts as a contact and stomach poison for the control of a variety of pest species. Grasshoppers, houseflies, jassids, lice, thrips and various caterpillars have all been reported to be controlled with crude extracts. Although not persistent, sabadilla has also been shown to be toxic to some soil bacteria and to honey bees. Mammals may also be affected by sabadilla, although its oral LD_{50}[2] is greater than 4000 mg/kg. It is still registered for use today in the USA for application to fruit crops for the control of thrips. Its rapid inactivation by sunlight means that crops can be harvested 24 h following treatment. This apart, the use of sabadilla as a pesticide has now been largely superseded by other crop protection chemicals. Sabadilla is still used by native peoples as an insecticide. Its chemical complexity means that it has not yet been synthesised.

Nicotine

The plant alkaloid nicotine (Figure 2.2) is the active insecticidal component in the tobacco plants *Nicotiana tabacum* and *N. rustica.* Like sabadilla, these plants were first discovered by Europeans during explorations of the

[2] The LD_{50} is the amount of a poison that will kill half of the individuals in a test population, i.e. it is the lethal dose for 50% of individuals assayed. See Chapter 3 and glossary for more details.

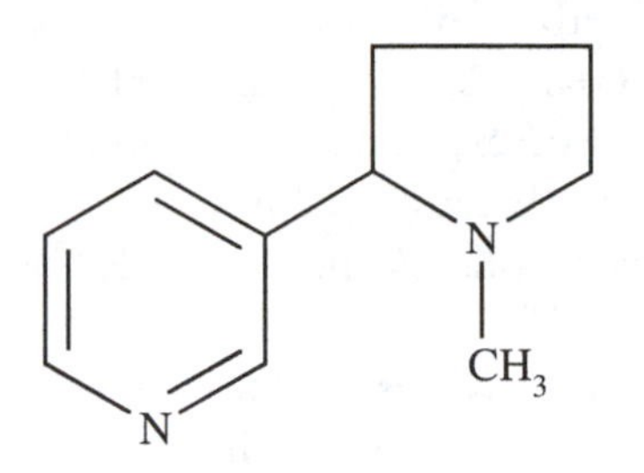

Fig. 2.2. Chemical structure of nicotine.

Americas in the sixteenth century. These plants however were native to the northern areas of the New World. By the late seventeenth century liquid tobacco extracts began to be widely used in Europe for crop protection. In the mid eighteenth century tobacco dust was used to protect seeds that were shipped from Europe to North America and by 1773 the French had produced a bellows that enabled tobacco to be used as a fumigant. The active ingredient within tobacco extracts was discovered and named nicotine in 1828. By 1843 a structural formula for the alkaloid had been proposed and in 1904 nicotine was synthesised. It is now known that the alkaloid fraction within tobacco plants comprises 12 different chemicals. Nicotine however accounts for 97% of the total alkaloid fraction. The nicotine content of plants varies from 5 to 10% in the leaves to trace amounts in the seeds.

Nicotine extracts have been used in a range of formulations. For example:

- as liquids, either as 40% free nicotine in water or as 40% nicotine sulphate;
- as dusts, where free nicotine or the salt is mixed with absorbent, inert or active carriers (e.g. talc, gypsum, hydrated lime, respectively);
- as fumigants, where concentrated extracts are volatilised by the action of heat;
- as fixed solutions such as nicotine–oil combinations.

Nicotine acts a rapid-contact insecticide, entering pest species by direct penetration of the integument. It acts directly on the central nervous system (CNS) causing excitation at low concentrations and paralysis at high concentrations. It works as an agonist (competitor) for acetylcholine receptors on postsynaptic membranes within the CNS. Build-up of acetylcholine within synapses causes the physical symptoms that are seen in poisoned species. Nicotine is not selective and is highly toxic to a range of other species, including bees, predatory invertebrates and earthworms. It is very toxic to mammals, with a mammalian oral LD_{50} of 50–60 mg/kg. Environmental degradation of nicotine typically occurs within 48 h of exposure to light and air. Within the UK six professional formulations remain available for the control of field and glasshouse pests. Nicotine formulations are not available for domestic use in the UK.

Originally nicotine was one of the cheapest botanical insecticides available. Formulations were prepared from waste or low-grade tobacco that was not suitable for the manufacture of cigarettes or cigars. With improvements in tobacco-processing technology however, and with the withdrawal of government subsidies for low-grade crops like nicotine, the costs of producing viable formulations became prohibitive. In the 1960s the costs associated with producing nicotine formulations on a commercial scale were six times as great as those associated with producing DDT. Because of the high costs of production and the high mammalian toxicity of nicotine formulations it is unlikely that this botanical insecticide will make any kind of large-scale return to use. However, nicotine is still widely used in some native cultures. For example, in China rice is protected against stemborers by immersing tobacco stalks in paddy fields at a density of 150–300 kg stalks per hectare and in Bolivia potatoes are protected from attack by aphids using nicotine sprays. In the UK nicotine formulations are primarily used in horticulture to protect against a range of pest species.

Quassia

Quassia was originally extracted from the Central American tree *Quassia amara*. However, in the eighteenth century the commercial extracts of quassia were obtained from the related shrub *Aeschrion excelsa*. These plants were grown extensively in Venezuela for export to Europe and North America. This export continued from the eighteenth century to the middle of the twentieth century. In the 1940s over 500 000 kg of quassia was exported to North America alone. Insecticidal formulations can be prepared from chipped trunk and branches.

The active component within extracts is called quassin (Figure 2.3), a water-soluble molecule that acts as a contact and stomach poison. Extracts are prepared by steeping wood chips in water for several hours prior to boiling to concentrate the liquid. The toxicity of extracts prepared this way has been found to vary enormously. However, good extracts have been found to be

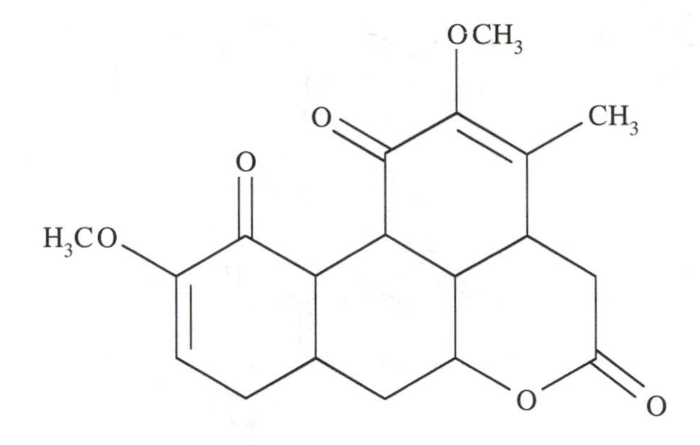

Fig. 2.3. Chemical structure of quassin.

potent aphicides and to be toxic to a number of lepidopterous pests. It has been reported that quassin can be absorbed by plant roots and that it is translocated within the vascular system of plants, i.e. that it has systemic activity. Quassin does not appear to be toxic to many other insect species. In the UK quassia is no longer used as an insecticide, mostly because of the high costs associated with the production and transportation of wood chips. Recently, quassia has been assessed in trials in Australia aimed at finding alternatives to synthetic insecticides for the control of pests of brassicas. Quassin has not yet been synthesised and its precise mode of action is still unknown. It was discovered recently that in India farmers use extracts from *Picrasma excelsa,* a shrub closely related to *Aeschrion excelsa*, for pest control. In the short term it is likely that quassia will continue to be used as an insecticide within native cultures.

Unsaturated isobutylamides

Unsaturated isobutylamides can be extracted from various South American plants in the families Compositae and Rutaceae. That crude extracts from these plants were used for pest control was discovered by explorers early in the nineteenth century. Native Mexicans burn the leaves of *Heliopsis longpipes*, releasing the active ingredients as fumigants within field crops. Extracts from the roots of this plant have also been used to control flies of medical and veterinary importance.

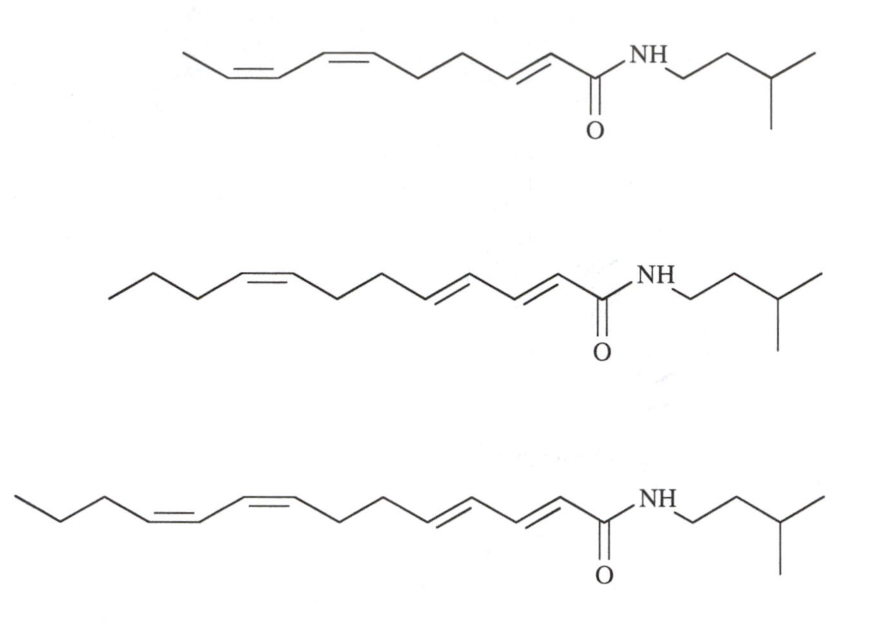

Fig. 2.4. Chemical structure of some isobutylamides.

The first of the unsaturated isobutylamides was discovered in 1834. The extract, from *Anacyclus pyrethrum*, was initially called pyrethrin. This caused confusion with pyrethrum that was extracted from *Chrysanthemum cinerariaefolium* and so in 1895 the compound was renamed pellitorine. This chemical was first synthesised in 1952 (Figure 2.4). Since then, many more isobutylamides have been discovered. By the early 1970s at least 11 different naturally occurring compounds had been named. Many of these compounds have now been synthesised and assayed for activity against pest species in the laboratory. These compounds are listed in Table 2.2.

All of the unsaturated isobutylamides are neurotoxins that impair or block voltage-dependent sodium channels on nerve axons (like DDT and the pyrethroids – see Chapter 3). They are extremely unstable molecules and, although toxic to a range of arthropod species, every field trial that has been undertaken with these chemicals (natural or synthesised) has given poor results. Most current research with these chemicals is concentrating upon improvements to their chemical stability. Few data are available on their environmental or mammalian toxicity, primarily because they are not yet commercially available. Like many of the other botanicals discussed, it is likely that these naturally occurring insecticides will continue to be used at a local level within native cultures.

Table 2.2. *Unsaturated isobutylamides: plant sources, active ingredients and test species assayed. All are toxic to* Musca domestica

Plant species	Plant part	Active	Species assayed
Anacyclus pyrethrum	Roots	Pellitorine	*Tenebrio molitor*
A. pyrethrum	Roots	Anacyclin	*T. molitor*
Spilanthes oleraceae	Leaves	Spilanthol	*Anopheles* spp.
Heliopsis longipes	Roots	Affinin	*Aedes aegypti*
H. scabra	Roots	Scabrin	*M. domestica*
H. scabra	Roots	Heliopsin	*M. domestica*
Echinacea augustifolia	Roots	Echinacein	*M. domestica*
Zanthoxylum piperitum	Fruit, bark	α-Sanshool	*Culex pipiens*
Z. piperitum	Bark	Sanshool-II	*M. domestica*
Z. clavaherculis	Bark	Herculin	*T. molitor*
Z. clavaherculis	Bark	Neoherculin	*T. molitor*

Ryanodine

Until the 1940s all the botanical insecticides that had been discovered by European and New World explorers were as a result of observations made on indigenous peoples. During the 1940s, however, a number of plant-screening programmes were initiated and this was how the alkaloid ryanodine (Figure 2.5) was discovered. This screening programme involved collaboration between Merck and Rutgers University in North America. The compound ryanodine comes from *Ryania speciosa,* a shrub native to South America, and is sometimes also referred to as ryania. It was known at the time that wood from the stem of this shrub was used by native Indians to make poison for arrows. However, these extracts had not yet been assayed for use in crop protection. Following the discovery of ryanodine as a crop protection chemical it began to be marketed for pest control from 1945 onwards.

Ryanodine is a contact and stomach poison that is more stable than many other botanical insecticides. Residual activity of ryanodine has been reported for over 1 week after application. Ryanodine works by binding to calcium channels in sarcoplasmic reticulum. This causes cells to flood with calcium and leads to death. Extracts are toxic to a number of lepidopterous and coleopterous pests. The mammalian oral LD_{50} for ryanodine is 750 mg/kg making it a moderately toxic molecule in terms of the World Health Organization pesticide toxicity categories (these categories are described in Chapter 3). It was last used as an insecticide in the UK in 1979. At present, it is sold for codling

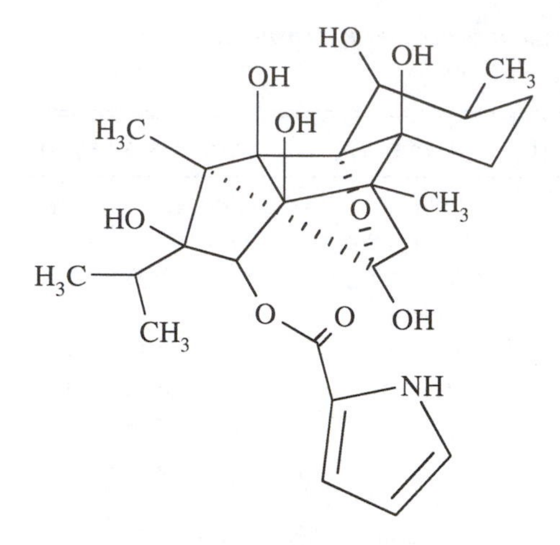

Fig. 2.5. Chemical structure of the alkaloid ryanodine.

moth control in organic orchards in the USA. There are some research data that suggest that ryanodine may also be useful for leafhopper control in orchards.

As with many other botanical insecticides, it has not been possible to synthesise the active ingredient chemically and the costs associated with large-scale production and processing of extracts are now becoming prohibitive. In native cultures however it is still used for pest control. Powdered roots, leaves and stalks can be mixed with an inert carrier and applied directly to crops or the mixture can be dissolved in water (30–40 g powder to 7–8 l of water) and then sprayed on to crops.

Naphthoquinones

The naphthoquinones are among the newest botanical insecticides to be discovered. Like ryanodine, these chemicals were discovered as a result of a plant-screening programme. The chemicals were discovered in *Calceolaria andina*, a plant native to the Andes, and the extraction and fractionation work was undertaken at Rothamsted in the UK in 1995. So far, they appear to be toxic to sap-feeding pest species and not toxic to beneficial species. Treated species die within minutes of exposure to these chemicals. Their precise mode of action is still not understood. Current research on these chemicals is

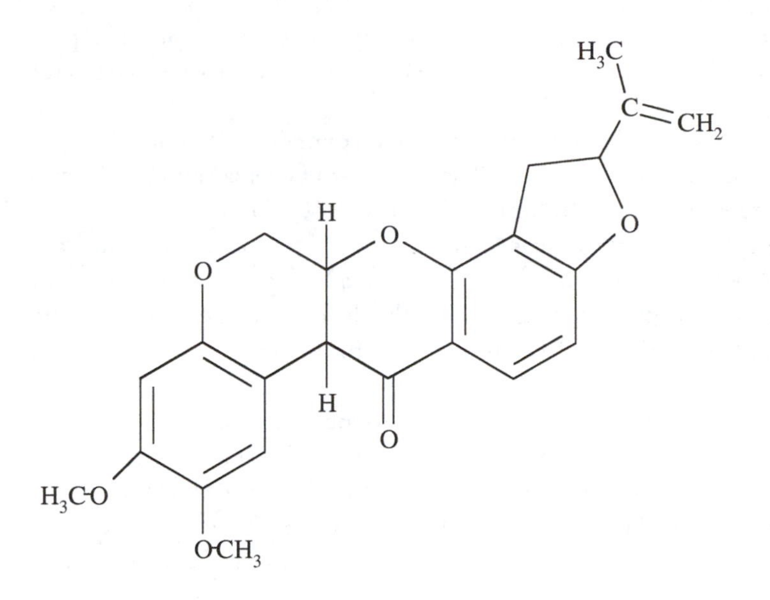

Fig. 2.6. Chemical structure of rotenone.

concerned with an investigation of their chemical structure and with their toxicological properties. At the time of writing, the results of this work were still a few years away (see Glossary for the latest information on this work).

Rotenone

Unlike the botanical insecticides discussed so far, rotenone (Figure 2.6) can be extracted from plants that are native to both the Americas (*Lonchocarpus* spp.) and to Asia (*Derris* spp.). Extracts were originally used by natives as fish poisons (macerated plants were thrown into water to paralyse fish that then floated to the surface). Their first use as a botanical insecticide was documented in 1848 when they were used against leaf-eating caterpillars. Since then, they have been used throughout Europe, Asia and the Americas for the control of a wide range of pest species. In 1902 rotenone was isolated for the first time and in 1932 its complete chemical structure was published. The name rotenone derives from 'roten' (native name for fish poison) and from 'one' (for ketone). Since then, 10 related chemical components have been identified. These are known collectively as rotenoids (rotenone, elliptone, sumatrol, malacol, toxicarol, deguelin, tephrosin, plus three others). So far, these chemicals have been identified in over 70 species of plant.

Most extracts are made from the roots which contain 5–10% active ingredient. None of the other rotenoids has proven to be as commercially successful as rotenone. Before the Second World War 50% of all rotenone came from *Derris* spp. grown in the Far East. However, after the start of the war alternative sources were sought and *Lonchocarpus* spp. grown in South America became the primary source. With the development of the synthetic chemical control market in 1945 the use of rotenone in crop protection declined.

Rotenone acts as an insecticide by binding to the mitochondrial respiratory chain and causing NADH dehydrogenase inhibition. This reduces oxygen uptake and leads to death. Pest species controlled by rotenone include aphids, caterpillars, beetles and various aquatic larvae. However, there are representative species in each of these groups that do not appear to be controlled. The estimated lethal dose for humans is 300–500 mg/kg and it is more toxic if inhaled than if ingested. Like most botanicals, rotenone is rapidly inactivated by exposure to light and air. The chemical also loses its toxicity when mixed with an alkali. Synergists such as piperonyl butoxide can be used to enhance the activity of rotenone.

In the UK rotenone is available as two professional and 12 amateur formulations. It is also formulated with other pesticides as combination products. Recently, there has been a suggestion that rotenone (like many other pesticides) may be withdrawn from use in the UK due to the high costs associated with its reregistration. Rotenone will continue to be used however in native cultures where simple homemade formulations can be concocted: 1 kg of derris dust (powdered roots) mixed with 500 g of soap will provide enough active material for 100 l of water. This formulation can be used to control a wide range of pests. Both homemade and commercially available formulations are highly toxic to fish.

Sweet flag

The leaves of sweet flag, *Acorus calamus,* were used in Europe to prepare the first botanical insecticide of Asian origin. Native to China and India, the leaves were first imported to Austria in 1574 and by the early 1600s the leaves were widely used in churches, castles and cottages to repel pest species, notably flies and lice. Since then, the range of species controlled by extracts has risen to include coleopterous and lepidopterous pests. Although sweet flag is still used by native peoples (both powdered rhizomes and simple extracts of essential oils), the active fractions have been found to contain the carcinogenic alkaloid asarone. Research to assess whether this alkaloid is necessary for activity is currently under way in western countries.

Marigolds

Another plant species that has been used for centuries by people in India and East Asia to control pests is the marigold (*Tagetes* spp.). These plants contain essential oils in the leaves and flowers that can be used to control flying pests. In Europe, marigolds have also been used to control pests by intercropping. The essential oils that *Tagetes* spp. contain are called thiophenes. Unlike many other botanical insecticides, these are light-stable. Research into the commercial applications of marigold extracts is currently an active field funded in the USA by the US Defense Department. The exact mode of action of these chemicals has not yet been determined. These plants will continue to be used at the local level for pest control.

Pyrethrum

Pyrethrum is probably the best-known botanical insecticide. Active extracts are derived from dried flowers of *Chrysanthemum cinerariaefolium*. These flower heads were exported to Europe for pest control from the beginning of the nineteenth century. It was not until 1851 however that the plant source was revealed. In 1828, the first commercial product was marketed for the control of lice and fleas. The main production region of chrysanthemums for export until World War I was Dalmatia.[3] Japan then took over as the major exporter to Europe until the start of World War II. The world's major producers of pyrethrum are now Kenya and Tanzania. The plants, which grow best at altitudes of 1500–3500 m, are harvested and processed for export on site. Kenyan flowers typically contain 1–2% pyrethrins.

Chemically, pyrethrum comprises six different active ingredients – pyrethrins I and II, cinerins I and II, and jasmolins I and II. All are esters comprising

[3] Dalmatia comprises a 320-km long and 50-km wide stretch of land lying in what is modern-day Croatia (the capital of Dalmatia is Split).

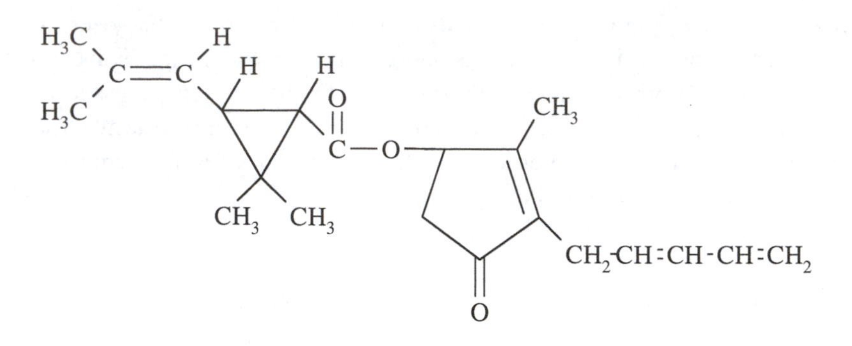

Fig. 2.7. Chemical structure of pyrethrin I.

an acid containing a three-carbon ring joined to an alcohol containing a five-carbon ring (Figure 2.7). These esters exist as an oily liquid that is not soluble in water. They are very unstable and are rapidly broken down following exposure to light and air. In aqueous environments the esters are rapidly hydrolysed which means that they are unable to kill insects as stomach poisons. Natural extracts are therefore used as fast-acting contact poisons. Most commercial formulations contain a synergist that enhances activity by slowing the rate at which the active ingredients are metabolised.

Pyrethrum extracts are neurotoxins (see also Chapter 3). They act by disrupting nerve conduction which leads to paralysis and death. Their toxicity is negatively correlated with temperature. This may be a function of the strength with which they are able to bind to receptor sites on nerve axons, i.e. as the temperature increases so does their dissociation. Their mammalian oral LD_{50} is *c.* 1600 mg/kg, which makes them very safe to mix and apply. They are however toxic to some fish and to beneficial species such as bees.

Natural pyrethrum extracts are used against an enormous range of pest species. Most of the insecticidal sprays sold for household use in western countries comprise pyrethrum extracts formulated as aerosols with synergists. In commercial situations extracts are used against livestock pests, in storage facilities to control grain pests and in field crop situations for the control of vegetable and fruit pests. Because they are broken down rapidly they can safely be used very close to harvest. Local formulations are still prepared by native people for use on crops. For example, a simple and effective formulation can be made by steeping 500 g of flower heads in 4 l of kerosene for half a day.

Pyrethrum was the first botanical insecticide from which synthetic analogues were developed. Studies on their chemistry began in the 1920s and in

1949 the first synthetic pyrethoid, allethrin, was produced. Since then, a large number of other synthetic pyrethroids have been produced (see Chapter 3). In the western world these synthetic pyrethroids have now become the most widely used insecticides. It has been this success that drives much of the current research concerning botanical insecticides.

Azadirachtin

Azadirachtin is the principal active ingredient in extracts derived from the neem tree *Azadirachta indica*. Neem trees are native to the Indian subcontinent but are distributed throughout South-East Asia, sub-Sahelian[4] Africa, eastern Asian islands and Central America. Neem trees thrive in a variety of climates and have modest soil requirements. Trees fruit when they are 4–5 years old, with an average yield of 30–50 kg of fruit per tree. The oil content of seeds, where the most potent extracts are found, is 35–45%. The leaves and the oil extracted from seeds have been used for thousands of years in Asia for the control of insect pests. Locally produced extracts are used to control lice and bugs in bedding and crude sprays can be simply prepared for crop protection. A basic formulation can be made by steeping crushed seeds in water at a concentration of 25–50 g/l.

Azadirachtin was first isolated from the neem tree in 1968. However, it took until 1985 for a full identification of this complex molecule to be published (Figure 2.8). Azadirachtin, perhaps not surprisingly, cannot yet be synthesised. Although azadirachtin is believed to be the most potent active ingredient within neem extracts, on its own it is less potent than neem. The other active components within extracts have yet to be identified.

Both neem and azadirachtin act as antifeedants and as disrupters of larval development. Their exact mode of action is still to be determined. Over 100 pest species can be controlled using these chemicals. So far, they do not appear to be toxic to beneficial hymenopterous species. The mammalian oral LD_{50} is >5000 mg/kg. To all intents and purpose therefore they are harmless to mammals. Like many other botanical insecticides though, they lose their potency in sunny conditions.

In the 1990s commercial formulations of neem extract began to be produced in both developed and developing countries. To overcome problems

[4] The Sahel is the transitional semiarid zone between the arid Sahara to the north and the humid savanna to the south. It extends across Africa from Senegal to Sudan.

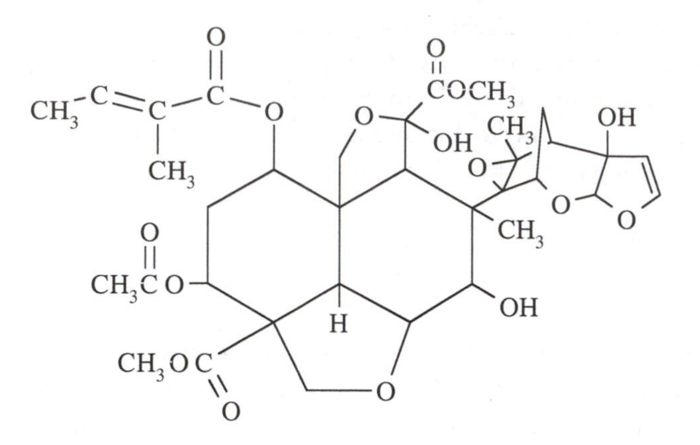

Fig. 2.8. Chemical structure of azadirachtin.

associated with photostability, sunscreens and antioxidants are often incorporated within these formulations. Local formulations of extract continue to be prepared. It is anticipated that the use of neem and azadirachtin as insecticides will expand and increase in the next 10–20 years.

Essential oils[5]

In addition to the botanical insecticides so far discussed, there are also a range of other extracts that have been shown to have insecticidal activity. These extracts are known as essential oils (Table 2.3). These oils can be extracted from the leaves, seeds and flowers of a range of plants and many have been used for thousands of years to control pest species. Most of these oils are highly volatile and are extracted in such small amounts that commercial formulations for pest control are uneconomic to produce. However, this does not preclude their use at a local level.

Botanical insecticides developed for use in the west

The use of botanical insecticides for pest control was widespread in western Europe and North America until the development of the modern synthetic insecticide industry. The one botanical insecticide that survived this

[5] Essential oils are volatile oils that exist in odoriferous plants. Oils are stored as microdroplets that diffuse out of glands, spread and then evaporate. They are called essential because they are believed to represent the essence of odour or flavour.

Table 2.3. *Essential oils with toxicity to arthropod pests*

Plant	Oil	Arthropod pest	Effect
Eucalyptus spp.	Eucalyptus	*Varroa* mites	Acaricidal
Eucalyptus spp.	Eucalyptus	*Pieris brassicae*	Antifeedant
Cymbopogen citratus	Lemongrass	*Aedes aegypti*	Larvicidal
Mentha piperita	Peppermint	*P. brassicae*	Antifeedant
Cymbopopon nardus	Citronella	*Dermataphagoides* spp.	Acaricidal
C. nardus	Citronella	Flying insects	Repellent
Laurus spp.	Laurel	*Dermataphagoides* spp.	Acaricidal

Source: Data from: Endersby, N.M. & Morgan, W.C. (1991). Alternatives to synthetic chemicals for use in crucifer crops. *Biological Agriculture & Horticulture*, **8**, 33–52; McDonald, L.G. & Tovey, E. (1993). The effectiveness of benzyl benzoate and some essential plant oils as laundry additives for killing house dust mites. *Journal of Allergy and Clinical Immunology*, **92**, 771–2. Singh, G. & Upadhyay, R.K. (1993). Essential oils – potent sources of natural pesticides. *Journal of Scientific and Industrial Research*, **52**, 676–83. Adams, S. (1995). Natural products show promise for controlling tracheal and varroa mites. *American Bee Journal*, **135**, 533–4.

development and is still widely used today is pyrethrum. However, in recent years there has been an upsurge of interest in the use of plant extracts for pest control, particularly among organic farmers (see Chapter 11) and an increase in the number of products that are marketed commercially has been observed. Neem extracts, in particular, are beginning to be more widely used and it is likely that this use will continue to expand. A list of the plant extracts that had been developed for commercial sale by 2001 is provided in Table 2.4. It is unlikely that the extracts that were once widely used (sabadilla, quassia, nicotine, etc.) will ever come back into large-scale production. This is primarily because there are still a large number of affordable alternative synthetic chemicals that are available. However, there may be niche markets, e.g. organic farming, where the use of these products may find widespread application.

At the local level in developing countries, however, where costly modern insecticides are not yet available or affordable, the use of botanical insecticides is likely to expand and increase. Western companies will keep looking for new sources of insecticidal material and so will continue to be interested in these pest control products from natural sources.

Table 2.4. *Plant-derived insecticides available on a commercial basis in 2002*

Compound	Source	Used against
Azadirachtin/ neem extract	*Azadirachta indica* (neem tree)	Variety of arthropod pests on diversity of crops
Canola oil	*Brassica napus* and *B. campestris* (rape plants)	Variety of insect pests primarily as a repellent
Capsaicin	Plants in genus *Capsicum* (chilli pepper)	Variety of insect pests primarily as a repellent
Citronella	Plants in genus *Cymbopopon*	As a repellent against mosquitoes and gnats
Eugenol[a]	Mainly *Laurus* spp. (laurel)	Variety of insect pests in fruit and vegetable crops
Garlic oil	*Allium sativum* (garlic)	As a repellent for a variety of insect pests
Jojoba oil	*Simmonsia californica* (jojoba)	Whiteflies in a range of crops
Maple lactone	Plants in genus *Acer*	As an attractant for cockroaches in traps
Nicotine	Plants in genus *Nicotiana* (tobacco)	Variety of arthropod pests in a diversity of crops
Pyrethrins	Powdered flowers of *Chrysanthemum cinerariaefolium*	Variety of arthropod pests in a diversity of crops
Rotenone	Plants in genus *Derris* and *Lonchocarpus*	Variety of arthropod pests in a diversity of crops
Ryania extract	Plants in genus *Ryania*	Codling moth, corn borer and thrips
Sabadilla powder	From *Sabadilla officinale*	Thrips in citrus and avocado

Note:
[a] Eugenol is the commonly used name, although it has no official status. The approved name for the chemical compound extracted from laurel is 4-allyl-2-methoxyphenol.

Source: Data from Copping, L.G. (ed.) (2001). *Biopesticide Manual*. Farnham, Surrey: BCPC Publications. This list only includes compounds that are actually extracted from plants, i.e. it does not include compounds that are found in plants but that are chemically synthesised for commercial use.

Table 2.5. *Examples of modern-day native uses of plants for pest control*

Region/area	Plant	Use for pest control
Nigeria	*Annona senegalensis* *Luffa aegyptica* *Hyptis spicigera*	Dried leaves of all these plants are mixed with grain to control postharvest pests
Zimbabwe	*Spirostachys africana*	Sap from bark used as a pesticide in granaries
Central America	*Mammea americana*	Leaves wrapped around new plantings to prevent pest attack
Philippines	*Gliricidia sepium*	Twigs inserted into paddy fields to control stemborers
Nepal	*Artemesia vulgaris*	Twigs inserted into paddy fields to control stemborers
South Asia	*Azadirachta indica*	Leaves mixed with grain for postharvest pest control
China	*Tripterygium wilfordii*	Aqueous extracts used against a variety of pests.

Source: Data modified from Ahmed, S. & Stoll, G. (1996). Biopesticides. In: *Biotechnology, Building on Farmers' Knowledge*, ed. J. Bunders, B. Haverkort & W. Heimstra, pp. 52–79. London: MacMillan Education.

Botanical insecticides in developing countries

At present there are an enormous number of plant species that are used in developing countries for the production of pest control formulations. Some estimates put the number of plant species in use as sources of botanical insecticides at over 2400. Despite this, it has been argued that resource-poor farmers in many developing countries could make far more use of these free, naturally occurring products. Table 2.5 lists some of the plants that are used for pest control today. Included in the list is the neem tree, mentioned above.

It has been argued that programmes are needed to familiarise farmers with the pesticidal properties of naturally occurring plants. Plants should be identified that can be used as sources of botanical insecticides and their positive qualities should then be emphasised. For example, Table 2.6 lists plants that can be used for pest control but that also have additional complementary uses. Complementary uses are often critical to convincing farmers to grow such species.

Table 2.6. *Plants with uses complementary to pest control that could be more extensively used on a local level*

Plant	Common name	Family
Acorus calamus	Sweet flag	Araceae
Allium cepa	Onion	Amaryllidaceae
A. sativum	Garlic	Amaryllidaceae
Annona reticulata	Custard apple	Annonaceae
Chrysanthemum cinerariaefolium	Chrysanthemum	Asteraceae
Derris eliptica	Derris	Fabaceae
Lantana camara	Common lantana	Verbenaceae
Mammea americana	Mammey tree	Clusiaceae
Ocimum sanctum	Holy basil	Lamiaceae
Piper nigrum	Black pepper	Piperaceae
Vitex negundo	Indian privet	Verbenaceae
Zingiber officinale	Common ginger	Zingiberaceae

Source: Data modified from Ahmed, S. & Stoll, G. (1996). Biopesticides. In *Biotechnology, Building on Farmers' Knowledge*, ed. J. Bunders, B. Haverkort & W. Hiemstra, pp. 52–79. London: MacMillan Education. All of these plants contain extracts that are toxic to a wide spectrum of pests. They also have a number of other complementary uses, i.e. these plants have other uses than as sources of insecticidal compounds.

Summary

In summary, research and development activities involving botanical insecticides are likely to expand in the twenty-first century. This expansion will come from two directions. First, large agrochemical companies will continue to screen plant-based extracts in their attempts to develop synthetic insecticides from plants that are known to be useful for pest control. Second, developing countries will develop their local exploitation of free, naturally occurring plant-based insecticides. It is likely that some of these extracts will become commercially marketable products, as has occurred with neem. For example, the establishment of Natural Plants, a company in Thailand committed to sustainable rural development, is an excellent model to follow here. This company markets a range of neem-based products for pest control. These insecticides have helped in the development of what is now a flourishing market for organic produce.

The market for botanical insecticides in the new millennium is therefore promising. At present, however, the majority of farmers in both developed and developing countries still rely on synthetic chemical insecticides for pest control. It is these products that are discussed in the next chapter.

Further reading

Arnason, J.T., Philogene, B.J.R. & Morand, P. (eds) (1989). *Insecticides of Plant Origin.* Washington, DC: American Chemical Society.

Benner, J.P. (1993). Pesticidal compounds from higher plants. *Pesticide Science*, **39**, 95–102.

Blunders, J., Haverkort, B. & Hiemstra, W. (eds) (1996). *Biotechnology: Building on Farmers' Knowledge.* London: Macmillan Education.

Casida, J.E. & Quistad, G.B. (1995). *Pyrethrum Flowers: Production, Chemistry, Toxicology, and Uses.* Oxford: Oxford University Press.

Dev, S. & Koul, O. (1997). *Insecticides of Natural Origin.* Amsterdam; Harwood Academic Publishers.

Endersby, N.M. & Morgan, W.C. (1991). Alternatives to synthetic chemicals for use in crucifer crops. *Biological Agriculture and Horticulture*, **8**, 33–52.

Grainge, M. & Ahmed, S. (1988). *Handbook of Plants with Pest Control Properties.* New York: Wiley Interscience.

Hedin, P.A. (ed.) (1997). *Phytochemicals for Pest Control.* Washington, DC: American Chemical Society.

Khambay, B.P.S. & Jewess, P. (2000). The potential of natural naphthoquinones as the basis for a new class of pest control agents – an overview of research at IACR-Rothamsted. *Crop Protection*, **19**, 597–601.

Schmutterer, H. (1990). Properties and potential of natural pesticides from the neem tree. *Annual Review of Entomology*, **35**, 271–97.

Singh, G. & Upadhyay, R.K. (1993). Essential oils – potent sources of natural pesticides. *Journal of Scientific and Industrial Research*, **52**, 676–83.

Stoll, G. (1992). *Natural Crop Protection.* Weikersheim, Germany: Verlag Josef Margraf.

Yang, R.Z. & Tang C.S. (1998). Plants used for pest control in China. *Economic Botany*, **42**, 376–406.

3

Modern synthetic insecticides

Introduction

Modern synthetic insecticides began to be used on a widespread basis in global crop protection following the end of World War II. Almost all of these chemicals belong to one of five chemical groups – organochlorines, organophosphates, pyrethroids, carbamates or neonicotinoids. Discussion of these five chemical groups comprises the majority of this chapter.

Prior to World War II, however, two other groups of organic insecticide had already been developed. These groups are the dinitrophenols and the alkyl thiocyanates. In this chapter these groups are briefly discussed first because examples from both groups are still in use today. The remainder of the chapter is then given over to: (1) discussion of crop losses and pesticide use; and (2) consideration of the five main chemical groups listed above.

Early synthetic insecticides

The first synthetic organic insecticide to be developed and marketed was the dinitrophenol compound potassium dinitro-o-cresylate. This insecticide was marketed in Germany in 1892 as a moth-proofing agent by Friedrich Bayer & Co. Further refinement of the molecule eventually led to the synthesis of dinitro-orthocresol (DNOC, Figure 3.1), an insecticide that is still in use today. This is a nonsystemic insecticide and acaricide that is toxic to moths, scale

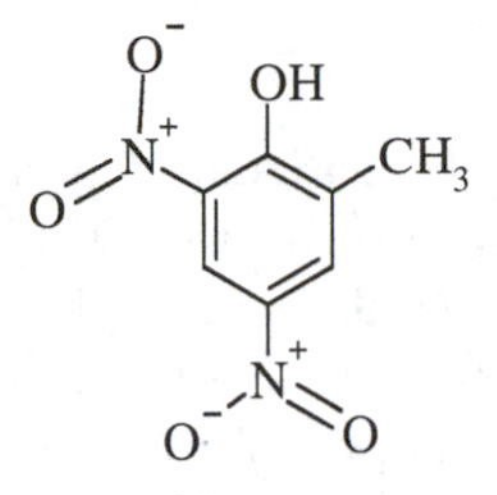

Fig. 3.1. Chemical structure of methyl dinitrophenol (DNOC: dinitro-*o*-cresylate).

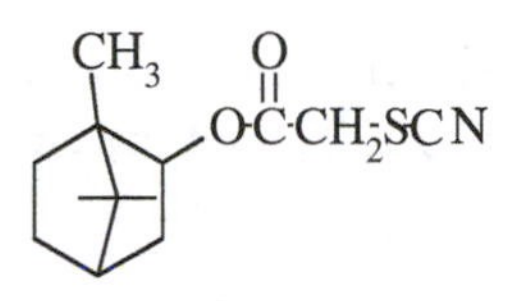

Fig. 3.2. Chemical structure of thanite (isobornyl thiocyanate).

insects and spider mites. Death of target species is caused by an uncoupling of oxidative phosphorylation in mitochondria. Most dinitrophenol compounds are toxic to a wide range of living organisms and they have also been used as herbicides and fungicides. In general, their use has been phased out because a variety of much safer insecticides are now available.

The second group of synthetic organic insecticides to be developed comprises the alkyl thiocyanates (Figure 3.2). These were developed in the 1930s and some are still in use today, although in most parts of the world they have been superseded by far safer compounds. It has been suggested that these chemicals were never fully investigated as insecticides because of the dramatic success achieved by the compounds that replaced them. They have a very rapid knock-down action and are thought to work by liberating the cyanide ion which then prevents respiration in target species.

Overall, the organic insecticides that were in use prior to the 1940s were expensive, difficult to work with and consequently dangerous to apply, often phytotoxic and not very effective by today's standards. The arrival of dichlorodiphenyltrichloroethane (DDT) in 1939 and the organophosphates somewhat later had a dramatic effect upon this situation. For the first time synthetic organic insecticides became available that were cheap, effective and, it was thought, safe.

Modern synthetic insecticides

Modern synthetic insecticides began to be widely used for agricultural pest control following the end of the Second World War. During the war, DDT had been used to protect troops from vectors of disease (malaria and typhus transmitted by mosquitoes, lice and fleas) and research had been undertaken that would eventually lead to the development of the organophosphate insecticides. Although synthetic insecticides are still widely used today (2002) for pest control, the heyday for synthetic insecticide application was to span the 30 years from 1945 to 1975. The end of this heyday is sometimes linked to the publication of Rachel Carson's book *Silent Spring*, which was published in 1962.[1] It was this book that brought to the attention of the general public some of the dangers associated with widespread and indiscriminate pesticide use. The five main insecticidal classes that were developed following the end of the Second World War are as follows:

- organochlorides (e.g. DDT – first commercially produced in 1942);
- organophosphates (e.g. tetraethylpyrophosphate (TEPP) – first commercially produced in 1943);
- pyrethroids (e.g. allethrin – first commercially produced in 1950);
- carbamates (e.g. carbaryl – first commercially produced in 1956);
- neonicotinoids (e.g. imidacloprid – first commercially produced in 1991).

All of these classes of insecticide (discussed individually below) have brought tremendous benefits to humanity in the control of vectors of disease and in the control of pests of crops and livestock. There is no reason to suppose that such benefits will not continue to be realised in the future.

The synthetic insecticide market

Over the period 1940–2000 there were a number of distinctive trends within the global synthetic insecticide market. These trends include: a decrease in the dose rate at which insecticides are used; a decrease in the rate of discovery of new insecticides; an increase in the costs associated with developing a new

[1] It should be noted that insecticide use globally increased even after the publication of Carson's book. The heyday is therefore identified not based upon sales but rather upon the time period during which pesticides were viewed as a complete solution to arthropod pest problems. From the mid-1960s onwards practitioners of pest control began to realise that there were a number of alternatives to synthetic pesticides (see also Chapters 1 and 12).

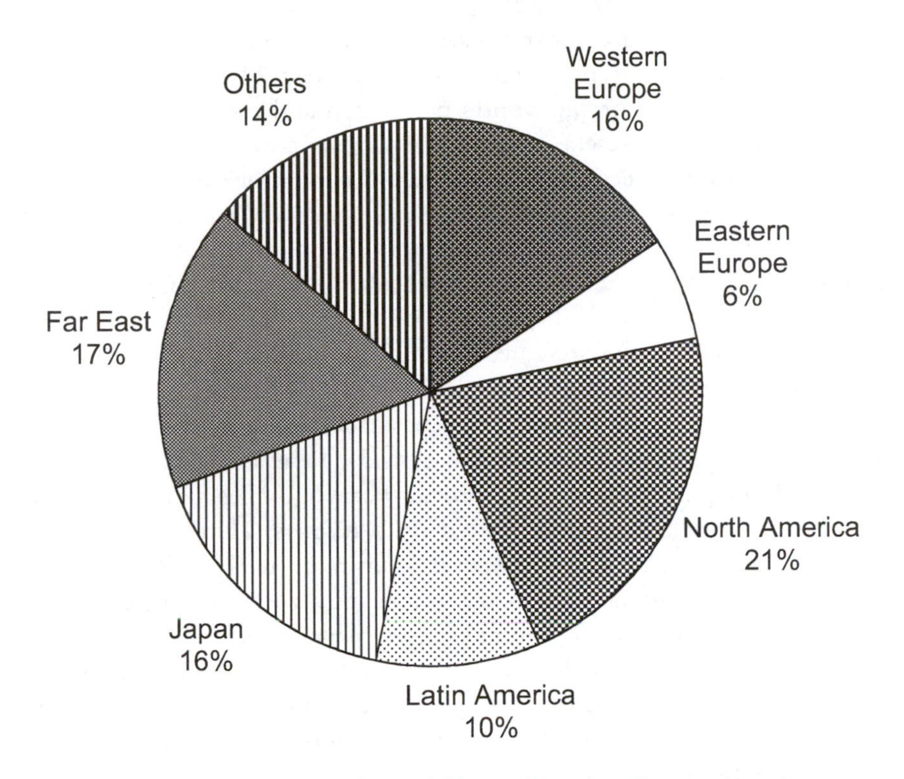

Fig. 3.3. Global breakdown of insecticide use by region. Data are based on regional market value in 1994.

insecticide; and an increase in the volume of data required to register a new insecticide. These trends were discussed in detail in Chapter 1.

Since the rapid expansion of the global synthetic insecticide market following the end of the Second World War, the use of these products has increased enormously. In 1998 total world consumption of pesticides was valued at *c.* £25–30 billion (insecticide use was valued at *c.* £10 billion). The global increase in insecticide use however has not been even. The greatest use of insecticides currently takes place in North America, Europe and Japan, which together use over 50% of the world's insecticides on only 25% of global cropland. By contrast, developing countries use about 20% of the world's insecticides (by value) on 55% of the world's cropland.[2] A regional breakdown for insecticide use is given in Figure 3.3. Given these data it is easy

[2] Because many developing countries use older and cheaper products, the actual differential may be greater than that presented here. A greater differential would only further strengthen the argument that is made.

to see why agrochemical companies expect to achieve expanded sales of their products within developing countries over the next 20 years. Developing countries house most of the world's population and contain most of the world's cropland. At present, however, developing countries also make the fewest insecticide applications. To illustrate this point we can contrast the Netherlands, which in the early 1990s was the world's highest user (by density) of insecticides, with an average of 21 kg/ha per year, with countries in sub-Saharan Africa where average insecticide use was less than 1 kg/ha per year.

In addition to the data presented above there are also a number of other reasons to expect an increase in insecticide use across Africa, Asia and South America. First, many governments view pesticides as the most viable way to achieve increased yields. Second, and as a result of the above, many governments in developing countries often provide heavy subsidies for pesticides. In many cases, pesticides are given away free to farmers. Third, agrochemical companies spend a great deal of money demonstrating the benefits associated with pesticide use to farmers in developing countries. Fourth, extension workers[3] in many developing countries often promote the use of pesticides because they are easy to apply, may produce immediate gains and are simple to explain. Fifth and last, alternative approaches to pesticide use, such as the development of integrated pest control programmes (discussed in Chapter 12) are often inadequately resourced and can sometimes appear too complex to get across to farmers. For example, Indonesia is a country that has been heavily involved in the development of integrated approaches towards pest control on crops and yet imports of chemical insecticides to this country continue to rise year on year.

The upshot then is that most of the growth in the chemically based global crop protection market that will take place in the next 20 years will be in developing countries. Whether this improves the food security of these countries remains to be seen. Certainly the experience in developed countries is that using pesticides is economic for farmers and can lead to increased yields. However, the damage caused by crop pests in these countries has, paradoxically, not declined. This paradox is explained in more detail in the next section.

[3] In many developing countries the Ministry of Agriculture operate systems that provide free sources of advice (and often free chemicals) to farmers. The individuals involved in these systems are referred to as extension workers. In essence, these workers will visit farms and provide free technical advice on crop protection and agronomy. These systems are often modelled on the extension service that is run by the US Department of Agriculture (USDA) in the USA. In the UK such individuals are often called crop consultants and their services have to be paid for!

Table 3.1. *Crop loss due to rice pests in various countries*

Pest	Country	Estimated losses
Stemborers	Bangladesh	30–70%
	India	3–95%
	Indonesia	Up to 95%
	Malaysia	33%
	Philippines	c. 7%
Leafhoppers and planthoppers	Bangladesh	50–80%
	India	1–33%
Rice bugs and gall midges	India	10–35%
	Vietnam	50–100%

Source: Data adapted from Yudelman, M., Ratta, A. & Nygaard, D. (1998). *Pest Management and Food Production*, p. 53. Washington, DC: International Food Policy Research Institute.

Global crop losses and insecticide use

It has been estimated that global pre- and postharvest crop losses vary between average levels of 10–30% in developed countries and 60–70% in developing countries. These losses occur despite the widespread use of agrochemicals and many researchers and analysts (especially those working for agrochemical companies!) suggest that these figures might double if agrochemicals were not used at all.

Analyses of actual crop loss data are always fraught with problems concerning the reliability of the data collected and with the associated data interpretation. For example, it is common for pest infestations to coincide with climatic changes. Both of these factors can cause yield changes and attributing crop losses to pests alone can therefore be misleading.

To emphasise this point, Table 3.1 gives data collected on losses caused by crop pests to rice crops around the world. Rice is mostly grown in developing countries and more pesticides are used on rice than any other food crop.[4] Since rice is a staple for about 2 billion people, crop losses will be critical in determining their food security. The table shows that estimated annual crop losses varied between 3% and 95%, that losses varied between countries and that losses varied between pests.

[4] When all crops are considered, cotton crops are the most heavily sprayed with insecticide worldwide. Some estimates have indicated that 25% of all insecticides (by value) are used on cotton crops.

Table 3.2. *Estimated percentage crop losses for eight major crops by pest and by region 1988–90*

	Pathogens	Insects	Weeds	Total
Africa	16	17	16	49
North America	10	10	11	31
Latin America	14	15	13	42
Asia	14	19	14	47
Europe	10	10	8	28
Former Soviet Union	15	13	13	41
Oceania	15	12	10	37

Source: Data taken from Oerke, E.C., Dehne, H.W., Schohnbeck, F. & Weber, A. (1995). *Crop Production and Crop Protection: Estimated Losses in Major Food and Cash Crops*. Amsterdam: Elsevier.

Overall, most researchers have concluded that a global average for total preharvest crop losses comes in at around 40% of potential production. Losses are attributed almost equally to pests, diseases and weeds. Another 10% of losses occurs postharvest so that the best global average would be somewhere around 50% of potential production. These losses are not divided equally among countries though. Developed countries typically have preharvest losses calculated at 25–30% of potential production while developing countries have losses of 40–50%. Table 3.2 gives summary data on calculated crop losses for eight major crops (barley, coffee, corn, cotton, potatoes, rice, soybeans and wheat) by region over the time period 1988–90. The table shows that crop losses are distributed almost equally between insect pests, pathogens and weeds, and that the greatest overall losses occur in developing countries.

What is most interesting though is that historical analyses of crop losses over the last 50 years show that the proportion of potential production lost to pests has increased. For example, analyses over the period 1940–90 indicate that total crop losses have increased by 30–37%, with damage caused by insects alone rising by 6–13%. And so we have the curious situation of rising pesticide use, rising crop losses and, most importantly, rising crop harvests. This situation has been referred to as the paradox of increased pesticide use.

It is this paradox that is really the key to understanding what is happening in global crop protection and it also explains why almost every economic analysis of pesticide use that is undertaken demonstrates that it is economically sensible to use pesticides for crop protection. The solution to the paradox of increased pesticide use and increased crop losses lies with plant breeding and intensive fertiliser use. Average yields for all of the major crops grown worldwide have increased as a result of research efforts by plant breeders and as a result of increased applications of synthetic fertilisers. Many of these high-yielding, nutrient-rich and attractive plant varieties are also highly attractive to crop pests. High-yielding varieties intensively treated with pesticides therefore suffer greater crop losses than low-yielding varieties that are treated with less pesticide. It is because of the differential in potential gross yield between these varieties that the greatest net gains are realised with high-yielding crops. To illustrate this, consider upland rice in the Philippines. Native varieties of rice suffer pest losses of around 2% while high-yielding modern varieties suffer losses in the region of 24%. However, native rice varieties produce a harvest of around 200 kg/ha while high-yielding varieties produce harvests of over 1000 kg/ha. The net gain is still greatest for the high-yielding rice with the associated heavier crop losses. The development of high-yielding crop varieties that are intensively treated with fertilisers explains why it is economically sensible to use these varieties knowing that they will suffer proportionately greater crop losses. It also explains why it is economically sensible to use pesticides on these crops in attempts to minimise losses. These data however only really apply in the short term.

In the long term it may be the case that the overall rise in pest species problems may ultimately make crop protection less and less sustainable. These issues are discussed more fully in Chapter 12. In the short term, though, it would seem that increasing pesticide use, increasing crop yields and increasing pest problems are the most economically sensible option. Just about every study that has been carried out has shown that using pesticides is economically beneficial for farmers, at least in the short term. For example, in 1960 it was estimated that for every $1 spent on pesticides in the USA, the return was *c.* $2.8. By 1997, the average return had been upgraded to $4 for every $1 spent. The trend towards increasing use of pesticides clearly makes economic sense. In the next five sections the chemistry and mode of action of the most commonly used insecticides worldwide are discussed.

Organochlorine insecticides

The organochlorines or chlorinated hydrocarbons were the first widely used synthetic organic insecticides. Chemically, all of the insecticides in this group contain carbon, hydrogen and chlorine atoms. They may also contain atoms of oxygen and sulphur. However, it is the stable covalent bonds that are formed between carbon, hydrogen and chlorine atoms that give compounds in this

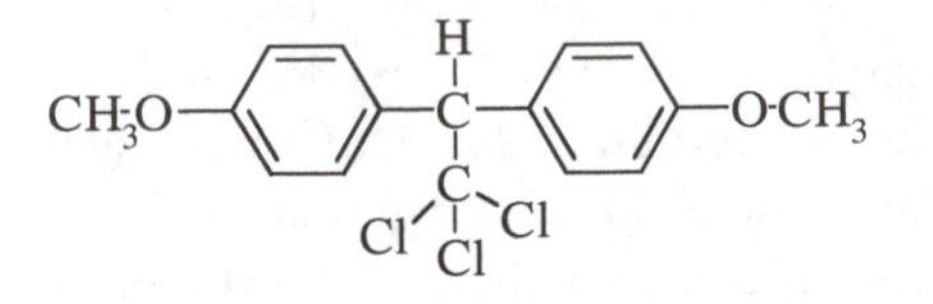

Fig. 3.4. Chemical structure of methoxychlor (the methoxy groups that replace the chlorine atoms that are found in dichlorodiphenyltrichloroethane (DDT) make this insecticide far less persistent in the environment).

group of insecticides their high degree of environmental persistence. For example, the half-life of DDT is measured in terms of years. For many pest control situations pesticide persistence is regarded as highly advantageous.

There are three main groups of organochlorine insecticide. These are:

- the DDT group;
- the hexachlorohexane (HCH) group;
- the cyclodiene group.

The DDT group

DDT is probably the best-known insecticide in the world. The full name of this chemical is dichlorodiphenyltrichloroethane (DDT). DDT was first prepared as a chemical in 1874 by a German graduate student called Othmar Zeidler. However, its powerful insecticidal properties were not discovered until 1939 when Paul Müller, a Swiss chemist working for the Geigy Corporation, came across it as part of a screening programme for new insecticides. Paul Müller was awarded the Nobel Prize for his work on DDT in 1948. In the years between 1939 and 1945 DDT was extensively used, with great success, in the control of vectors of diseases such as malaria and typhus (mosquitoes and lice). After the war, DDT became widely used for the control of numerous agricultural pests.

The main advantages associated with the use of DDT were its stability, persistence of insecticidal toxicity, low cost of manufacture and its moderate mammalian toxicity. Other insecticides in the DDT group include dicofol and methoxychlor. The chemical structure of methoxychlor is shown in Figure 3.4. This compound is not so fat-soluble as DDT and has therefore found more widespread acceptance. DDT was banned for use as an insecticide in the USA

Table 3.3. *Biomagnification of dichlorodiphenyltrichloroethane (DDT)*[a]

Species group	DDT concentration	Concentration factor[a]
Water	0.000003 ppm	Not applicable
Zooplankton	0.04 ppm	c. 13 000
Small fish	0.5 ppm	c. 17 000
Large fish	2 ppm	c. 700 000
Fish-eating birds	25 ppm	c. 8 million

Notes:
[a] Concentration from amount in water.
Source: Data collated from measurements made at Long Island Sound in the USA.

in 1973 and in the UK in 1984. It has also been banned for use in a number of other countries. The primary problem is its chemical persistence. It is a very fat-soluble compound (very lipophilic) and accumulates in the body fat of animals that consume it. This leads to biomagnification (bioaccumulation or bioconcentration) of the pesticide within ecosystems. As a result of this process, organisms that are at the top of food chains can accumulate levels of DDT that are toxic. An example of biomagnification is given in Table 3.3.

In addition to biomagnification, DDT also appears to affect calcium metabolism in many predatory birds. This can lead to egg shell-thinning and thus reproductive failures for species such as falcons, eagles and ospreys. The American national bird, the bald eagle, was particularly hard-hit by exposure to DDT and in 1967 it was officially classified as endangered in all states south of the 40th parallel.[5] Despite being banned in many countries DDT is still widely used for the control of vectors of disease (particularly mosquitoes) in many developing countries. For example, DDT was recently excluded from an international list of chemicals that should not be used worldwide because of its utility in suppressing mosquitoes that act as a vector for malaria. DDT is currently (2002) used extensively in South America and Africa for mosquito control.

Mode of action of DDT DDT, like many modern synthetic pesticides, is a neurotoxic molecule. It is toxic to target species' nervous systems. Symptoms of poisoning include tremors, loss of movement, convulsions and death. It

[5] It was not until July 2000 that the US Fish and Wildlife Service were able to declare this species fully recovered – 27 years after DDT was banned from use in the USA!

appears to exert its toxic effects by upsetting the flow of ions across axonal membranes. More specifically, DDT appears to act on sodium channels within axonal membranes, disrupting the normal flow of sodium ions. This is similar to the mode of action of many pyrethroid insecticides (see later). That said, the exact mode of action of DDT has never been clearly established.

It has been noted that resistance to DDT does not necessarily lead to resistance to other organochlorine compounds, presumably because of the differences that exist between their modes of action (see later). It displays a negative toxicity correlation with temperature (like pyrethroids), possibly caused by lower levels of dissociation (or greater blocking of sodium channels on axons) at lower temperatures. Biologically, DDT acts as a contact and feeding poison; it is not systemic. It has a low vapour pressure and so does not act as a respiratory poison. Penetration through the insect cuticle occurs slowly, so insecticidal action is also relatively slow. It is generally agreed that DDT has been responsible for saving millions of lives worldwide (through vector control) and that the molecule has had an undeserved and disproportionate amount of negative popular publicity.[6]

The HCH group

The insecticidal properties of HCH[7] were discovered by British and French scientists in 1942. Chemically, there are five isomers of this compound, although only the gamma isomer is toxic to arthropod pests. This is the isomer that is manufactured and sold for pest control. The common name of the gamma isomer is lindane, taken from the name of Van der Linden, who first isolated the pure compound in 1912. Its chemical structure is shown in Figure 3.5. Lindane has greater volatility and solubility that DDT. This means that it can act as a contact, feeding and respiratory poison. The higher volatility, though, means that its persistence is reduced in comparison to DDT. It is very heat-stable and can also be used as a fumigant. Superficially at least, its mode of action appears similar to that of DDT in that it causes tremors, paralysis and death in treated species. These behavioural changes occur much more rapidly than those observed following exposure to DDT, but are not caused by the disruption of sodium channel activity. Rather, it appears that lindane (and cyclodiene organochlorines, see later) exerts its toxic effects by

[6] Paul Müller was awarded the Nobel Prize for Medicine in 1948 in recognition of his life-saving discovery of the toxic properties of DDT.

[7] HCH is also often referred to as benzenehexachloride (BHC).

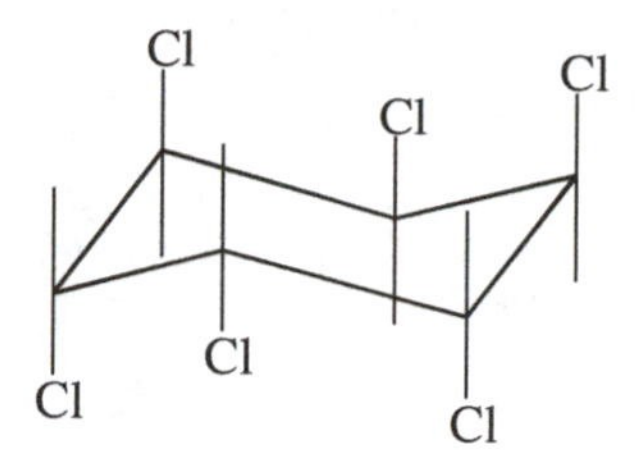

Fig. 3.5. Chemical structure of lindane (hexachlorohexane).

antagonism of the inhibitory neurotransmitter gamma-aminobutyric acid (GABA). This neurotransmitter hyperpolarises postsynaptic membranes by binding to receptors containing chloride channels, i.e. when GABA binds to postsynaptic receptors, chloride channels open. Lindane binds to these chloride channels and so blocks their activation by GABA. This results in hyperexcitation of nerve cells within the central nervous system (CNS) and therefore the behavioural changes observed. Unlike DDT, lindane is still widely used as an insecticide in agriculture and in the control of both public health pests and animal ectoparasites, particularly in many developing countries. In Europe, the use of lindane for pest control has largely been phased out.

The cyclodiene group

The insecticidal acitivities of this group were discovered in 1945. The group includes aldrin, dieldrin, endrin and endosulfan, amongst others. These are all relatively persistent chemicals that have been widely used to control soil pests and termites. For example, wood treated with aldrin or dieldrin during the 1950s for termite control is still free from damage today (2002). They all have higher mammalian toxicities than DDT and care should be taken when handling them. They are prepared using the Diels–Alder reaction. Their mode of action appears similar to that of lindane (see earlier), i.e. they bind to chloride channels on postsynaptic receptors with the CNS and so act as antagonists to GABA. However, unlike lindane and DDT they show a positive temperature correlation, i.e. they get more toxic with increases in ambient temperature. Their exact mode of action is still not fully understood.

Examples of these compounds are given in Figure 3.6. Like DDT, most of these insecticides are no longer used. This is because of their high mammalian toxicity, their persistence (and so problems with residues in food), and

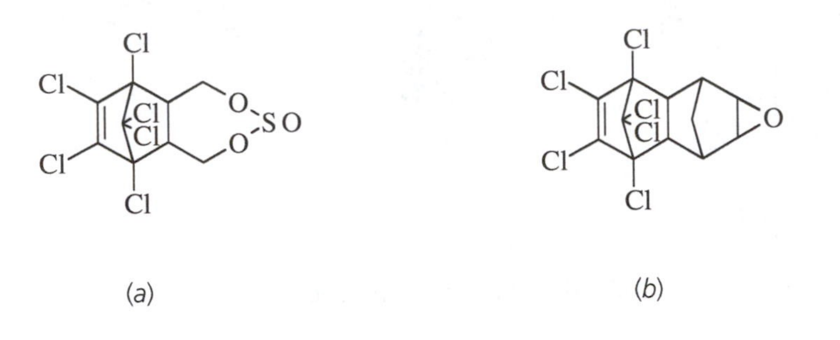

Fig. 3.6. Chemical structures of (*a*) endosulfan and (*b*) dieldrin.

because of the development of pest species resistance. The Environmental Protection Agency (EPA) in the USA cancelled the registrations for agricultural use of cyclodienes between 1975 and 1980. However, one cyclodiene that is still widely used is the insecticide endosulfan. This insecticide is used for the control of crop and public health pests, particularly in many developing countries.

Organophosphate insecticides

The organophorus insecticides were discovered as a result of research carried out during World War II that was concerned with developing compounds as potential nerve gases (chemical warfare agents). In Germany, Schrader synthesised the highly toxic nerve gases tabun and sarin (Figures 3.7 and 3.8), which, although highly insecticidal, were also extremely toxic to mammalian species. The first recognised organophosphorus insecticide was called schradan. However, even this compound had a high mammalian toxicity and it has now been replaced by less toxic compounds. As an insecticidal group the organophosphates are still the most toxic of all pesticides to vertebrates.

The organophosphates are all derived from phosphoric acid. Unlike the organochlorines, they break down rapidly (days) and some are systemic (they are translocated within a plant's vascular system). Systemic compounds have the advantage that the insecticide is redistributed within the plant and can reach parts not directly exposed to the active ingredient, such as roots and new foliage. Systemic compounds are also excellent for controlling small hemipterous sucking pests such as aphids.

```
CH3  O
   \  ||
    N–P–CN
   /  |
CH3  O–CH2CH3
```

Fig. 3.7. Chemical structure of tabun.

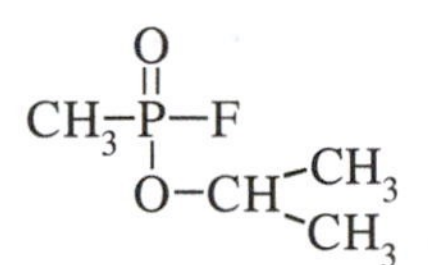

Fig. 3.8. Chemical structure of sarin.

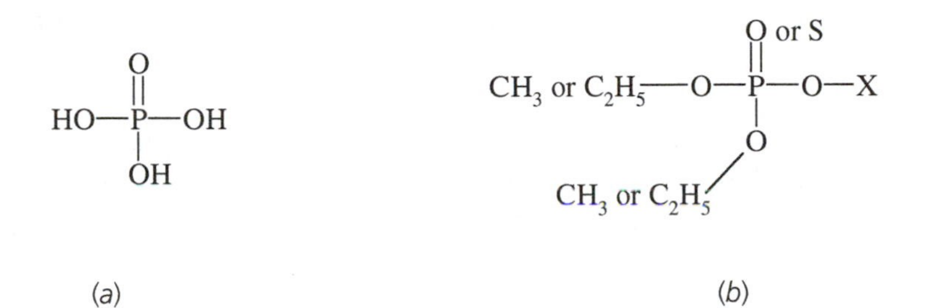

Fig. 3.9. Chemical structures of (*a*) phosphoric acid and (*b*) a generalised organophosphate, in which 'X' represents the leaving group.

The generalised structure for all organophosphorus insecticides is given in Figure 3.9. The leaving group or 'X' in Figure 3.9 is the chemical group that splits from the phosphate group after the insecticide binds at its site of action. This is discussed in more detail below. However, it is the manipulation of the leaving group that has been responsible for the synthesis of over 100 different active ingredients. These active ingredients are available in thousands of different formulations, making the organophosphates by far the most successful group of insecticides ever to be developed. We can distinguish three groups of organophosphorus compound: aliphatic (straight-chain) compounds, phenyl (ring) compounds and heterocyclic (modified ring) compounds. Each group will be discussed in turn.

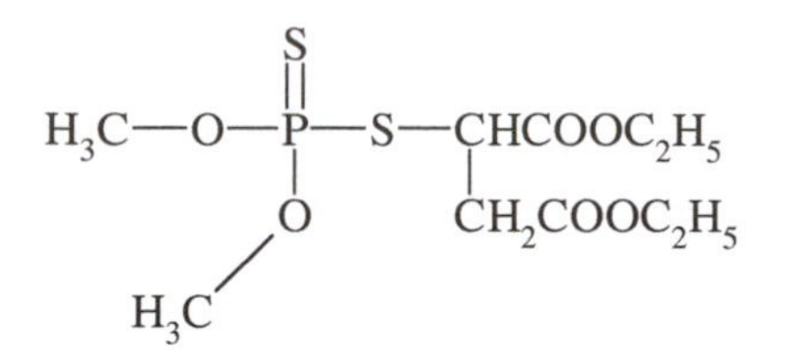

Fig. 3.10. Chemical structure of malathion. In insects, the thion sulphur atom is replaced by oxygen which produces the compound malaoxon, an organophosphate with much higher toxicity.

Aliphatic compounds

Aliphatic compounds are those with straight carbon chains. The very first organophosphate to be used in agriculture (TEPP) belongs to this group.[8] The most widely used aliphatic organophosphate is malathion. Malathion is primarily a contact insecticide that is used in agriculture and in the control of medical and veterinary pests.

Another fairly widely used contact aliphatic organophosphate is dichlorvos. The aliphatic group also includes a number of systemic compounds, for example, acephate, dimethoate, disulfoton, phorate, methamidophos and mevinphos. The chemical structure of malathion is shown in Figure 3.10.

Phenyl compounds

These organosphophates, as the name suggests, have a phenyl ring in their structure. One of the oldest and most widely used compounds in this group is parathion. This was first introduced in the ethyl form but the use of this compound has now diminished in preference to the less toxic methyl form. Parathion is a contact, feeding and respiratory poison with limited systemic activity in plants. It has been used against a wide variety of pests. The chemical structure of methyl parathion is shown in Figure 3.11. Other phenyl organophosphates include fenitrothion and fenthion. Fenthion is used as an animal systemic for protection against grubs while fenitrothion is a non-systemic broad-spectrum insecticide with a range of uses in crop protection.

[8] TEPP (tetraethylpyrophosphate) is no longer manufactured or registered for use as an insecticide.

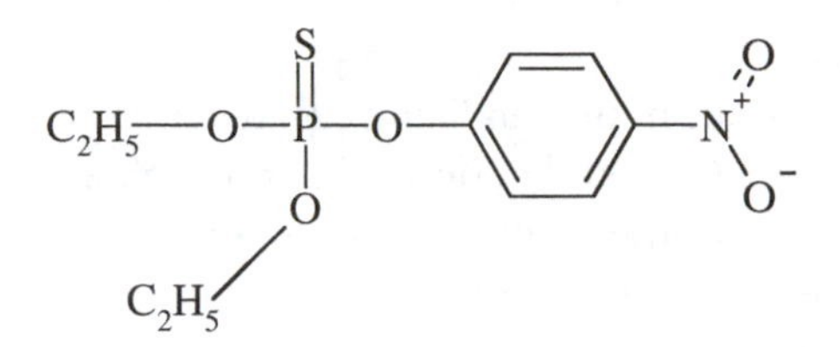

Fig. 3.11. Chemical structure of methyl parathion.

For example, fenitrothion has been fairly widely used to control the desert locust *Schistocerca gregaria* and in Oman fenitrothion has been used to control the hemipterous pest of date palm, the dubas bug *Ommatissus lybicus*.

In general, the stability of phenyl compounds is greater than that of the aliphatics so that they last longer in the environment. This stability is still measured in days rather than weeks or months, as is the case with organochlorine insecticides.

Heterocyclic compounds

These compounds, like phenyl organophosphates, have ring structures. However, one or more of the carbon atoms within the ring is displaced by O, N or S. Rings in this group may have three, five or six atoms. Some of the most widely used compounds in this group are diazinon, azinphosmethyl and chlorpyrifos. These have all been used against a wide range of insect pests. All of these are contact and feeding insecticides with little systemic action in plants.

Mode of action of organophosphates

Organophosphates, like organochlorines, are neurotoxic. The symptoms of poisoning by organochlorines and organophosphates are similar and include tremors, convulsions and paralysis prior to death (see also earlier). However, the mode of action of organophosphates is different from that of the organochlorines. While organochlorine insecticides disrupt ion movement across axonal cell membranes (e.g. DDT at sodium channels) or at postsynaptic cell membranes (e.g. cyclodienes or HCH at chloride channels), organophosphates interfere with normal nerve cell activity by inhibiting the activity of the enzyme responsible for the breakdown of the neurotransmitter acetylcholine. This is explained in more detail below.

One of the most widespread neurotransmitters in living organisms is acetylcholine. This chemical is found in the CNS and peripheral nervous system of vertebrates and invertebrates. The neurotransmitter mediates communication between nerve cells and between nerve and muscle cells at synapses. Once the neurotransmitter has completed its function it is broken down into its constituent molecules (acetic acid and choline) by the enzyme acetylcholinesterase. The normal process in which the enzyme combines with the neurotransmitter is called acetylation.

Organophosphates interfere with this process by inhibiting the enzyme. In essence, the enzyme, which would normally be acetylated prior to the release of choline and acetic acid, is phosphorylated by binding with the insecticide. This occurs because organophosphate molecules are structurally similar to acetylcholine, at least as far as enzymatic binding goes. During phosphorylation the enzyme is blocked for about 1 million times longer than it would be under normal circumstances such as acetylation. The enzyme is not destroyed – it is just inhibited. The length of time it takes for phosphorylation of the enzyme to happen depends greatly on the rate of dissociation of the leaving group 'X' (see earlier) from the phosphate group. The leaving group of organophosphate molecules is analogous to the choline molecule that is released by the enzyme following acetylation. The rate of dissociation of the leaving group (and hence the time it takes for phosphorylation to occur) in turn depends to a large extent upon the electron-attracting ability of the leaving group (its electrophilicity). Leaving groups with a (comparatively) high electrophilicity are more stable and so are quicker to form. This can mean that the enzyme is phosphorylated faster, i.e. the toxicant works quickly. Variation in the chemical structure of leaving groups of organophosphates can also have an impact upon the toxicity of the insecticide. Some of the organophosphates with the most stable leaving groups are also the most toxic.[9]

The result of the enzyme inhibition is a build-up of the neurotransmitter within synapses, which is why poisoned organisms display the symptoms mentioned above. The mode of action is exactly the same in both insects and mammals as both of these groups utilise the same enzyme to break down acetylcholine. However, in insects the effects of organophosphates are largely

[9] The toxicity of different organophosphates is determined not just by the electrophilicity of the leaving group. For example, structural features of molecules will be important. The actual formulation used will also impact upon product toxicity, most notably if synergists are present.

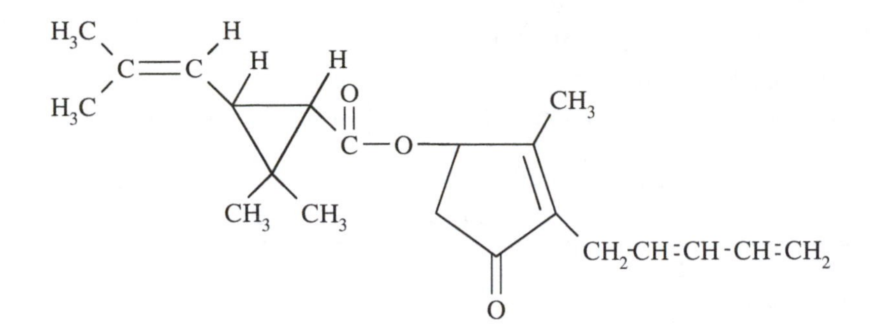

Fig. 3.12. Chemical structure of pyrethrin I (natural pyrethroid).

confined to the CNS simply because this is where most of the cholinergic receptors are. When humans are poisoned with organophosphates they are usually given an injection of the acetylcholine inhibitor atropine. In addition to the known neurotoxic effects of organophosphates, there is also some evidence that these insecticides may upset water regulation in insects.

Pyrethroid insecticides

The pyrethroids (pyrethrum-like insecticides) include some of the most recently developed insecticides (the most recently developed group comprises the neonicotinoids, see later). The very first pyrethroid to be developed was allethrin in 1949. This insecticide is simply a synthetic version of pyrethrin I with a couple of carbon atoms removed. Compare the two structures given in Figures 3.12 and 3.13. Since 1949, approximately 30 different active ingredients have been developed and marketed. These insecticides are all contact and feeding poisons; they have no systemic activity. They have been successfully used to control a wide range of pests of agricultural and medical significance.

Synthetic pyrethroids offer a number of advantages for pest control over natural pyrethrins. First, natural pyrethrins are not photostable, while many synthetic pyrethroids are. Second, natural pyrethrins are relatively expensive while synthetic pyrethroids are comparatively cheap. Third, synthetic pyrethroids are more toxic to insects than natural pyrethrins and they can be applied at very low dose rates. For example, the synthetic pyrethroid deltamethrin can be applied at dose rates as low as 2.5 g/ha. Similarities between

Table 3.4. *World Heath Organization (WHO) and US Environmental Protection Agency (EPA) toxicity classification categories for pesticides based on the amount of active ingredient needed (mg/kg of body weight) to kill 50% of a group of rats that feed on the toxicant, i.e. the oral LD_{50}*[a]

WHO class	Rat (oral) LD_{50}	EPA class	Rat (oral) LD_{50}
Ia: Extremely hazardous	<5	I	<50
Ib: Highly hazardous	5–50	II	51–500
II: Moderately hazardous	51–500	III	501–5000
III: Slightly hazardous	>501	IV	>5001

Source: [a] Data collated from Tomlin, C. (ed.) (2000). *The Pesticide Manual*, 12th edn. Farnham, Surrey: BCPC Publications.

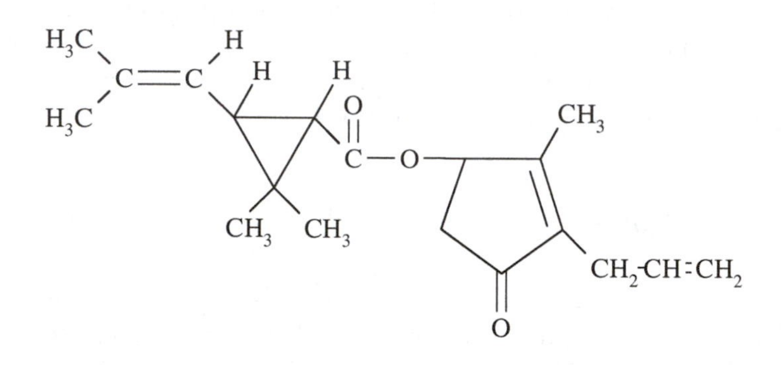

Fig. 3.13. Chemical structure of allethrin (first synthetic pyrethroid).

synthetic and natural pyrethroids include relatively high mammalian LD_{50}[10] and a lack of selectivity. Many pyrethroids are toxic to beneficial species, as well as pest species. Throughout the development of the synthetic pyrethroids there has been a general trend to increase their environmental stability and to decrease the rate at which they are applied (i.e. to increase their intrinsic toxicity). The trend to increase the environmental stability of this group of insecticides has been made possible because of chemical changes at both the alcohol and acid end of the molecule.[11]

[10] The dose required to kill 50% of species in a toxicological assay. Pesticides are often compared toxicologically in terms of their LD_{50}. Table 3.4 gives the current World Health Organization and US Environmental Protection Agency classification categories for pesticides with different toxicities.

[11] Chemically the pyrethroids are composed of an acid containing a three-carbon ring linked to an alcohol.

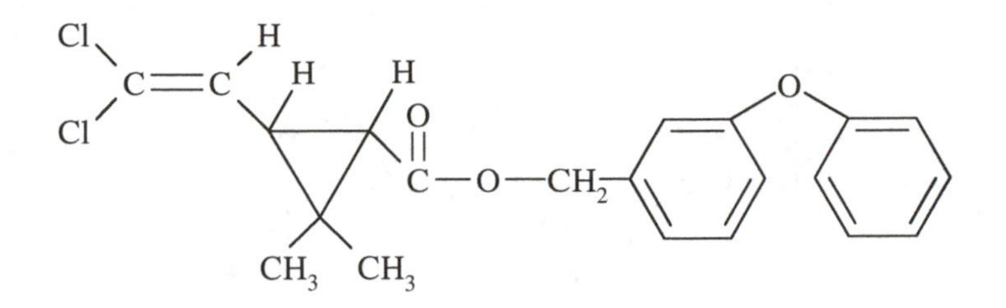

Fig. 3.14 Chemical structure of permethrin.

At the alcohol end the changes have largely comprised modifications to the aliphatic (straight-chain group) group such that sites of autooxidative attack (unsaturated sites) are removed and more stable groups (e.g. benzene rings) have been added. At the acid end of the molecule the changes have mostly comprised the removal of methyl groups and the addition of more stable halide molecules. These changes were seen in the development of allethrin from pyrethrin and can also be seen in the development of permethrin, which was introduced for use in 1973 (Figure 3.14). In permethrin the aliphatic side group of pyrethrin has been replaced by benzene rings and chlorine atoms have been added at the acid end of the molecule.

These changes meant that the insecticide could be applied at rates as low as 50 g/ha and that the environmental stability of the molecule was such that permethrin was the first pyrethroid that could be used as an effective seed treatment. One of the next pyrethroids to be introduced (in 1976) took these changes even further. This chemical was called deltamethrin. In deltamethrin bromine replaces the chlorine atoms and a cyano group has been added in the centre of the molecule. These changes have meant that this pyrethroid can be used at dose rates of 10–20 g/ha. The chemical structure of deltamethrin is shown in Figure 3.15. Finally, one of the most modern pyrethroid insecticides to be developed is tefluthrin. This chemical compound hardly resembles the original pyrethrin. It has been changed so much that it now has an environmental stability that is measured in months. The chemical structure of tefluthrin is shown in Figure 3.16. In summary, throughout the development of the synthetic pyrethroids there has been a trend to increase their environmental stability and a trend to decrease the rate at which these insecticides can be applied. These points are summarized in Table 3.5.

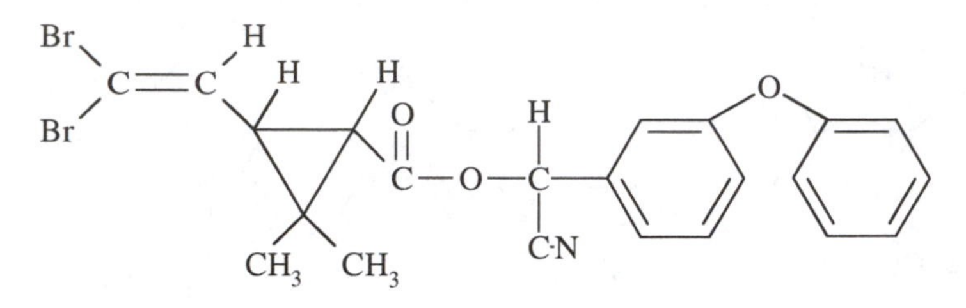

Fig. 3.15. Chemical structure of deltamethrin.

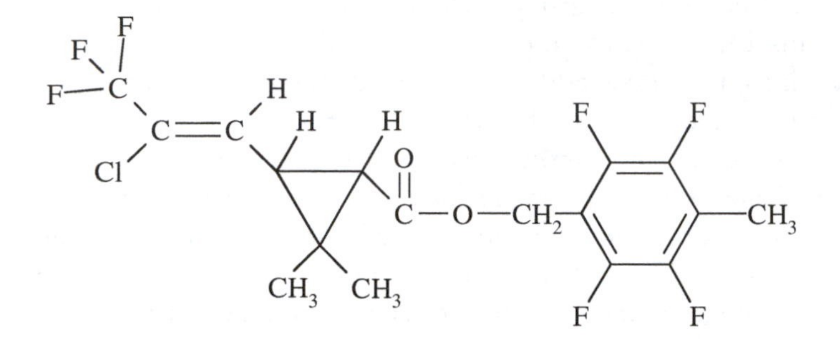

Fig. 3.16. Chemical structure of tefluthrin.

Mode of action of pyrethroids

All of the pyrethroids work as contact and feeding insecticides. Pyrethroids do not readily penetrate plants and have no systemic activity. They are neurotoxic and function toxicologically by interfering with axonal transmission of nerve impulses, like DDT. Because pyrethroids have a rapid knock-down effect on target species it is thought that their toxic effects are first manifest in the peripheral nervous system of poisoned species. Insects exposed to pyrethroids may display symptoms of poisoning (tremors and/or hyperexcitability and/or incoordination) within minutes – a process that is called knock-down. Thereafter, it is likely that they also poison target species' CNSs. It is known that sodium and potassium ion movements are altered in poisoned nerves. In particular, it is known that pyrethroid insecticides disrupt the flow of sodium ions by interfering with sodium channels (like DDT). It

Table 3.5. *Toxicity, recommended application rate and stability for some synthetic pyrethroids*

Chemical	Housefly Toxicity	Dose rate	Stability[a]
Pyrethrin	0.3 μg/kg	NA	Hours
Allethrin (1949)	0.1 μg/kg	NA	1–2 days
Permethrin (1973)	NA	50–200 g/ha	3–5 days
Deltamethrin (1976)	0.003 μg/kg	10–20 mg/ha	2–7 days
Tefluthrin (1980s)	NA	10–150 mg/ha	1–3 months

Notes: Data collated from Cremlyn, R.J. (1991). *Agrochemicals: Their Preparation and Mode of Action*. Chichester: John Wiley.
[a] Average stability in terms of time to molecular degradation and inactivation. NA, data not available.

is now known that there are two types of pyrethroid, referred to as type I and type II. Both types affect the movement of sodium ions. However, the former modify sodium currents for tens or hundreds of milliseconds while the latter modify currents for minutes or longer. In poisoned species, type I pyrethroids produce tremors and/or hyperexcitability while type II pyrethroids produce incoordination. Type I pyrethroids become less toxic as the ambient temperature increases (like DDT) while type II pyrethroids become more toxic with temperature (like cyclodienes). Examples of type I pyrethroids are allethrin and tetramethrin while examples of type II pyrethroids are fenvalerate and fluvalinate.

Carbamate insecticides

Carbamate insecticides were first developed by the Geigy Corporation in 1951 but were not made commercially available until 1956. Research into carbamates began following a search for compounds other than organophosphates that had anticholinesterase activity. One of the known compounds in this group was the alkaloid physostigmine (Figure 3.17), found in the calabar bean (*Physostigma venenosum*), which had been used for trial by ordeal in its native West Africa.[12] This compound is highly water-soluble, but its

[12] Native to West Africa, but introduced elsewhere, the seeds (beans) of this leguminous plant have been used in what is called trial by ordeal. Accused individuals were given a mixture of pounded seeds infused with water. If the accused vomited, and survived, he or she was pronounced innocent. If the accused

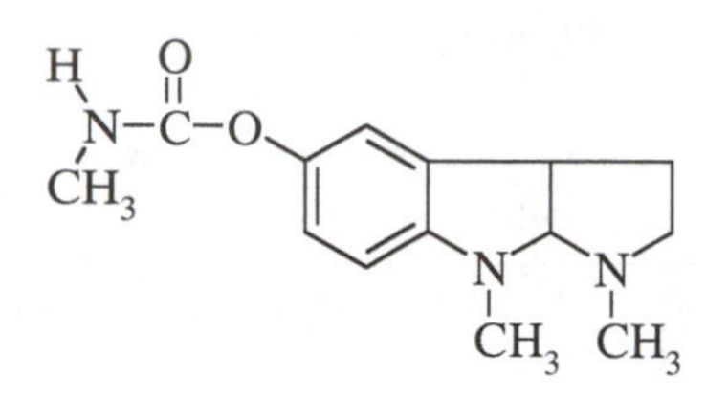

Fig. 3.17. Chemical structure of the naturally occurring carbamate physostigmine.

identification led to the development of the much more lipophilic carbamates. The carbamate insecticides, which include a number of systemic compounds, have been used to control a wide variety of pests of agricultural, veterinary and medical importance. Some of the most important carbamates to have been developed include carbaryl, carbofuran, methiocarb, pirimicarb, propoxur and aldicarb. Overall, about 30 different active ingredients have been developed since 1956. Chemically the carbamates can be divided into three distinct groups based on whether the leaving group attached to the *N*-methyl carbamate group is aliphatic, carbocyclic or heterocyclic. These three subgroups of the carbamates are therefore exactly the same as those of the organophosphates.

Carbocyclic carbamates

The first carbamate to be used was carbaryl.[13] This was introduced for use in crop protection in 1956. Its low mammalian toxicity has meant that it has been widely used around the home as well as in agricultural production. Carbaryl is widely used in orchards both as an insecticide and as a fruit-thinning agent,[14] and although its use in the European Union has declined in recent years, more carbaryl has been used worldwide than all

footnote 12 (*cont.*)

died, then he or she was found guilty! In West Africa seeds are referred to as *esere*, from which the alternative name for the active ingredient, eserine, is derived. Medically, this compound slows the circulation, raises blood pressure and depresses the CNS. Treatment for poisoning comprises stomach evacuation (vomiting or by pumping) and administration of atropine (see also Chapter 1).

[13] In 1984 a pesticide factory in Bhopal, India, that produced carbaryl released a cloud of methyl isocyanate gas that killed 3800 people and injured at least 10000 others. This remains the world's worst industrial accident (see also Chapter 4).

[14] Carbaryl can also act as a plant growth regulator by causing fruit-thinning in orchards.

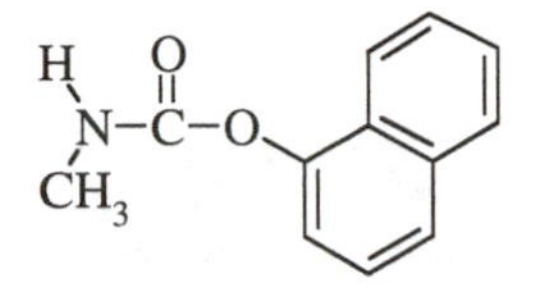

Fig. 3.18. Chemical structure of carbaryl.

the remaining carbamates combined. It is a contact and feeding poison that is mainly active against biting pests. It is not systemic. Its chemical structure is shown in Figure 3.18. Other carbocyclic carbamates include propoxur and methiocarb.

Heterocyclic carbamates

In contrast to carbaryl, carbofuran, another widely used carbamate, is systemic. This insecticide has found a wide variety of uses in pest control. Unfortunately, it has a high mammalian toxicity and so must be handled carefully. The other widely used heterocyclic carbamate insecticide is pirimicarb. This is also a systemic compound and it is marketed as being useful for integrated pest control programmes because it is not toxic to bees and other beneficial insects when applied at recommended doses.

Aliphatic carbamates

Examples of aliphatic carbamate insecticides include aldicarb, aldoxycarb, methomyl and oxamyl. All of these insecticides have systemic activity and have found a variety of uses in pest control. Unfortunately, they are also extremely toxic molecules and so care must be taken in handling these insecticides. Aldicarb has the distinction of being one of the most toxic insecticides ever developed. The mammalian oral LD_{50} for aldicarb is 1 mg/kg, i.e. only 70–80 mg would be enough to kill a person of average weight. As a result of this, in the UK this active ingredient is only sold as granules. Liquid formulations would present too much of a hazard to use. The chemical structure of aldicarb is shown in Figure 3.19.

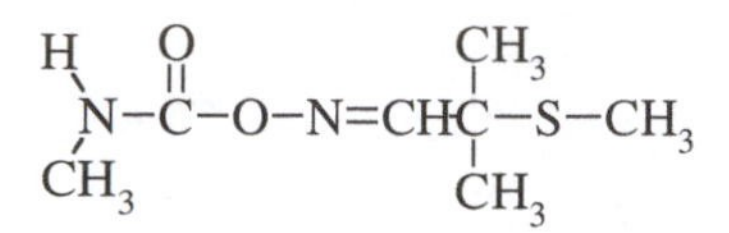

Fig. 3.19. Chemical structure of aldicarb.

Mode of action of carbamates

Like the organophosphates the carbamates have limited environmental persistence. They are produced from carbamic acid and have a similar mode of action to organophosphates in that they also block the enzyme acetylcholinesterase. This process of enzyme inhibition is called carbamylation. Unlike the situation with organophosphates, the rate at which enzymatic inhibition takes place is determined by the rate of carbamylation. With organophosphates the rate at which enzymatic inhibition takes place is largely determined by the rate of dissociation of the leaving group from the enzyme (see earlier discussion). In comparison to a phosphorylated enzyme complex, the carbamylation process is relatively less stable, i.e. the enzyme is not blocked for so long. Like many insecticides the symptoms of poisoning are similar. Tremors are followed by paralysis and death. Like organophosphates, atropine can be administered in cases of accidental mammalian poisoning. Unfortunately, many carbamates are fairly toxic to Hymenoptera (including parasitoids and pollinators), so do not integrate well with integrated pest management programmes. The active ingredient, pirimicarb, is the one exception to this rule.

Neonicotinoids

The neonicotinoids represent the newest group of synthetic insecticide to be developed. They are so called because their development was based on the chemical structure of the alkaloid nicotine (Chapter 2). The first commercial neonicotinoid, imidacloprid (Figure 3.20), was discovered in 1984, and was commercially marketed for the first time in 1991 for the control of sucking pests. Since 1991 three further neonicotinoids have been commercially developed. These are nitenpyram (1995), acetamiprid (1996) and thiamethoxam (2001; Figure 3.21). All of these insecticides (including imidacloprid) have contact and systemic activity. In addition to controlling sucking pests, these

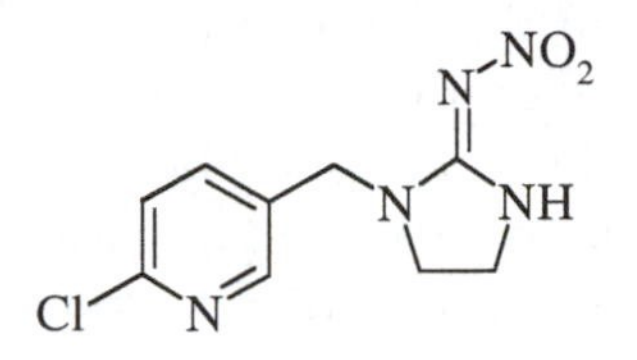

Fig. 3.20. Chemical structure of imidacloprid.

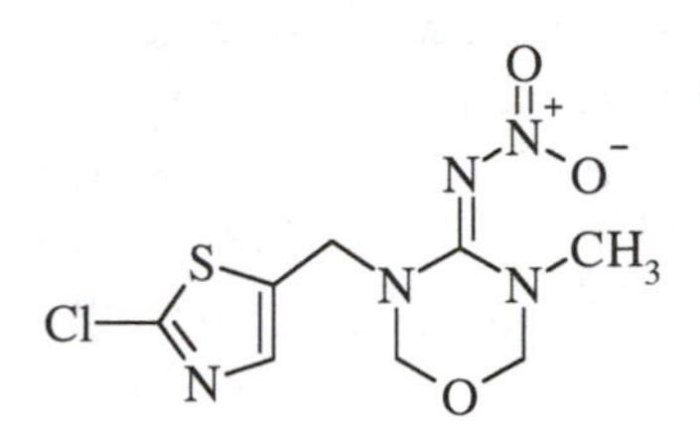

Fig. 3.21. Chemical structure of thiamethoxam.

later-generation neonicotinoids can also control a number of chewing and/or biting pests. Neonicotinoids are available in a variety of formulations and are registered for use as foliar treatments, as soil treatments and as seed treatments. In addition to the above, there are a number of lead chemical structures that are based on neonicotinoid chemistry, and it is highly likely that the number of commercially available neonicotinoids will increase in the next 10 years. Globally, in terms of tonnage and extent of use, the neonicotinoids had overtaken the organochlorines and were about to overtake the carbamates at the time of writing this book. One author even claims that, by volume, imidacloprid may be the most extensively used insecticide worldwide.[15]

Mode of action of neonicotinoids

Like nicotine, the neonicotinoids act as agonists to acetylcholine by binding to postsynaptic nicotinic acetylcholine receptors.[16] This binding is persistent

[15] See Ware, G.W. (1999). *An Introduction to Insecticides*. http://ipmworld.umn.edu/chapetrs/ware.htm.

[16] Acetylcholine receptors can be classified as nicotinic or muscarinic, with further subdivisions of these two categories. It is not within the scope of this book to discuss these neurotransmitter receptor sites. See glossary for more details.

because the molecules are insensitive to the action of the enzyme acetylcholinesterase. Persistent stimulation of acetylcholine receptors then results in hyperexcitation, tremors, paralysis and death, i.e. the symptoms of poisoning are similar to those observed with almost all other insecticidal groups. At least one reason for their toxicity to arthropod pests is that insects make more extensive use of nicotinic receptor sites than mammals.

Other insecticidal chemical groups

Chemical pest control in the year 2002 is still dominated on a worldwide basis by the use of the five insecticidal classes discussed above. Other types of insecticide have been developed. For example, insecticides that are classed as amidines, benzilates, nonester pyrethroids, phenyl ethers and phenylpyrazoles, amongst others, have all been developed. To give one example, fipronil is phenylpyrazole that was commercialised for use in the 1990s. This insecticide binds to GABA-gated chloride channels and is effective against a range of pest species. However, the total number of active ingredients within all of these additional chemical groups numbers less than 20, and it is outside the scope of this book to discuss all of these products individually. For further information the reader is referred to the list of texts given at the end of this chapter.

Apart from the neonicotinoids, the only other truly successful class of insecticide to be developed in more recent years comprises the benzoylphenylureas. The development of this group began in 1975 with diflubenzuron and there are now nine different compounds sold within this group. All of these compounds are chitin synthesis inhibitors. They are discussed in detail in Chapter 8.

Summary

In summary, the use of synthetic chemical insecticides to control pest species has expanded greatly since the end of the Second World War. At present, there are in the region of 160 different active ingredients (in thousands of different formulations) that are available for pest control. Almost all of these active ingredients can be placed in one of five chemical groups – organochlorides, organophosphates, pyrethroids, carbamates and neonicotinoids. Table 3.6 lists the active ingredients by chemical group that were available for pest control in 2000. By far the most successful group has been the organophosphates. All of these synthetic chemicals are neurotoxic. DDT and the pyrethroids seem to

Table 3.6. *Insecticides, acaricides and nematicides registered for pest control globally in 2000 by major chemical group (name of active ingredient given)*

Organochlorides	Organophosphates		Pyrethroids	Carbamates	Neonicotinoids
Chlordane	Acephate	Methidathion	Acrinathrin	Alanycarb	Acetamiprid
DDT	Azamethiphos	Mevinphos	Allethrin	Aldicarb	Clothianidin[a]
Dicofol	Azinphos-ethyl	Monocrotophos	Bifenthrin	Bendiocarb	Dinotefuran[a]
Endosulfan	Azinphos-methyl	Naled	Bioallethrin	Benfuracarb	Imidacloprid
Lindane	Cadusafos	Omethoate	Bioresmethrin	Betocarboxim	Nitenpyram
Heptachlor	Chlorethoxyfos	Oxydemeton-methyl	Cycloprothrin	Butoxycarboxim	Thiamethoxam
Methoxychlor	Chlorfenvinphos	Parathion	Cyfluthrin	Carbaryl	
	Chlormephos	Parathion-methyl	β-cyfluthrin	Carbofuran	
	Chlorpyrifos	Phenthoate	Cyhalothrin	Carbosulfan	
	Chlorpyrifos-methyl	Phorate	λ-Cyhalothrin	Ethiofencarb	
	Coumaphos	Phosalone	Cypermethrin	Fenobucarb	
	Cyanophos	Phosmet	α-Cypermethrin	Fenoxycarb	
	Demeton-*S*-methyl	Phosphamidon	β-Cypermethrin	Formetanate	
	Diazinon	Phoxim	ζ-Cypermethrin	Furathiocarb	
	Dichlorvos	Pirimiphos-methyl	θ-Cypermethrin	Isoprocarb	
	Dicrotophos	Profenphos	Cyphenothrin	Methiocarb	
	Dimethoate	Propaphos	Deltamethrin	Methomyl	
	Dimethylvinphos	Propetamphos	Empenthrin	Metolcarb	
	Disulfoton	Prothiofos	Esfenvalerate	Oxamyl	
	EPN	Pyraclofos	Fenvalerate	Pirimicarb	
	Ethion	Pyridaphenthion	Fenpropathrin	Propoxur	
	Ethoprophos	Quinalphos	Flucythrinate	Thiodicarb	

Table 3.6. (*cont.*)

Organochlorides	Organophosphates		Pyrethroids	Carbamates	Neonicotinoids
	Famphur	Sulfotep	Flumethrin	Thiofanox	
	Fenamiphos	Sulprofos	Imiprothrin	Trimethacarb	
	Fenitrothion	Tebupirimfos	Phenothrin	XMC	
	Fenthion	Temephos	Prallethrin	Xylylcarb	
	Fosthiazate	Terbufos	Resemethrin		
	Heptenophos	Tetrachlorvinphos	RU15525		
	Isofenphos	Thiometon	Tefluthrin		
	Isopropyl *O*-salicylate	Triazophos	Tau-fluvalinate		
	Isoxathion	Trichlorfon	Tetramethrin		
	Malathion	Vamidothion	Tralomethrin		
	Mecarbam		Transfluthrin		
			ZX18901[a]		
Total 7		65	34	26	4

Notes
[a] Products listed in the manual that were under commercial development at the time of writing.
DDT, dichlorodiphenyltrichloroethane.

Source: Data collated from Tomlin, C. (ed.) (2000). *The Pesticide Manual*, 12th edn. Farnham, Surrey: British Crop Protection Council Publications. This list only covers the major insecticidal classes, i.e. it does not include products that do not belong to one of the classes listed.

Table 3.7. *Summary data for the five main chemical groups of insecticide*

	Environmental stability	Exposure and toxicity	Number of AIs
Organochlorides	Months to years	Contact and feeding poisons	c. 10–20
Organophosphates	Days to weeks	Contact, feeding and systemic poisons	c. 80–100
Pyrethroids	Days to weeks	Contact and feeding poisons	c. 30–40
Carbamates	Days to weeks	Contact, feeding and systemic poisons	c. 30–40
Neonicotinoids	Days to weeks	Contact, feeding and systemic poisons	c. 5

Note:
AI, active ingredient. Most active ingredients are available in a series of different formulations (see also Chapter 4).

mediate their toxic effects by interfering with axonal transmission of nerve impulses. The organochlorines, HCH and the cyclodienes, along with the organophosphates, the carbamates and the neonicotinoids, by contrast seem to mediate their toxic effects by interfering with synaptic transmission of nerve impulses. All of these insecticides are contact and feeding poisons; the organophosphates, carbamates and neonicotinoids also include insecticides that have systemic activity. A summary of these main points is given in Table 3.7.

There can be absolutely no doubt that these insecticides have had a dramatic effect upon disease control and on food production in the last 50 years. However, the deliberate release of chemicals that are known to be poisons has not been without problems. The release of these chemicals (their application) and some of the problems that have subsequently occurred form the subject matter of the next chapter of this book.

Further reading

Bloomquist, J.R. (1993). Toxicology, mode of action, and target site mediated resistance to insecticides acting on chloride channels. *Comparative Biochemistry and Physiology*, **106**, 301–14.

Bloomquist, J.R. (1996). Ion channels as targets for insecticides. *Annual Review of Entomology*, **41**, 163–90.

Cremlyn, R.J. (1991). *Agrochemicals: Preparation and Mode of Action*. Chichester: John Wiley.

Curtis, C.F. & Davies, A.R. (2001). Present use of pesticides for vector and allergen control and future requirements. *Medical and Veterinary Entomology*, **15**, 231–5.

Krieger, R.I. (2001). *Handbook of Pesticide Toxicology*, 2nd edn. San Diego, CA: Academic Press.

Maienfisch, P., Angst, M., Brandl, F. *et al.* (2001). Chemistry and biology of thiamethoxam: a second generation neonicotinoid. *Pest Management Science*, **57**, 906–13.

Müller, F. (ed.) (2000). *Agrochemicals*. Weinheim, Germany: John Wiley.

Oerke, E.C., Dehne, H.W., Schohnbeck, F. & Weber, A. (1995). *Crop Production and Crop Protection: Estimated Losses in Major Food and Cash Crops*. Amsterdam: Elsevier.

Thompson, W.T. (1998) *Agricultural Chemicals – Book I – Insecticides*. Fresno, CA: Thomson Publications.

Tomlin, C. (ed.) (2000). *The Pesticide Manual*, 12th edn. Farnham, Surrey: British Crop Protection Council Publications.

Vais, H., Williamson, M.S., Devonshire, A.L. & Usherwood, P.N.R. (2001). The molecular interactions of pyrethroid insecticides with insect and mammalian sodium channels. *Pest Management Science*, **57**, 877–88.

Ware, G.W. (2000). *The Pesticide Book*, 5th edn. Fresno, CA: Thompson Publications.

4

Formulation, application and the direct and indirect side-effects of insecticides

Introduction

Chapter 3 introduced the notion that when modern synthetic insecticides are used for pest control, we are deliberately releasing into the environment chemicals that are known to be toxic to a diversity of life. We do this because we think that the benefits accrued from the release far exceed any of the costs involved in our purposeful contamination of the environment.

For example, with most modern insecticides any contamination (including death to nontarget species) is usually short-lived. As a result, many scientists will argue that when the death of a small number of individuals from nontarget species does occur, it is irrelevant to the long-term survival of those organisms. As for human concerns, many people (especially regulators and industry representatives) will tell you that these are more than adequately catered for in the legislation that exists to regulate pesticide use, particularly in developed countries.[1]

Like the risks associated with insecticide use, it is also easy to demonstrate and/or explain the benefits that are associated with using modern synthetic insecticides. For example, Chapter 3 showed how in simple economic terms most insecticide applications more than pay for themselves, especially over the short term. The results of insecticide use can also be shown to benefit humanity in general and not just individual farmers, particularly in relation to increased yields and to disease control. For example, Table 4.1 shows the

[1] The current rapid developments in organic farming (Chapter 11) would suggest that many people are not assured that all is well with human exposure to pesticides.

Table 4.1. *Malaria cases in KwaZulu-Natal. Cases by year 1990–2000 and by month for 2000. Spraying of dichlorodiphenyltrichloroethane (DDT) stopped in 1995 and resumed in May 2000*

Year	Number	Month (2000)	Number
1990	2758	January	4350
1991	1933	February	5759
1992	599	March	6161
1993	6118	April	4926
1994	5209	May	6225
1995	4117	June	4553
1996	10535	July	4154
1997	11425	August	2155
1998	14575	September	763
1999	27238	October	937
2000 (first 10 months)	39983		

Source: KwaZulu Department of Health.

benefits associated with the use of dichlorodiphenyltrichloroethane (DDT) for controlling anopheline mosquitoes that act as a vector for malaria in KwaZulu-Natal. When DDT was withdrawn from use in 1995 the number of cases of malaria began to increase immediately. Reintroduction of this pesticide in 2000 has now begun to reduce the number of cases. The arguments for insecticide use in many cases therefore can seem almost irrefutable. In fact, many commentators on insecticide use spend a great deal of time and energy explaining how widespread organic farming (without insecticide use) would lead to widespread food shortages and to concomitant substantial increases in the cost of food.

Despite seemingly irrefutable arguments, however, we are still left with the fact that an insecticide application constitutes the deliberate release into the environment of a poisonous molecule that is harmful to life. Because this is so, it would be prudent to use this chemical with the aim of minimising its adverse environmental impact. To achieve this we need to look at the formulation of the insecticide and at the method that is used to apply it. Both of these factors have been, and can be, manipulated to minimise environmental damage. These are the subjects that form the first two topics of this chapter. The remainder of the chapter is then given over to the main problems that are associated with insecticide use. Despite the benefits that insecticides may have, in a large number of cases it could be argued that the costs associated with such use have now become too great. The chapter ends with a brief consideration of the legislation that exists to regulate insecticide use, particularly in developed countries.

Formulation of insecticides

It is very rare for active ingredients to be applied in pure form to control pest species. This is not only because it is usually dangerous and often impractical to do so, but also because with most modern pesticides a very small amount of active ingredient must be distributed over a very large area. Consider, for example, the pyrethroid insecticide deltamethrin. This active ingredient can be used at rates as low as 10 g/ha or 1 mg/m^2. Even if this insecticide were completely harmless to nontarget species it would be very difficult to apply it neat at such low dose rates.

In most cases therefore pesticides must be formulated – a process that involves adding a number of other chemicals or substances to the active ingredient[2]. It is the formulation, and not pure active ingredient, that is purchased by pesticide users. The chemicals that are added to the active ingredient include solvents, diluents, synergists, surfactants, stickers, penetrants, dispersion aids, safeners, deodorants, antifoam agents, buffers and thickeners. A more complete list of these chemicals and their associated putative effects upon the final formulation is given in Table 4.2.

Overall, the aim of the formulation process is to make the active ingredient easy to use and to ensure that it is stable during transport and storage. In many cases the most important aim of a formulation chemist is to achieve an adequate shelf-life for the formulation. For most pesticides a satisfactory shelf-life would be anything greater than 2 years. However, it is also important that a formulation is safe to use and is simple to apply. The following section of this chapter briefly reviews the types of insecticide formulation that are available on a commercial basis.

Categories of formulation

Pesticide formulations can be classified according to whether they are wet, dry or gaseous. Most pesticides belong to the first category. Wet formulations are

[2] Pure active ingredient often exists as a solid and is referred to as technical-grade material.

Table 4.2. *Chemicals that may be used in pesticide formulations and their putative functional properties*

Chemical	Putative function
Acidifier	Lowers pH
Activator	Increases biological efficacy
Antievaporant	Inhibits foam formation
Antifoam agent	Reduces evaporation of pesticide
Buffer	Causes resistance to change in pH
Compatibility agent	Permits mixing of agrochemicals
Defoaming agent	Suppresses foam in spray tank
Deodorants	Improve pesticide odour
Deposition agent	Improves deposition on targets
Diluents	Diluting active ingredient
Drift control agent	Reduces spray drift
Emulsifier	Stabilises emulsions
Extender	Increases life of active ingredient following application
Foaming agent	Increases stability of a foam
Humectant	Increases drying time of a deposit
Mineral oil	NA
Penetrating agent	Aids penetration of pesticides
Solvents	Dissolving technical-grade material
Spreader	Increases spread on target
Sticker	Increases adhesion to targets
Surfactant	Surface-active agent
Synergists	Increase pesticide toxicity
Vegetable oil	NA
Wetter	Increases wetting – same as a spreader
Water conditioner	Reduces harmful effects of hard water

Note:
NA, not applicable.

Source: Data collated from the European Adjuvant Association 1996 (an organisation that is now defunct).

referred to as emulsifiable concentrates, suspension concentrates (flowables), solutions or microencapsulated pesticides. Dry formulations exist as granules, dusts, etc. Gaseous formulations exist for those insecticides that are vapour-active. Table 4.3 provides an overview of the most commonly used pesticide formulations.

Table 4.3. *Common pesticide formulations*

Wet formulations	Dry formulations
Emulsifiable concentrates	Dusts
Suspension concentrates (flowables)	Granules
Solutions (liquid concentrates)	Wettable powders
Ultra-low-volume concentrates	Soluble powders
Microencapsulated pesticides	Baits

Wet formulations

The most extensively used insecticide formulations are emulsifiable concentrates. These formulations comprise the active ingredient dissolved in a petroleum organic solvent (xylene, kerosene, etc.) and mixed with an emulsifying agent. The emulsifier is a surface-active agent (a surfactant) that ensures a stable emulsion is produced following dilution with water. These emulsions are very stable and should not require constant agitation to prevent the emulsion breaking. Typical concentrations of insecticide within an emulsifiable concentrate will vary between 30 and 50%. Typical dilution rates for a formulation will be 1:200–500 of formulation-to-water.

In suspension concentrates the active ingredient is blended with an inert dust diluent (talc,clay, etc.) and mixed with a small amount of a liquid carrier such as water or oil. The typical particle size of the inert diluent will be less than 3 μm in diameter and this will form 50–80% of the formulation depending upon the concentration of the insecticide. Various surfactants and dispersion agents keep the semisolid formulation stable and ensure that a stable emulsion is produced following dilution with water. Because inert ingredients exist in the formulation the spray tank mixture must be constantly agitated to prevent the active ingredient from settling to the bottom and so precipitating out of solution. Suspension concentrates are also sometimes referred to as 'flowables'.

Solutions are also sometimes called liquid concentrates and exist as the active ingredient dissolved in an organic solvent. Some solutions may be further diluted with oil before use while others may be used directly. For example, ultra-low-volume concentrates are usually applied while they are 95% neat. Ultra-low-volume concentrates, although dangerous to handle, have been successfully used where it is impractical to use high-volume sprays, for example, in many African countries where diluting agents (water- or oil-based) are either costly or unavailable.

Finally, microencapsulated pesticides are a relatively newer development and consist of the active ingredient enclosed in a microscopic polymeric sphere that is 1–10 μm in diameter. These spheres are formulated as concentrated aqueous liquids that contain dispersion and emulsifying agents. The concentrated liquid is diluted with water before application. The spheres provide a slow-release quality for the active ingredient, which diffuses out through time following application. Increasing the thickness of the sphere will slow the release rate and so increase the residual activity of the pesticide.[3]

Dry formulations

Dry formulations include dusts, granules, wettable powders, soluble powders and baits. Dusts are among the oldest pesticide formulations and comprise mechanically pulverised active ingredient with a particle size of 3–30 μm in diameter. Usually the active ingredient is diluted using an organic flour or a finely ground inert mineral. However, some insecticides may be used as dusts without further dilution. For example, ground pyrethrum flowers can be applied as dusts without any further dilution. The primary use for dusts is in enclosed, small areas. Dusts are highly sensitive to wind and do not deposit well on exposed vegetation. They also present a safety hazard to applicators through inhalation.

Granules are prepared by applying liquid pesticides to coarse, porous particles of inert or organic origin. These granules then release the pesticide over time. Granules are typically 0.25–1.00 mm in diameter and so are much safer to handle because they cannot be inhaled. Aldicarb, one of the most toxic insecticides ever developed (rat oral LD_{50} *c.* 1 mg/kg – see also Chapter 3), is sold as a granular formulation. Because of obvious problems that would be encountered with coverage on target surfaces, most granular formulations contain systemic active ingredients. Granular formulations are also often used to control soil pests and are incorporated with seeds or seedlings at planting.

Wettable powder formulations are dusts containing surfactants that enable mixing and dilution with water. In essence, these formulations are a dry version of a suspension concentrate. Like suspension concentrates, frequent agitation of the liquid mixture is required to prevent separation and keep the insecticide in suspension. Wettable powder formulations contain much more active ingredient than dusts (15–95% active versus 1–10% active in dusts)

[3] Microencapsulation and the use of slow-release formulations are also discussed in Chapter 7 in relation to pheromone use.

because they are designed for further dilution. They should never be used as dusts, as this would be extremely hazardous.

Soluble powders, unlike wettable powders, dissolve fully in water and form a true solution. It is not usually necessary to agitate these solutions. Very few pesticides exist as soluble powders because most active ingredients are not soluble in water.

Baits combine edible or attractive substances with the active ingredient. Pests are then exposed to the pesticide following consumption of the bait or by simply walking across it. Most baits contain *c.* 5% active ingredient and they are widely used to control domestic pests such as ants and cockroaches as well as vertebrate pests such as birds and rodents. In agricultural situations baits have been extensively used for slug control.

Gaseous formulations

The final category of formulation comprises those pesticides with a vapour action. These are fumigants and aerosols. Fumigants are usually active ingredients that exist in a gaseous state at normal temperatures. For example, the naphthalene in moth balls slowly evaporates from solid crystals at room temperature. However, some pesticides may be used as fumigants by vaporising the active ingredient with specialised pesticide application equipment. Fumigants are used in enclosed spaces such as greenhouses and storage facilities, e.g. grain stores. They are also used at ports and for treatment of ships and aeroplanes.

Aerosols are widely used in household pest control. The active ingredient is dissolved in a petroleum solvent and pressurised in a container with a propellant gas such as carbon dioxide. When sprayed, the petroleum solvent quickly evaporates to leave a fog of microscopic droplets that are suspended in air. The droplets produced are typically less than 10 μm in diameter and, because there is a danger associated with inhaling these droplets, most aerosols have a very low concentration of active ingredient. This makes aerosols expensive to use. In agricultural situations, fogging machines can be used to produce aerosols, by using thermal energy to vaporise active ingredients. For these to work successfully the active ingredient must obviously be heat-stable.

Formulation summary

In summary, most insecticides are available in a diversity of formulations. Which formulation is used for any particular situation will usually depend on availability and price, as well as a user's history of success with the product. The majority of insecticide formulations are prepared by the manufacturer either for immediate use or for dilution with water. In recent years, however, there has been increasing interest in the use of chemicals called adjuvants that can be added to formulations. The aim of using these chemicals is to enhance the effectiveness of the insecticide. Figure 4.1 gives summary data on the types of adjuvant that are available for use with insecticides. At least one of the reasons why these chemicals are not premixed with the formulation is that they allow applicators to customise formulations depending on the prevailing environmental circumstances. Whichever formulation is finally used for an insecticide will also depend on the application equipment available. This equipment is discussed in more detail in the next section of this chapter.

Application of insecticides

Pesticides can be applied to control pest species using a range of methods and equipment. For liquid applications (*c.* 75% of all pesticides are applied as liquids) it has been suggested that we can categorise at least six different types of sprayer on the basis of the energy used to create a spray. These categories are hydraulic, gaseous, centrifugal, kinetic, thermal and electrical energy sprayers. Many of the sprays that are produced using these energy types can also benefit from air assistance in terms of moving the spray droplets towards their target. Air assistance is not discussed in detail here. However, a brief summary of the energy categories that are used to produce liquid sprays is provided below. Because of space constraints, application equipment for dry formulations is not discussed. The reader is referred to the reading list at the end of the chapter for more information on pesticide application.

Hydraulic sprayers

Hydraulic sprayers are probably the most widely used sprayers throughout the world (Figure 4.2, colour plate). In essence, these sprayers use hydraulic energy to force a large amount of liquid through a very small hole in a short space of time. The forces involved in this process are sufficient to create a spray. The hydraulic pressure may be generated manually (as in knapsack and pneumatic sprayers) or by using compressed-air cylinders or engine-driven pump units. The spray produced will typically have a wide spectrum of droplet sizes from 10 to 1000 μm in diameter. The actual spectrum produced will

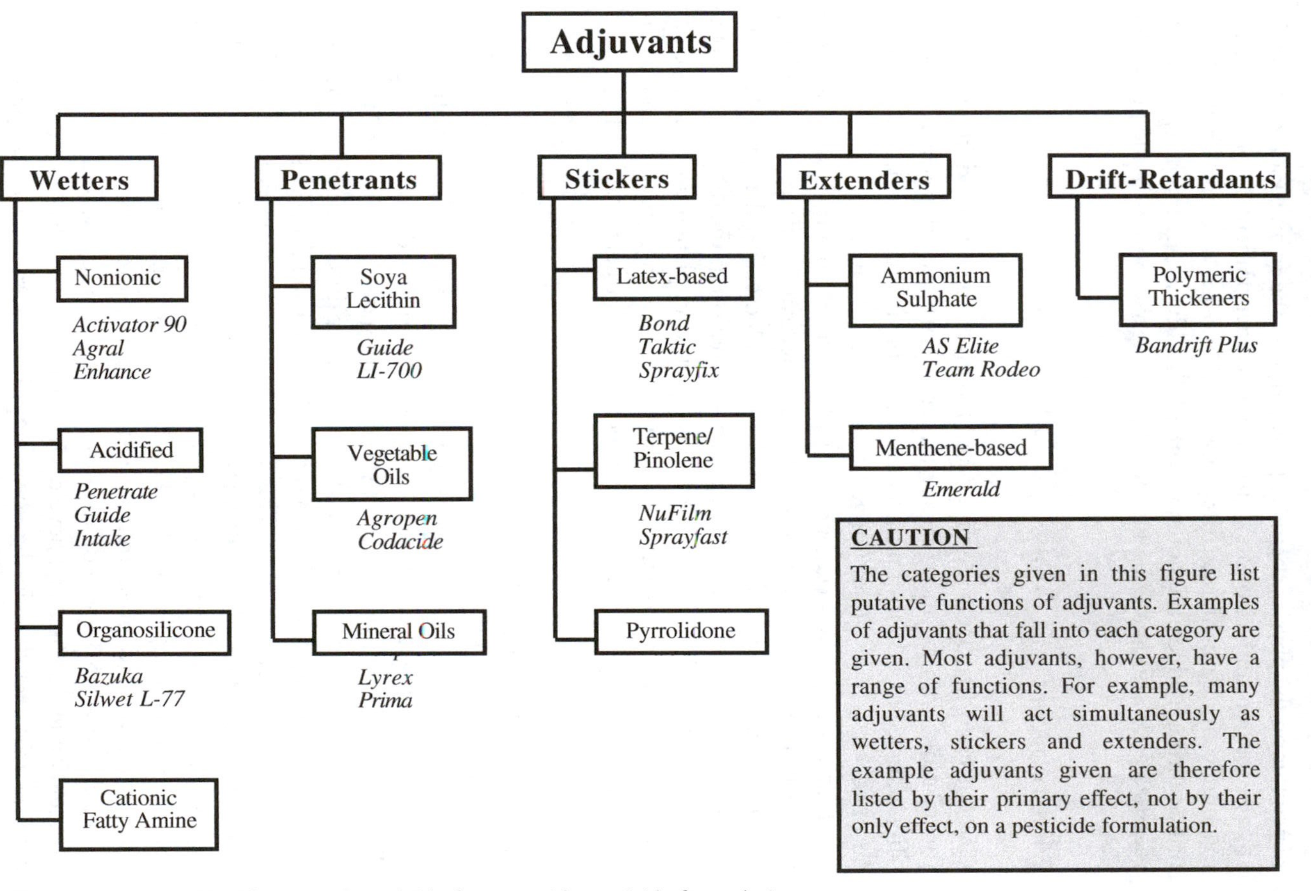

Fig. 4.1. Categories of adjuvant available for use with pesticide formulations.

depend upon the physicochemical properties of the formulation (principally surface tension and viscosity), the ambient environmental conditions, the magnitude of the hydraulic forces involved and the size and shape of the orifice through which the liquid is forced. Worldwide, hydraulic sprayers are regarded as reliable and effective tools for the application of liquid pesticides.

Mist-blowers

Gaseous energy sprayers atomise liquid by impacting one fluid (the pesticide formulation) with another (usually an air stream). For this reason they are often referred to as twin-fluid systems. Probably the most widely used sprayers of this type today are motorised mist-blowers (Figure 4.3, colour plate). A jet of air, produced by an engine, is supplied with the liquid formulation via a restrictor. The liquid is then atomised and a spectrum of droplets is produced within the size range 20–500 μm in diameter. Decreasing the liquid feed rate and increasing the air speed both cause a reduction in the droplet-size spectrum produced. These sprayers are widely used in plantation crops such as cocoa where it is impossible to use machine-mounted sprayers and where good coverage of the spray on a complex target (a tree) is required.

Spinning disc sprayers

Centrifugal energy sprayers atomise liquids by throwing droplets from the edge of a spinning disc. These sprayers are also referred to as spinning discs (Figure 4.4, colour plate). Typically, the rotating disc has grooves and teeth that promote the formation of droplets. The primary difference between these sprayers and those discussed already is that the droplet-size spectrum produced is much narrower. This has permitted the development of so-called controlled-droplet application where sprays with monosize droplets can be produced for specific crop and pest situations. Typical monosize sprays comprise droplets of 50, 100, 200 or 300 μm in diameter. The droplet size produced will depend on the physicochemical properties of the formulation, the rotational speed of the disc, the size of the disc and the flow rate of liquid on to the disc. The development of spinning-disc sprayers has permitted a vast reduction in the volume application rate for many pesticides. For example, most hydraulic systems apply pesticides at volume application rates of 200–500 l/ha, whereas spinning-disc sprayers can operate at rates as low as 5 l/ha. Not surprisingly, spinning-disc sprayers have proven to be a great success for pesticide applications in countries where water is at a premium.

Watering cans

Kinetic-energy nozzles rely simply on the principle that liquid falling through a small hole forms a filament that can be shaken to produce droplets. The simplest pesticide applicator, and one which uses this principle, is therefore a watering can. These sprayers produce relatively large droplets (>500 μm in diameter) and work at high-volume application rates. They are most suitable for the home gardener.

Fogging machines

Thermal-energy nozzles use a stream of hot gas to vaporise a liquid pesticide which then condenses on leaving the sprayer to form a cloud of aerosol droplets. When these droplets are less than 15 μm in diameter a fog is produced. These sprayers (which are really a variation on the gaseous-energy sprayer theme) are also called fogging machines. The hot gas is the exhaust from a petrol engine that is expelled through a long pipe (0.5–1 m in length). The liquid pesticide is fed into the hot gas stream at a rate controlled by a restrictor. Increasing the flow rate will decrease the droplet size produced. Only pesticides that are heat-stable are suitable for use with fogging machines. These machines have found widespread use for pesticide application in enclosed environments like greenhouses, food storage facilities and buildings. They have also been used successfully in plantation crops with very dense canopies. Fogging machines have also been used to sample the diverse fauna of tropical rainforest canopies. Pyrethroid insecticides are sprayed into the rainforest canopy and the arthropods that fall out are collected on sheets placed at the base of treated trees.

Electrostatic sprayers

The final type of energy that can be used in the application of insecticides is electrical energy. This energy has been used to charge droplets that are produced conventionally using hydraulic or centrifugal forces and it has also been used to atomise sprays directly. The overall aim of providing droplets with an electrical charge is to improve deposition on plant targets. There are three important forces that can work to achieve this goal with charged sprays. These have been called the nozzle effect, the space cloud effect and the induced-field effect. The nozzle effect derives from the electrical field created between a charged nozzle above a crop and earthed object (plant) below it. Droplets that have the same electrical potential as the nozzle are accelerated in this field towards the target object, thus, it is hoped, improving coverage. The space cloud effect is created in dense sprays with droplets of similar charge. Droplets are repelled from one another, so increasing the overall volume of the droplet cloud and, it is thought, improving coverage on the target plant. The induced-field effect is created by charged droplets inducing an opposite charge on the target plant which can then enhance deposition and coverage. Because there is an inverse-square relationship between force and distance, these forces only operate over very small distances.

While it has proven possible to charge pesticide sprays, the magnitude of improvements in deposition that have been recorded for most systems has not been great. The problem with most systems has been the amount of charge which droplets have been able to pick up. This is known as the charge-to-mass ratio. With hydraulic systems the electrical charge imparted to droplets has not been great enough to modify the behaviour of droplets that are also exposed to gravitational and wind-induced forces. These latter forces simply dominate in terms of droplet behaviour. With centrifugal systems the means of droplet production (horizontally from the disc) means that the droplets are unable to take advantage of the electrical forces discussed above.

The one electrical system that has proven to be relatively successful is one which uses electrical energy to produce the spray in the first place. This process is called electrohydrodynamic atomisation. It is based on the principle that a droplet will break up into smaller droplets in order to increase its overall surface area once it has taken on sufficient charge. This is known as the Rayleigh limit. In these sprayers liquid is fed through a charged orifice that causes the creation of a spray. The droplets produced are small (20–100 μm in diameter) and so have a relatively high charge-to-mass ratio with the upshot that improved coverage on target plants can clearly be seen. Because the droplets are small and the coverage high, these systems can be used at very-low-volume application rates. Like spinning-disc sprayers, they have been most widely used where conventional diluents like water are either unavailable or prohibitively expensive.

Unfortunately, there are a number of problems associated with electrohydrodynamic sprayers and some of these explain why these sprayers have not been more widely used. First, the atomisation process requires liquids with specific conductivity properties. Not all pesticides can be atomised this way. Second, charged droplets have often been found to deposit preferentially on the exterior of crop canopies. This has meant that penetration of the pesticide into the crop canopy has worsened. Third, it has been observed that charged droplets do not deposit well on pointed leaves (like many cereal crops) because of a process called corona discharge. Finally, and perhaps most importantly, the benefits from using these systems have not been perceived to be sufficiently great when a comparison is made with the equipment that users already possess.

Application summary

In summary, pesticides can be applied using a range of equipment. Hydraulic systems are by far the most extensively used, though centrifugal energy systems that can work at very low dose rates have been extensively used in many developing countries. Fogging machines are used in enclosed environments and mist-blowers are used in plantation crops on difficult terrain. Electrostatic sprayers remain to find widespread application in crop protection. In recent years many application systems have benefited from the use of air assistance and hybrid sprayers have also been produced. For example, Micron Sprayers has now produced a spinning disc attachment that can be fitted to a mist-blower (Figure 4.5, colour plate). With this sprayer the air stream is used to turn the disc from which droplets are produced.

Unfortunately, and in spite of the diversity of application equipment available for pest control, much of this equipment is extremely inefficient.[4] This inefficiency relates the proportion of active ingredient that is released in comparison to the amount that would be needed if pesticide targeting was 100% effective. Most arthropod pests are small and so are difficult to target with a pesticide. Since most estimates of pesticide application efficiency are below 1%, a large amount of pesticide has to be released to ensure that target species contact enough active material for mortality to occur. This is one reason why pesticides can have a detrimental impact on nontarget species. This, and other side-effects associated with pesticide use, is discussed in the remainder of this chapter.

[4] It should be noted that inefficiency relates to the waste of pesticide. Pesticides *per se* and application equipment both usually function well for the intended purpose. It is just that a lot of waste occurs because pest species are so difficult to target.

Direct and indirect side-effects of insecticides

Despite the enormous benefits that accrue from using insecticides to control arthropod pests, their use is not without both direct and indirect costs. These problems were first brought to the attention of the general public with the publication in 1962 of Rachel Carson's *Silent Spring*. Since then, widespread prophylactic use of insecticides has generally declined, particularly in developed countries, and attempts to use insecticides far more judiciously have increased. However, problems do still exist and these will be briefly discussed in the next few pages. We can partition problems associated with insecticide use into the following categories.

- the development of pest species resistance to insecticides;
- the development of secondary pest problems following insecticide applications;
- the development of resurgent pest populations;
- mortality of nontarget species following insecticide applications;
- contamination of aquatic and terrestrial environments by insecticides;
- contamination of food products destined for human and/or animal consumption by pesticides;
- the effects of occupational exposure to insecticides (as a user of pesticides and as a manufacturer), particularly cumulative, long-term exposure.

These problems are not always mutually exclusive. For example, mortality of nontarget predatory invertebrate species may lead to the development of resurgent or secondary pest populations.

Resistance to insecticides

The development of resistance is a natural phenomenon that results from the selective pressure exerted upon a pest population by exposure to an insecticide. It has been defined as an inheritable change in the susceptibility of a pest population to a chemical. Resistance is not new. The first documented case of resistance was in 1914 when the San José scale became resistant to the fumigant inorganic insecticide hydrogen cyanide. It was not until after the introduction of modern synthetic insecticides that resistance became an increasingly serious problem. In 1938 seven insect and mite species were known to be resistant to at least one pesticide; by 1950 this number had increased to 11. In 1969, *c.* 156 species were known to be resistant and by 1984 this number had increased to 477. By the end of the 1990s the number of resistant insects and mites had risen to *c.* 600 species and by 2001 it had increased to over 700 species. These data are shown graphically in Figure 4.6. There is no reason to expect that this trend will not continue well into the twenty-first century.

The impact of the development of resistant pest species has varied worldwide. More often than not, its impact has been greatest in tropical climates in developing countries where pest species reproduce rapidly and have numerous generations per year. Some of the best-known examples of resistance development come from the cotton-growing regions of the world. Cotton is attacked by a complex of pest species and is often treated with numerous insecticide applications. It should come as no surprise to learn that resistance to insecticides developed rapidly for pest species attacking cotton crops. For example, the development of resistance to insecticides by cotton pests led to the near-destruction of cotton growing in Russia, to make cotton growing completely uneconomical in the Yellow River valley in China, to a reduction in the cotton hectarage in Mexico in the 1960s from 280000 ha to 400 ha, and to a decline in cotton yields in countries as diverse as Nicaragua, Peru, Egypt and India. In almost all of these cases a similar pattern was observed. Insecticides selected for resistant individuals and caused mortality to nontarget predatory species. This meant that farmers had to apply insecticides more frequently and at higher rates. This situation escalated until yields began to decline and it became completely uneconomic to try to protect crops from cotton pests. The only solution was to stop using insecticides and to try alternative approaches towards pest control or to grow a different crop altogether. It is perhaps not surprising that integrated pest management as a philosophy for pest control developed in part as a result of these problems in the cotton agroecosystem and that cotton crops were among the first to be managed using an integrated approach. This is discussed in more detail in Chapter 12.

Resistance management

Notwithstanding the development of integrated approaches towards pest control, the techniques that have been developed to try to control resistance development can be partitioned into three categories. All these tactics are part of what has been called insecticide resistance management. The management techniques can be categorised as moderation, saturation and multiple attack.

Moderation of insecticide use involves using insecticides less frequently, making only localised applications, using higher thresholds for applications,

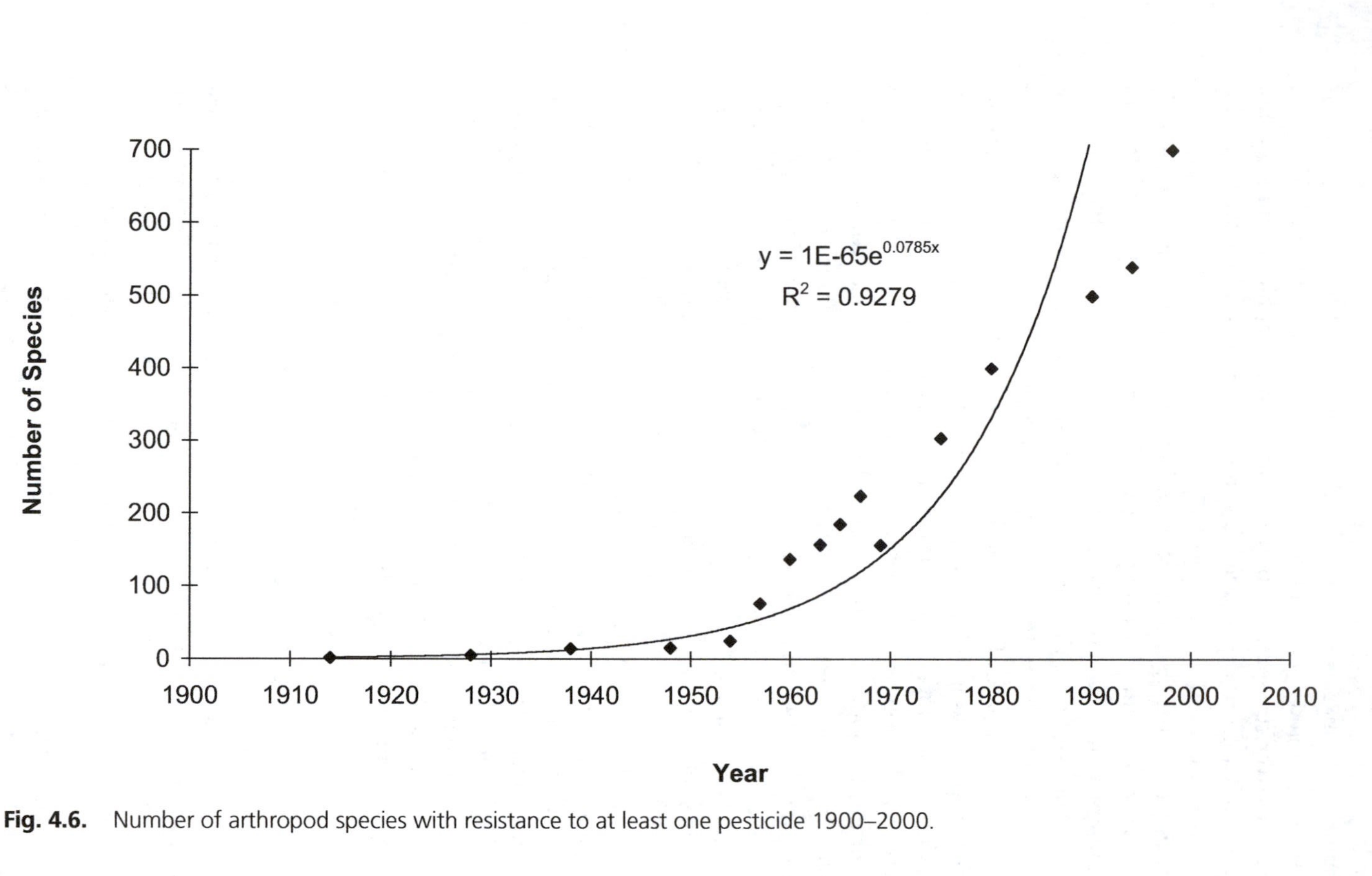

Fig. 4.6. Number of arthropod species with resistance to at least one pesticide 1900–2000.

using reduced-rate applications and integrating applications with other management practices. In essence, moderation is part and parcel of an integrated approach towards insecticide use and pest management.

Saturation, by contrast, takes a far harder line and was the approach adopted up until the 1960s when it became clear that ultimately it was unsustainable. The idea was to use higher doses of pesticide in order to kill all the resistant genotypes. Once this had been achieved, a switch to another product could be made. It rapidly became clear that the costs and environmental impacts associated with using very high doses of insecticide were unacceptable. Insecticide users rapidly became locked into a pesticide treadmill in which more and more insecticide was required to achieve the same level of control. Unfortunately, it is a practice which still continues in some countries where only a limited range of products are available and/or governments provide subsidies for particular products.

The final approach towards resistance management is to use a multiple-attack strategy. This means using mixtures of insecticides, using mosaic patterns for application and rotating the products that are applied. It is an approach that has had some success and is one that is favoured by the agrochemical industry because it does not mean using less insecticide. For many years, individuals within the agrochemical industry argued strongly that moderation, and reduced-rate insecticide applications in particular, could lead to a more rapid development of resistance within a pest population.

Although we can categorise strategies for resistance management, historically chemical pest control has relied on the development of new products as the ultimate solution to this problem. So far, this approach has been successful and it has been claimed that no insecticide has ever been withdrawn from use because of resistance. Insecticides have only been withdrawn because they were environmentally harmful or because newer and better products were available. The problem at the start of the twenty-first century is that the rate of discovery of these new products has declined significantly (see Chapter 1). Many researchers have claimed that the upshot of this will be that in the future pest control will become less and less reliant upon insecticidal inputs. Whether this is the case remains to be seen.

Resistance mechanisms

One of the main problems with many of the insecticides used today is that they have similar modes of action. For example, organophosphates and carbamates interfere with synaptic transmission of nerve impulses while some organochlorine and all pyrethroid insecticides disrupt the axonal transmission of nerve impulses (see also Chapter 3). This means that it is a relatively easy process for species that show single-compound resistance to develop multiple-compound resistance[5] and cross-resistance to compounds in another insecticidal class with a similar mode of action. The mechanisms of resistance in many cases are now also well understood.

In general, resistance is often controlled by a single major gene (monogenic resistance) that can spread rapidly within a pest population. The resistance mechanisms that have so far been identified include elevated levels of detoxification enzymes, a reduced sensitivity at target sites, an increased cuticular thickening preventing toxin penetration to the target site and various behavioural adaptations that lead pests to avoid exposure to the insecticide. Some of these changes highlight the fact that resistance does not necessarily entail immunity to an insecticide. Rather, these resistance mechanisms can ensure that pest species live sufficiently longer to produce enough offspring to cause the target pest population to rebound. Pesticides do not produce resistance – they only select for it and all of these resistance mechanisms would have pre-existed in the pest population at large. There is virtually no evidence that pesticides can cause genetic mutations that lead to the development of resistance.

Resistance ecology

It is because resistance is often monogenic that it is able to spread rapidly throughout pest populations. However, there are also a number of other pest species characteristics that can influence the rate of resistance development. Paramount among these are the pest species fecundity, mobility, pest status and diet. Species with a high fecundity are more likely to become resistant more rapidly than species with a lower fecundity, all other things being equal. It should come as no surprise that mites are especially liable to develop resistance and that it has been reported that over the period 1952–73 some 20 different acaricides were used in American apple orchards. Mobility will affect resistance development by spreading resistant genes (resistant individuals) or by diluting resistant populations (susceptible individuals). Since resistant individuals are usually less fit (biologically) than susceptible individuals, in the absence of the selective pressure, dispersal of resistant individuals does not usually make a significant contribution to the development of resistance in a

[5] See glossary for a definition of these terms.

larger population. Dilution of resistant populations by contrast is very important and it has been shown that leaving untreated areas (untreated pest populations) can slow the rate of resistance development. This is a strategy that is mandated for some genetically modified crops that express toxins continuously (see Chapter 13). Pest status and diet are important in relation to the number of pesticide applications to which species are likely to be exposed. Key pests that are vectors of disease or feed on high-value crops and pests that are polyphagous and widespread are likely to be more exposed to pesticides than those that feed on low-value crops and are monophagous. Frequency of exposure to pesticides is known to be critical to driving resistance development.

Crop characteristics that will enhance the development of resistance in pest species include the relative value of the crop and the area given over to it. A high-value crop is likely to be heavily treated with pesticides because thresholds for pesticide applications will be low. If it is planted over a large area, then a large number of pests will be exposed to these pesticides. Both of these factors will speed up the development of resistance.

Although the number of pest species that are resistant to insecticides continues to increase, there are only a small number of species that give cause for major concern at present. Some of the species that are of current (2002) major concern are listed in Table 4.4. Not surprisingly, the table lists species that show most of the characteristics already discussed. What will happen concerning the control of these species can only be guessed at. However, there are certainly some researchers who believe that the use of chemicals to control pest species will inevitably decline if catastrophic situations are to be minimised.

Resistance and transgenic crops

The impact of developments in molecular biology on insect pest control will be discussed in detail in Chapter 13. However, it is worth noting here some of the likely impacts upon resistance development. A number of crops have now been genetically engineered to be resistant to insect pests. The resistance mechanism employed in the plants that are commercially available involves expression of a monogenic toxin derived from a bacterium. The problem with these plants is that the toxin is continually expressed. This means that the pest population is continually exposed to a selective pressure to develop resistance. It is the author's belief that use of these plants will lead to rapid and

Table 4.4. *Major arthropod pests that have been recorded as resistant to at least one insecticide in more than one country*

Pest species	Common name
Anopheles spp.	Mosquitoes (vectors of malaria)
Anthonomus grandis	Cotton boll weevil
Aphis gossypii	Cotton aphid
Bemisia tabaci	Whitefly
Chilo suppressalis	Rice stemborer
Cydia pomonella	Codling moth
Frankliniella occidentalis	Western flower thrip
Heliocoverpa, Heliothis spp.	Bollworms, budworms
Musca domestica	Housefly
Myzus persicae	Peach-potato aphid
Nephotettix spp.	Rice leafhoppers
Nilaparvata lugens	Rice brown planthopper
Panonychus citri	Citrus red mite
P. ulmi	Fruit tree red spider mite
Plutella xylostella	Diamondback moth
Spodoptera spp.	Armyworms
Tetranychus urticae	Twospotted spider mite

Source: Data partially collated from Insecticide Resistance Action Group at http://plantprotection.org/IRAC/.

widespread resistance development to the expressed toxins. Insect pest species will simply develop to overcome the selective pressure exerted.

Resurgent pests, secondary pests and nontarget mortality

The second major problem with many modern insecticides concerns their effects on nontarget beneficial species. Most insecticides have a broad spectrum of activity and are toxic to both pest and beneficial species alike. A broad spectrum of insecticide activity tends to be favoured by both manufacturers and growers since the same product can then be used against a number of different pests. The most extensively applied insecticides therefore also tend to have a broad spectrum of activity. Because of economies of scale, this often makes these products the cheapest to purchase and so the cycle continues. The result of a broad spectrum of activity is that beneficial predators and parasitoids are often killed along with the target pests following an insecticide application. This nontarget mortality may have an insignificant impact upon the overall survival of the species affected but it may have a dramatic effect upon the functional impact that these species make. The elimination of this free or natural pest control has frequently led to reinvasion and/or reproduction by the original pest as well as reinvasion and/or reproduction by additional pests. These pests are referred to as resurgent pests and secondary pests, respectively.

Resurgent and secondary pest outbreaks often occur rapidly following a pesticide application because many of these species are typically highly fecund and highly dispersive (i.e. invasive). Biologically, many of these species display life-history characteristics that are consistent with the *r*-end of the *r*–*K* continuum[6] (see Chapter 1). Reinvasion and reproduction by predators and parasitoids, by contrast, usually takes longer. Removing any endemic natural control (with a pesticide application) therefore presents pest species with an ideal opportunity to increase in number rapidly, i.e. because they can colonise, feed and reproduce in a food-rich environment that is predator-free. In 1995 it was estimated that the extra pesticide applications that are made to control these additional pests cost the USA alone about $520 million per year.

The situation where the number of pesticide applications that has to be made increases year after year has been called the pesticide treadmill. In a few cases it has led to the complete collapse of the agroecosystem. Figure 4.7 details the outbreak of the whitefly *Bemisia tabaci* that occurred following the introduction of pyrethroids in the Imperial Valley, California, in 1975. Pyrethroid insecticides were applied to control pink bollworm and tobacco budworm populations. In addition to controlling these pests, the beneficial parasitoids *Encarsia* spp. and *Eretomocerus* spp. were wiped out. The upshot then was the development of whitefly as a secondary pest. Other examples of resurgent and secondary pest population development are given below.

In north-west Mexico in the 1970s excessive pesticide use in cotton crops led to a minor pest (*Heliothis zea*) achieving major pest status following destruction of natural enemies. This problem eventually became so great that the cotton industry collapsed. In Egypt, pesticides used on cotton to control bollworms led to the development of severe whitefly problems. In North America, one of the most widely studied examples of a secondary pest

[6] Species that are *r*-selected are ideally suited to exploit temporary agricultural crops. These are species that have evolved to exploit ephemeral habitats rapidly (see also Chapter 1). It should also be noted that the trend to use nonpersistent pesticides (see Chapter 3) can enhance the rate at which resurgent and/or secondary pest outbreaks occur.

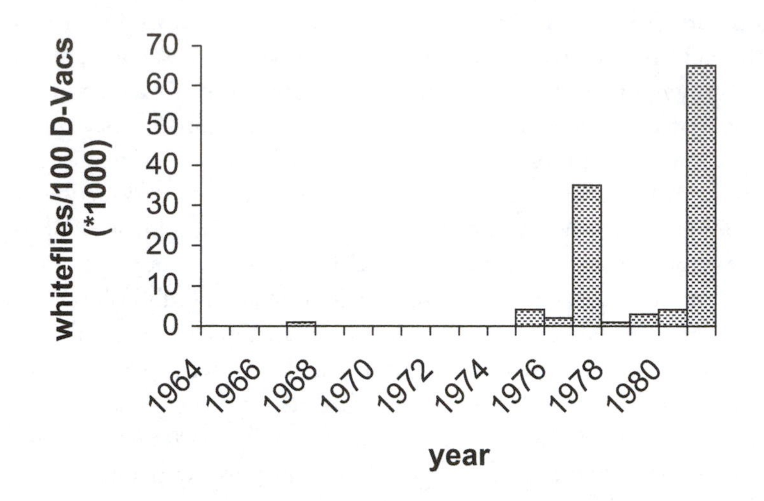

Fig. 4.7. Outbreak of the whitefly *Bemisia tabaci* in the Imperial Valley, California following the introduction of synthetic pyrethroids in 1975. Reproduced with permission from Metcalf, R.L. and Luckmann, W.H. (1994). *Introduction to Insect Pest Management*, 3rd edn. New York: John Wiley. Copyright © 1994 by John Wiley & Sons, Inc.

outbreak concerns spider mites in fruit orchards. These phytophagous mites are typically controlled by a complex of predatory species that includes other mites and beetles. However, when pesticides are applied, destruction of natural enemies has frequently led to spider mite outbreaks. The result is that many growers are now being far more careful concerning the products they use and the timing of their pesticide applications. It has been estimated that on a global basis *c.* 50% of the pests that are controlled today using pesticides became pests as a direct result of pesticide use: before pesticides were extensively used, these organisms were not considered important enough to merit control measures.

Ecosystem side-effects associated with insecticide use

In addition to problems associated with resurgent and secondary pests there are a number of other ecosystem problems that can develop as a result of pesticide use. These problems are often interrelated and include a general contamination of terrestrial and aquatic environments, a disruption of microbial activity in soils, a decline in food availability for species at higher trophic levels and a disruption in community structure. Of these, perhaps one of the

Table 4.5. *Biomagnification of DDD in Clear Lake, California during the 1950s. The lake was treated at a rate of 20 ppb to control larvae of the gnat* Chaeborus astictopus

Source	Concentration	Magnification factor[a]
Lake water	0.02 ppm (20 ppb)	NA
Plankton	5 ppm	250
Frogs	40 ppm	2000
Sunfish	240 ppm	12000
Western grebes	1600 ppm	80000

Note:
[a] In relation to original concentration applied.
NA, not applicable.

Source: Data modified from Hunt, E.G. & Bischoff, A.I. (1960). Inimical effects on wildlife of periodic DDD applications to Clear Lake. *California Fish and Game*, **46**, 91–106.

best-documented problems comprises that of biomagnification of DDT. This process (also referred to as bioconcentration) occurs because DDT is a stable, lipophilic molecule that accumulates in fatty tissues in organisms that consume it. The build-up of DDD[7] (an insecticide that is structurally similar to DDT – sometimes referred to as rothane) in Clear Lake, California, following its application for midge control in the mid-1950s is shown in Table 4.5. DDT, although now banned in most developed countries, is still widely used for vector control in developing countries. Its continued use and its stability have meant that residues of the molecule can now be found worldwide, east to west and north to south.

Community-level problems that have occurred following pesticide use have been related to mortality of species that are used as food by species at higher trophic levels. For example, the decline in numbers of the grey partridge within the UK over the past 30 years has been attributed to the indirect effects of widespread pesticide use. The Game Conservancy, in particular, has argued vociferously that pesticide use is one of the factors that has been

[7] DDD is produced by removing one chlorine atom from the trichloromethyl group of DDT (see Chapters 1 and 3). It is also a breakdown product of DDT. In North America it is often referred to as tetrachlorodiphenylethane (TDE). It is no longer used for pest control. See also Chapters 1 and 3 for discussion of biomagnification.

responsible for the decline in numbers of this species. Pesticides have killed invertebrates that form a large part of the diet of young chicks, leading to a smaller number of young being successfully fledged.[8]

Insecticides may also affect communities by interfering with microbial activity in soils. It has now been well documented that some soils have high levels of pesticide-detoxifying bacteria. While this may seem to be a good thing, it has in fact led to a complete breakdown of pesticide activity in some cases. The best-known example is carbofuran and the control of soil pests on corn crops in the USA. In some states (e.g. Iowa) soils have become so aggressive that the pesticide is no longer effective.

In summary, the range of adverse effects that pesticides can have upon ecosystems is extensive. Only a few have been touched upon here. For example, contamination of water supplies – a critical issue for many countries – has not been discussed. The list is extensive because pesticides are toxic molecules that are purposely released into the environment. Once in the environment, they can spread and accumulate and exert effects many kilometres from their point of release. The purpose of this section was to illustrate some of these adverse effects. Further examples can be found in the selected reading at the end of the chapter.

Human problems associated with exposure to insecticides

The final category that needs to be discussed in relation to pesticide side-effects concerns their adverse effects upon the human population. Humans can be exposed to pesticides by eating contaminated food or drinking contaminated water, by inhaling pesticide droplets from the atmosphere, by coming into contact with treated vegetation or through involvement in their manufacture or application.

It is known that pesticides can cause certain cancers, birth defects, sterility problems, genetic mutations and behavioural changes. In recent years researchers have also begun to investigate the effects of pesticides on the human endocrine system because of evidence that some pesticides may be responsible for altering the gender of species.

Despite this extensive list, the general public in either developed or

[8] Despite the decline in numbers of the grey partridge, this organism has never been at risk of extinction, i.e. its long-term survival has never been threatened. The issue has been the supply of a high-enough bird population for individuals in the shooting community to pay to kill!

Box 4.1

Instructions for farmers using pesticides in developing countries and associated problems. Instructions abstracted from Bull, D. (1982). A Growing Problem. Oxford: Oxfam[a]

1. 'Store in a cool, dark, dry place' – usually impossible for rural farmers in developing countries.
2. 'Use rubber gloves and a face mask' – usually unavailable.
3. 'In case of accident call a doctor' – a doctor may be more than a day away.

Note:

[a] Instructions are rarely in local dialects and many rural workers cannot read anyway!

developing countries are not at particular danger from exposure to pesticides. It is the workers involved in applying pesticides who are most at risk. This is especially the case in developing tropical countries where pesticides may be applied by hand and protective clothing may be unavailable or unaffordable. Box 4.1 highlights some of the problems faced by farmers in developing countries who are exposed to pesticides. One of the most dramatic examples of human poisoning comprises that of 1500 Costa Rican banana plantation workers who became sterile after repeated exposure to the nematicide dibromochloropropane. Overall, data that have been collected from India, Central America, Malaysia, Uganda, Brazil and parts of the former Soviet Union indicate that growers who have experienced prolonged exposure to pesticides have developed eye, skin, pulmonary, neurological and renal problems. At the end of the 1990s in the UK, sheep farmers claimed that exposure to organophosphorus sheep dips had caused a range of neurological problems. Similar claims were also being made by veterans of the 1991 Gulf War. At the time of writing these matters still had to be resolved, although a condition called organophosphate-induced delayed polyneuropathy (OPIDP) had been identified. This condition presents as memory loss and mood swings and is caused by irreversible neurological defects following exposure to organophosphates.

The number of people involved in pesticide poisoning varies enormously depending upon the data collected. For example, a World Health Organization (WHO) report for 1989 concluded that about 1 million people were poisoned annually by pesticides and that about 20000 deaths occurred. The International Labour Organisation, by contrast, calculated in 1994 that about 5 million people are poisoned each year, 40000 of them fatally. Finally, a World Health Statistics report from 1987 concluded that about 3 million people were poisoned by pesticides each year, with 20000 fatalities. These figures, although different, are not that far from each other and are probably a fairly accurate reflection of the consequences of human exposure to pesticides. Part of the problem with the data is that reports of poisonings will tend to be overestimated in developed countries and underestimated in many developing countries. What all researchers agree on is that 80–95% of pesticide-related fatalities happen in developing countries – countries that are responsible for only 20% of global pesticide use.

Industrial production of pesticides – Bhopal

In addition to user exposure to insecticides, the industrial manufacture of pesticides may also cause problems for human health. While data for single events are normally unavailable, one exception is the data that resulted from the world's worst-ever industrial accident. In 1984, a pesticide-manufacturing plant in Bhopal, India, accidentally released a cloud of toxic methyl isocyanate. The plant was owned by Union Carbide and was involved in the industrial production of the insecticides carbaryl and aldicarb. Of the 800000 people living in Bhopal at the time, 2000 died immediately, 300000 were injured and up to 10000 have died since. Compensation for the victims of the tragedy was eventually agreed in 1989 when $470 million dollars was paid in a full and final settlement. The incident led to the establishment of a responsible care programme by the Chemical Manufacturing Association that has the aim of improving community awareness, emergency response procedures and employee health and safety. Since 1984 no tragedies on a similar scale have occurred that involve the industrial manufacture of pesticides. Rather, industrial problems with pesticides have been associated with the release (leakage) of pesticides from storage and/or disposal facilities.

Consumption of food

Apart from direct exposure to sprays, the other main way in which the human population is exposed to pesticides is through consumption of contaminated food. Except for obvious cases where people have died from eating poisoned food, the real human concern here surrounds prolonged exposure to low levels of pesticides. You would probably never be able to prove that eating food contaminated with pesticides in miniscule (safe) amounts could cause illness 30 years later in life. This, however, is a real concern for people. We know that pesticides are present in most of the food products we eat and that in some cases the residues exceed official limits. It is known that official (safe) levels for pesticides are often exceeded in many developing countries. This occurs for a variety of reasons, including neglect of instructions and lack of enforcement of abuses associated with pesticides, notably in critical areas like harvest intervals.[9] Most developing countries do not have the money to carry out this policing role. For chemical companies, of course, overuse of their products means higher sales. Fortunately, in developed countries, the wealth does exist to police abuses associated with pesticide misuse. This and other legislation associated with pesticide use are discussed briefly in the final section of this chapter.

Legislation and pesticides

In developed countries the methods that are used to protect the public from abuses associated with pesticides are largely legislative. Such laws cover the following areas.

- legislation associated with pesticide manufacture, transport, storage, purchase and use;
- legislation associated with permissible residues in food for human consumption;
- legislation associated with the export or import of food items both in relation to pest control and in relation to residues.

In the UK it is the Department for Environment, Food and Rural Affairs (DEFRA, formerly the Ministry of Agriculture, Fisheries and Food) that is largely responsible for ensuring that pesticide-related legislation is adhered to, while in the USA it is the Environmental Protection Agency. The number of separate articles of legislation that exist in relation to pesticides are too numerous to be adequately discussed here (see Glossary for further information). However, in the UK some examples of the legislation that exist include the following:

- Health and Safety at Work Act 1974 – ensures employers and employees take reasonable care for their own safety and for that of others.
- Food and Environment Protection Act 1985 – designed to safeguard the environment to ensure humane pest control.

[9] All pesticide manufacturers provide information on pesticides that is concerned with the minimum amount of time between product application and crop harvest. This is the harvest interval. These intervals are designed to minimise human exposure to pesticides and to ensure that the levels to which we are exposed are safe.

- Food Safety Act 1990 – relates to provision of safe food.
- Control of Pesticide Regulations 1986 – relates to training in pesticide use and to the supply and storage of pesticides.
- Control of Substances Hazardous to Health 1994 – relates to protection from exposure to toxic substances during work.

At the time of writing, European Union countries were attempting to harmonise much of their pesticide-related legislation. Some progress had been made on this, particularly in relation to residues permitted in water and some processed foods. It is anticipated that this trend to harmonisation will continue, not least because it will save manufacturers an enormous amount of money.

Overall, the aims of most legislation in developed countries are threefold. The first aim is to protect humans, other creatures and plants from pesticides. The second is to protect the environment from the harmful effects of pesticides. The third and final aim is to secure safe, efficient and humane pest control. For the most part the legislation that is in place achieves these aims excellently. The same cannot be said of the situation in the developing world.

While most countries use International Phytosanitary Certificates as part of their quarantine procedures to help ensure that pests are not spread around the globe, less attention is given to pesticides. It is frequently the case that pesticides will be applied at the wrong rate (usually higher than needed), without safety equipment (because it is not available) and that the correct harvest interval will not be used (because it is difficult to police this). Farmers in rural areas often have additional problems with pesticides because they find alternative uses for pesticide containers once they are empty. Governments and some companies do not help farmers either because they often provide free or subsidised pesticides that are of an older, more dangerous chemical composition. Some countries also base their product registration procedures on the spurious notion that if chemicals are registered for use in a developed country then they must be 'OK' for a developing one – a mindset that totally fails to acknowledge that climatic conditions are likely to have a dramatic effect upon the toxicity of a pesticide. This situation is slowly beginning to change. However, it is still the case that the developing world uses the least pesticide and has the most problems associated with human exposure to these chemicals (see also earlier comments).

Summary

In summary, the development of modern synthetic insecticides has had an enormous impact on lives all around the globe. The number of lives saved as a result of the development of DDT as an insecticide has been claimed to exceed the numbers saved as a result of penicillin use. If these data are correct then this must surely make it one of the most important chemicals on our planet. Although banned in most developed countries, DDT continues to be used successfully for the control of vectors of disease in developing countries. Although DDT is a very stable molecule and is liable to bioconcentrate in fatty tissues within animals, it is also very safe to humanity.

Although perhaps difficult to appreciate, the cost of food in many developed countries is a fraction of what it was before the Second World War. This change is due in part to more effective chemical pest control. It is also the reason why many claim that food prices would double or treble if pesticides were not used at all. The importance of this change in terms of quality of life cannot be overestimated. On a global basis, in 1970 it was estimated that one in three people on the planet faced food shortages (because of cost and availability) while by 1995 this number had declined to one in five, despite a concomitant increase in the world population. This is a highly significant development in which pest control using chemicals has undoubtedly played a part.[10]

At present, the world population is expected to reach around 7 billion by about 2010[11] and it has been calculated that global food output will need to rise by about 70% if mass starvation is to be avoided. Since most of the world's suitable agricultural land is already in production, this will necessitate more intensive farming because expanding the area farmed to increase output will not be an option. To farm more intensively requires a high level of pesticide use.

Despite their benefits, pesticides have also come in for a lot of public criticism. Such criticism tends to emphasise public health and environmental impacts. Because of such criticisms these areas are now beginning to be taken more seriously by chemical companies. There is no reason to expect this trend to change.

In the twenty-first century it is highly likely that chemicals will continue to play a critical role in crop protection. It is unlikely that their role will decline, especially once the general public fully considers the importance of the need for a global increase in food production. However, it is also likely that other techniques for pest control will receive greater attention in the next decade, particularly with developments in molecular biology. Many of these

[10] While there is no doubt that the price of food (in general) has declined worldwide, improved pest control is only part of the story. Clearly there are also political and socioeconomic reasons for this change.

[11] The current (2002) world population is *c.* 6 billion. The estimate of 7 billion by 2010 is provided by both the United Nations and the World Bank.

techniques will become complementary and will be part of an integrated approach towards pest control. Integrated pest management is discussed in Chapter 12. These other techniques will include an expanded use of biological control agents and an increased use of genetically manipulated pests, predators and plants. The next two chapters in this book discuss the use of biological control agents. Chapter 5 considers predators and parasitoids while Chapter 6 discusses pathogens and their utility for arthropod pest control.

Further reading

Brooks, G.T. & Roberts, T.R. (1999). *Pesticide Chemistry and Bioscience*. Cambridge: Royal Society of Chemistry.

Bull, D. (1982). *A Growing Problem – Pesticides and the Third World Poor*. Oxford: Oxfam.

Carson, R. (1962). *Silent Spring*. London: Hamish Hamilton.

DeLorenzo, M.E., Scott, G.I. & Ross, P.E. (2001). Toxicity of pesticides to aquatic microorganisms: a review. *Environmental Toxicology and Chemistry*, **20**, 84–98.

Hislop, E.C. (1991). Air-assisted crop spraying: an introductory review. *British Crop Protection Council Monograph*, **46**, 3–14.

Hunt, E.G. & Bischoff, A.I. (1960). Inimical effects on wildlife of periodic DDD applications to Clear Lake. *California Fish and Game*, **46**, 91–106.

Johnsen, K., Jacobsen, C.S., Torsvik, V. & Sorensen, J. (2001). Pesticide effects on bacterial diversity in agricultural soils – a review. *Biology and Fertility of Soils*, **33**, 443–53.

Landis, W.G. & Yu, M.H. (1995). *Introduction to Environmental Toxicology: Impacts of Chemicals on Ecological Systems*. Boca Raton, Florida: Lewis.

Marrs, T.C. (2000). Organophosphates and human health. In *Proceedings of British Crop Protection Conference – Pests and Diseases*, vol. 2, pp. 1107–16. Farnham, Surrey: British Crop Protection Council.

Matthews, G.A. (2000). *Pesticide Application Methods*, 3rd edn. Oxford: Blackwell Science.

Metcalf, R.L. & Luckmann, W.H. (1994). *Introduction to Insect Pest Management*, 3rd edn. New York: John Wiley.

Pimental, D. (1996). *Techniques for Reducing Pesticide use*. New York: John Wiley.

Ragnarsdottir, K.V. (2000). Environmental fate and toxicology of organophosphate pesticides. *Journal of the Geological Society*, **157**, 859–76.

Ragoucy-Sengler, C., Tracqui, A., Chavonnet, A. *et al.* (2000). Aldicarb poisoning. *Human and Experimental Toxicology*, **19**, 657–62.

Sanchez-Hernandez, J.C. (2001). Wildlife exposure to organophosphorus insecticides. *Reviews of Environmental Contamination and Toxicology*, **172**, 21–63.

Stephens, R., Spurgeon, A., Calvert, I.A. *et al.* (1995). Neuropsychological effects of exposure to organophosphates in sheep dip. *Lancet*, **345**, 1135–39.

Trewavas, A. (2001). Urban myths of organic farming. *Nature*, **410**, 409–10.
Wiktelius, S. & Edwards, C.A. (1997). Organochlorine residues in the African terrestrial and aquatic fauna 1970–1995. *Reviews of Environmental Contamination and Toxicology*, **151**, 1–38.

5

Biological control agents

Introduction

Biological control agents are organisms that use pest species as trophic resources. They are often referred to as natural enemies, beneficials or biocontrol agents. In most situations these beneficial species are so efficient that pest populations are held below economically damaging levels. Some estimates have suggested that *c.* 99% of all potential pests are controlled by natural enemies. Indeed, it is the destruction of the relationship between naturally occurring biocontrol agents and pests that is often responsible for pest outbreaks. This can occur when pest species colonise new geographic areas without their natural enemies, when natural enemies are destroyed by pesticide applications (see Chapter 4), or when the habitat that pest and predator occupy is modified to the benefit of the pest, for example, when a monoculture is planted over a very large area. Even in situations where pests do cause economic damage it will usually be the case that this damage would be even greater were it not for the activity of beneficial natural enemies. Figure 5.1 gives a simple example of how easy it is to produce a pest outbreak artificially by simply applying a broad-spectrum pesticide to control a pest. The pesticide severely curtails predatory activity, with the result that the pest population rapidly resurges. The process of pest resurgence was explained in detail in Chapter 4.

Within pest control, the aim of using biological control agents is to restore and/or enhance the relationship between pest and biocontrol agent either by reintroductions and/or by creating the habitat conditions under which a relationship will be strengthened or will form naturally. Whether the aim is to restore the relationship on a temporary or permanent basis will depend on

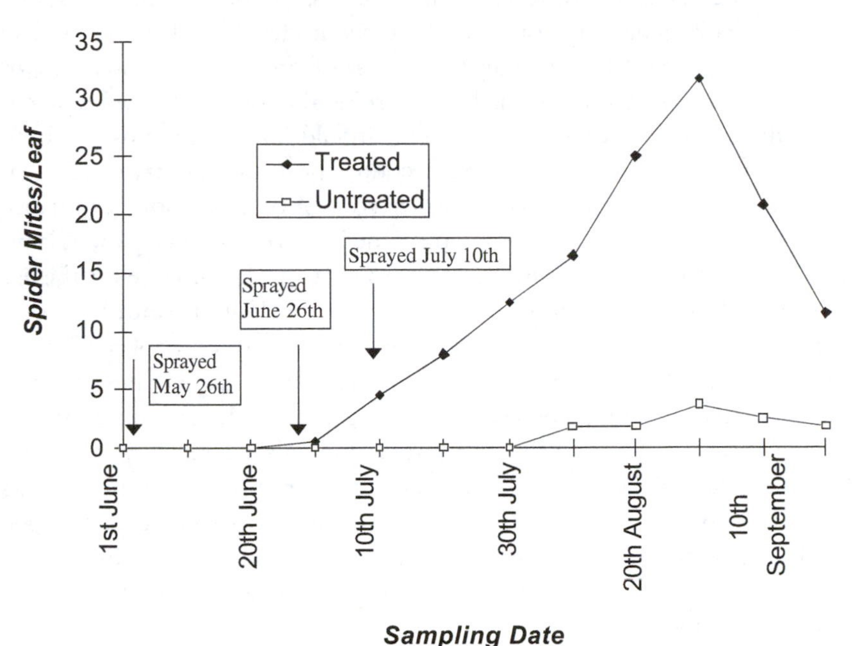

Fig. 5.1. An experimental demonstration of the increase in Pacific mite *Tetranychus pacificus* following applications of carbaryl. Despite the pesticide applications, the pest population rapidly resurges because of natural enemy mortality. Reproduced with permission from Debach, P. & Rosen, D. (1991). *Biological Control by Natural Enemies*. Cambridge: Cambridge University Press.

circumstances. For stable plantation agroecosystems it is likely that long-term control (over more than one season) would be desirable. For temporary (or unstable) annual crops, however, it may be enough that the control lasts only until harvest. Within the context of biological control it is important to recognise that the word 'control' does not mean eradication. Rather, it is meant to reflect population suppression to a point where the pest is no longer economically damaging.

As a control technique, biocontrol is most suited to pest species with a relatively high economic injury level (EIL; see Chapters 1 and 12). This is because a minimum prey density will usually be required to support a permanent predatory population. Where a permanent association is not the aim it will still be necessary to have a relatively high EIL since predators are rarely able to exploit prey to the point of extinction. The species that act as biocontrol agents and use pests as trophic resources include viruses, bacteria, fungi, protozoa, nematodes, arthropods, amphibians, reptiles, fish, birds and mammals. Pathogenic organisms are discussed in detail in Chapter 6 and so the bulk of this chapter is given over to the use of predatory invertebrates for pest control. Brief mention is made concerning the use of vertebrate predators at the end of the chapter. The reader should note that the use of both invertebrate and vertebrate predators to control pests may or may not be synonymous with what is called biological pest control. Some authors regard biological control as including all pest control methods that use living organisms, such as host-plant resistance and various genetic techniques, as well as the use of microbial organisms. Other authors consider biological control to comprise only the use of predatory invertebrates to control pest species. In the present chapter I focus upon the use of predatory invertebrates for pest control. I will leave it up to the reader to decide whether this is a subset of biological control or whether this constitutes biological control *in toto*. This chapter is structured so that characteristics of invertebrate biocontrol agents are considered first. Different procedures for using these organisms are then described and specific examples of successful biocontrol programmes are provided. The chapter concludes with a brief review of the results from biocontrol programmes worldwide.

Predatory invertebrates

Predatory invertebrates that feed on insects are known as entomophages. It has been estimated that only 15% of all entomophages have so far been identified. This is not surprising. Most insects are not pests, because they are controlled by predatory species. Most research on entomophages relates to the small proportion of insects that are of economic importance, i.e. natural control that is working does not receive funding and/or associated research attention.

The predatory invertebrates that are of interest to biocontrol can be characterised as belonging to one of two groups: predators or parasitoids. In general, the former are external feeders that consume prey items rapidly (seconds to minutes) while the latter are external or internal feeders that consume prey items more slowly (hours to days). The word parasitoid means 'parasite-like' and is used to describe species that kill their host. Most parasites do not kill their hosts, at least in the short term. Most parasitoids, by contrast, complete their juvenile development within a host (which is killed) but, as adults, are free-living.

Of predators and parasitoids, it is the latter that have been most extensively studied for biological control. Table 5.1 lists the species that were

Table 5.1. *Commercially available predators and parasitoids*

Common name (family)	Latin name	Primary target
Predatory beetle (Coccinellidae)	*Adalia bipunctata*	Aphids
Predatory mite (Phytoseiidae)	*Amblyseius barkeri*	Thrips
Predatory mite (Phytoseiidae)	*Amblyseius californicus*	Spider mites
Predatory mite (Phytoseiidae)	*Amblyseius cucumeris*	Thrips
Predatory mite (Phytoseiidae)	*Amblyseius degenerans*	Thrips
Predatory mite (Phytoseiidae)	*Amblyseius fallacis*	Spider mites
Predatory mite (Phytoseiidae)	*Amblyseius montdorensis*	Thrips
Parasitic wasp (Mymaridae)	*Anagrus atomus*	Leafhoppers
Parasitic wasp (Mymaridae)	*Anagrus pseudococci*	Mealybugs
Parasitic wasp (Mymaridae)	*Anaphes iole*	Lygus bugs
Predatory bug (Anthocoridae)	*Anthocoris nemoralis*	Aphids
Parasitic wasp (Aphelinidae)	*Aphelinus abdominalis*	Aphids
Parasitic wasp (Aphidiidae)	*Aphidius colemani*	Aphids
Parasitic wasp (Aphidiidae)	*Aphidius ervi*	Aphids
Parasitic wasp (Aphidiidae)	*Aphidius matricariae*	Aphids
Predatory midge (Cecidomyiidae)	*Aphidoletes aphidimyza*	Aphids
Parasitic wasp (Aphelinidae)	*Aphytis lignanensis*	Scale insects
Parasitic wasp (Aphelinidae)	*Aphytis melinus*	Scale insects
Parasitic wasp (Braconidae)	*Bracon hebetor*	Grain moths
Predatory lacewing (Chrysopidae)	*Chrysoperla carnea*	Homoptera
Predatory lacewing (Chrysopidae)	*Chrysoperla rufilabris*	Homoptera
Parasitic wasp (Aphidiidae)	*Cotesia* spp.	Caterpillars
Predatory beetle (Coccinellidae)	*Crytolaemus montrouzieri*	Mealybugs
Predatory beetle (Cybocephalidae)	*Cybocephalus nipponicus*	Scale insects
Parasitic wasp (Braconidae)	*Dacnusa sibrica*	Leafminers
Predatory beetle (Coccinellidae)	*Delphastus pusillus*	Whitefly
Predatory bug (Anthocoridae)	*Deraeocoris brevis*	Homoptera
Parasitic wasp (Eulophidae)	*Diglyphus isaea*	Leafminers
Parasitic wasp (Aphelinidae)	*Encarsia formosa*	Whitefly
Parasitic wasp (Aphelinidae)	*Eretmocerus* spp.	Whitefly
Predatory midge (Cecidomyiidae)	*Feltiella acarisuga*	Spider mites
Predatory mite (Phytoseiidae)	*Galendromus annectens*	Spider mites
Predatory mite (Phytoseiidae)	*Galendromus occidentalis*	Spider mites
Predatory beetle (Coccinellidae)	*Harmonia axyridis*	Aphids
Predatory beetle (Coccinellidae)	*Hippodamia convergens*	Homoptera
Predatory mite (Phytoseiidae)	*Hypoaspis aculeifer*	Fungus gnats
Predatory mite (Phytoseiidae)	*Hypoaspis miles*	Fungus gnats
Parasitic wasp (Encyrtidae)	*Leptomastix dactylopii*	Mealybugs
Predatory bug (Miridae)	*Macrolophus caliginosus*	Whitefly

Table 5.1. (*cont.*)

Common name (family)	Latin name	Primary target
Parasitic wasp (Encyrtidae)	*Metaphycus bartletti*	Black scale
Parasitic wasp (Encyrtidae)	*Metaphycus helvolus*	Soft scales
Predatory bug (Anthocoridae)	*Orius albidipennis*	Thrips
Predatory bug (Anthocoridae)	*Orius insidiosus*	Thrips
Predatory bug (Anthocoridae)	*Orius laevigatus*	Thrips
Predatory bug (Anthocoridae)	*Orius majusculus*	Thrips
Parasitic wasp (Eulophidae)	*Pediobius foveolatus*	Bean beetles
Predatory mite (Phytoseiidae)	*Phytoseiulus persimilis*	Spider mites
Predatory bug (Pentatomidae)	*Podisus maculiventris*	Lepidoptera
Predatory beetle (Coccinellidae)	*Rhyzobius lophantae*	Scale insects
Predatory beetle (Coccinellidae)	*Stethorus punctillum*	Spider mites
Parasitic wasp (Trichogrammatidae)	*Trichogramma brassicae*	Lepidoptera
Parasitic wasp (Trichogrammatidae)	*Trichogramma evanescens*	Lepidoptera
Predatory mite (Phytoseiidae)	*Typhlodromus occidentalis*	Spider mites
Predatory mite (Phytoseiidae)	*Typhlodromus pyri*	Spider Mites

Source: Data collated from Copping, L.G. (ed.) (2001). *Biopesticide Manual.* Farnham, Surrey: BCPC Publications. This list is not complete and does not include some of the organisms that have been used in government-sponsored biological control programmes.

commercially available worldwide for pest control in 2001. Of the 54 species available, 21 are parasitoid wasps, 13 are predatory mites, eight are predatory bugs, eight are predatory beetles, two are predatory midges and two are predatory lacewings. In addition to these commercially available biocontrol agents, there are also numerous examples of successful biocontrol projects that have been government-funded. Some of these government-funded projects are discussed later in the chapter. The biological characteristics of parasitoids are described in the next section.

Parasitoids

Parasitoids can be classified depending upon whether they are external feeders (ectoparasites) or internal feeders (endoparasites), and by whether they feed on eggs, larvae, pupae or adults. These categories can overlap. For example,

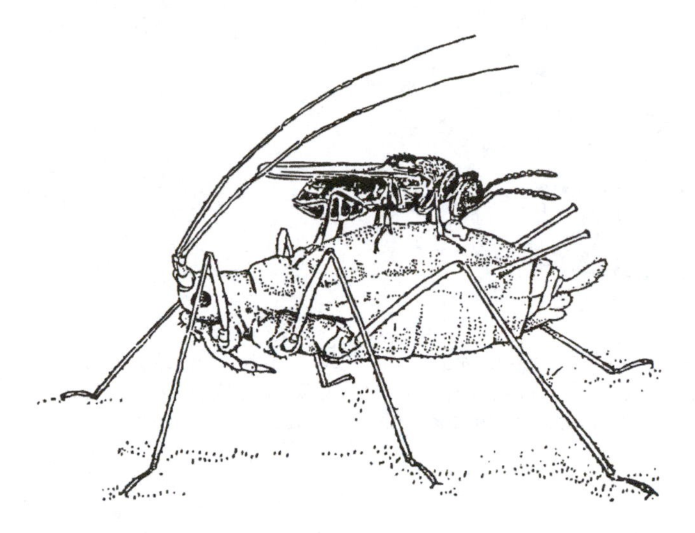

Fig. 5.2. Predatory host-feeding by *Aphidencyrtus aphidivorus*. Reproduced with permission from Debach, P. & Rosen, D. (1991). *Biological Control by Natural Enemies*. Cambridge: Cambridge University Press.

the larvae of the parasitic wasp *Scutellista cyanea* feed on the eggs of soft-scale insects. However, if no eggs are available then the larvae will switch and feed on the scale insect itself. With most parasitoids, it is the larvae that attack the host. However, there are also cases where adults will host-feed on prey, much like spiders. In these cases it is probably more correct to refer to parasitoid larvae and predatory adults. The process of host-feeding is shown in Figure 5.2. Where adult parasitoids do not host-feed (most cases) they generally feed on nectar and pollen from flowers. The importance of providing adequate food for adult parasitoids is often overlooked and will be discussed in more detail later.

Ectoparasitic species are usually found in situations where the host is protected, for example, where the host prey items are leafminers or scale insects. In these situations the adult may simply deposit an egg inside a feeding tunnel (e.g. leafmine) or near the host without needing to contact the host directly. Endoparasitism is much more common with hosts that are not protected, such as caterpillars or aphids. In these circumstances adults use their ovipositors to deposit an egg directly inside the host. The main difficulty that this then presents for the parasitic larvae is respiration in a liquid

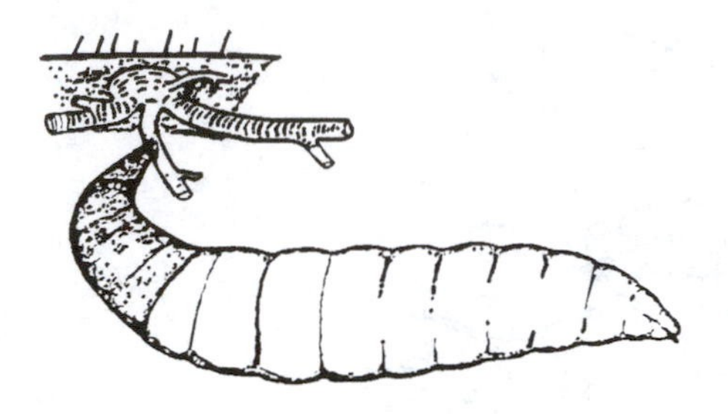

Fig. 5.3. Larva of tachinid fly *Prosena siberita* attached to the host's trachea. Reproduced with permission from Debach, P. & Rosen, D. (1991). *Biological Control by Natural Enemies*. Cambridge: Cambridge University Press.

environment. To overcome this, larvae may simply breathe oxygen directly from the host's body fluid, they may link their spiracles to the host's tracheal system or they may construct a tube that passes through the host and connects to the external air. An example of connection to the host's tracheal system is shown in Figure 5.3.

Both ectoparasites and endoparasites may be solitary or gregarious (more than one larva per host). Some species are polyembryonic, in which a single egg gives rise to many embryos. Other parasitoids display superparasitism, in which the host is attacked by more than one adult of the same species, and multiparasitism, where more than one species attacks the same host. Finally, parasites may themselves be attacked by hyperparasites. Hyperparasitism is of particular concern when species are imported into a new country for release. In most cases, imported parasites will be reared through more than one generation in quarantine before release to check for any potentially harmful hyperparasites.

The most important parasitoids belong to the insect order Hymenoptera. About 66% of all successful biocontrol programmes have involved parasitoids that belong to this insect order. At least one of the reasons for the success of these species in biocontrol is their possession of a modified ovipositor that can be used to lay eggs, to secrete venom to paralyse prey, to secrete feeding tubes and to locate prey via sensitive neural receptors. Some of the most important hymenopterous parasitoids that have so far been described are listed in Table 5.2 by superfamily and family.

Table 5.2. *Simple classification of some important hymenopterous parasitoids*

Superfamily[a]	Family[a]	Example
Chalcidoidea	Aphelinidae	*Aphytis* spp. for control of scale insects
	Encyrtidae	*Epidinocarsis* spp. for mealybug control
	Eulophidae	*Chrysocharis* spp. for control of leafminers
	Mymaridae	*Anagrus* spp. for control of leafhoppers
	Trichogrammatidae	*Trichogramma* spp. for control of Lepidoptera
Ichneumonoidea	Aphidiidae	*Aphidius* spp. for control of aphids
	Braconidae	*Dacnusa* spp. for control of leafminers
	Ichneumonidae	*Agrypon* spp. for the control of Lepidoptera
Proctotrupidoidea	Platygasteridae	*Amitus* spp. for control of whitefly
	Scelionidae	*Trissolcus* spp. for control of Homoptera

Note:
[a] These lists are not complete. There are 13 superfamilies within the Hymenoptera and substantial numbers of families. This list just gives examples of some of the more important groups that have been of use to biological control so far.

In addition to hymenopterous parasitoids, the other important group comprises dipterous parasitoids in the family Tachnidae. In general these parasitoids appear to be less host-specific. For example, some species of tachinid fly have been found to attack more than 100 different host insects. As yet, tachinids have not been used extensively in biological control programmes. There have however been a few successes with government-sponsored programmes. For example, the winter moth *Operophthora brumata* was controlled in Canada in 1957 following combined releases of the tachinid *Cyzenis albicans* and the ichneumonid *Agrypon flaveolatum*. However, there are still no commercially

Table 5.3. *Some common predatory species that contribute towards arthropod pest control*[a]

Order	Family	Examples
Coleoptera	Coccinellidae	Ladybirds feeding on Homoptera
	Staphylinidae	Polyphagous rove beetles
	Carabidae	Polyphagous ground beetles
Neuroptera	Chrysopidae	Green lacewings on Homoptera
	Hemerobiidae	Brown lacewings on Homoptera
Hymenoptera	Formicidae	Ants feeding on Lepidoptera in citrus
	Vespidae	Wasps feeding Lepidoptera to larvae
Diptera	Syrphidae	Hoverflies feeding on aphids
	Asilidae	Robberfly larvae on soil insects
Hemiptera	Anthocoridae	Polyphagous bugs
Dictyoptera	Phasmida	Polyphagous preying mantis
Odonata	'Many'	Dragonflies and damselflies feed on mosquitoes as larvae and as adults
Acari	Phytoseiidae	Predatory mites feeding on phytophagous mites and thrips
Araneae	'Many'	Polyphagous spiders

Note:
[a] This list is by no means exhaustive. It is intended to illustrate the diversity of predatory species that feed on arthropod pests.

available parasitoid formulations that incorporate tachinid flies. Figures 5.4–5.7 (colour plates) show some of the parasitoids that have been useful for biological control programmes.

Predators

In addition to parasitoids, almost all pest species are attacked by a number of different predators. Some of these predators prey on hosts as both larvae and adults (e.g. ladybirds), but many are predatory only as larvae (e.g. hoverflies). A list of the best-known predatory species is given in Table 5.3. From a commercial point of view, by far the most successful group comprises predatory mites. Predatory mites are, at present, used on a relatively large scale by farmers, by nursery managers and by home gardeners to control phytophagous mites and thrips. Figure 5.8–5.11 (colour plates) show some of the predators that can be used in biological control programmes. Biocontrol agents (whether predators or parasitoids) can be used in three different ways. These techniques are not mutually exclusive. However for simplicity, each will be discussed in turn.

Methods for using biocontrol agents

The three ways in which biocontrol agents can be used have been called: (1) classical biological control; (2) augmentation/inoculation; and (3) habitat manipulation. In many situations a combination of these techniques can be used. For example, habitats may be manipulated to assist with the survival of predatory species that are released with the aim of controlling a pest outbreak. The first two techniques involve rearing and/or collection of predators prior to their release; the last technique does not. The first and last techniques may be more or less permanent, once effective pest control is established. The differences between these techniques have been described as a continuum from products (predators) to techniques (habitat manipulation; Figure 5.12). In real-life situations most biological control programmes will involve a blend of products and techniques, as stated earlier.

The primary difference between the classical approach and an augmentative/inoculative release is that the former attempts to set up a permanent relationship between pest and predator while the latter seeks to use the predator as a biological pesticide. Augmentation is used where a predator population already exists, albeit at ineffectual (for pest control purposes) levels. Inoculation is carried out in situations where natural enemies are absent.

Potentially, all of these techniques are environmentally friendly and safe to human health.[1] Unlike the use of pesticides, none of these techniques is designed to be eradicative. The aim is to maintain pest populations below levels that are economically damaging. In most cases it will in fact be essential that some pests survive as a resource for the natural enemy population.

Because the first two techniques involve the collection/rearing and release of natural enemies, this means that techniques need to be developed to produce, store, distribute and apply these organisms while ensuring that they remain viable.[2] Production (or rearing) is often expensive because

[1] Unless the biocontrol programme goes wrong and introduced animals begin to attack nontarget species.

[2] Quality control of mass-reared organisms is often critical to their success once released. This can add substantially to the costs associated with breeding effective biocontrol agents.

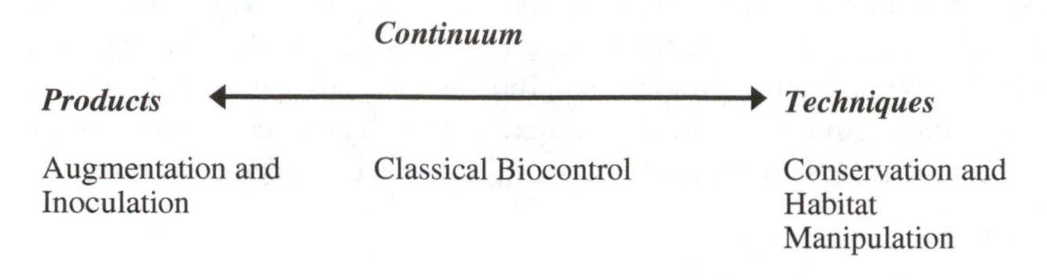

Fig. 5.12. Classification of biological control methods. Adapted from Gurr, G.M. & Wratten, S.D. (1999). 'Integrated biological control': a proposal for enhancing success in biological control. *International Journal of Pest Management*, **45**, 81–4.

tritrophic[3] culturing systems are required in which plants are grown to support pest populations that can then support predator populations. It is also important that species that are mass-reared for release suffer no loss of fitness as a result of laboratory culturing. Storage is also usually expensive because it is much harder to keep a living organism alive than it is to keep a pesticide viable for a satisfactorily commercial shelf-life of about 2 years (see Chapter 4). Storage is obviously more critical for augmentative/inoculative programmes where the biological pesticide is designed to be used like a chemical. In the following three sections we will look at each of these biocontrol methods in more detail.

Classical biological control

Although the use of biological control agents can be dated back at least 2000 years to the Chinese, who used ants to protect their crops from insect pests, the first well-documented case of biological control can be found in California in the USA, and is an example of what is now called classical biological control. The basic approach with this technique is to identify natural enemies that control a pest in its home location and to introduce these enemies in the pest's new location. It is therefore usually used to control species that have achieved pest status following invasion of a new geographical location. Two specific examples are provided below.

Following escalating outbreaks of the invasive alien pest, the cottony

[3] Tritrophic pertains to three 'trophic' or feeding levels. At the lowest level, plants operate as autotrophs or 'self-feeders'. Pests and predators then act as heterotrophs, feeding on the plant and the pest, respectively.

cushion scale *Icerya purchasi,* in citrus orchards in California during the nineteenth century, a predatory beetle was eventually found and released to control these pests in 1888. This scale insect had been introduced and had become established in the Los Angeles area some time before 1886 and by 1888 had already destroyed thousands of citrus trees. It was known that the pest was of Australian origin but that it was not a pest there. Preliminary studies in Australia identified three putative biocontrol agents – a parasitic fly *Cryptochetum iceryae*, a beetle *Rodolia cardinalis* and a lacewing *Chrysoperla carnea.* Of these species, it was the parasitoid and the beetle that were to prove useful in the control of the scale pest. From initial releases of about 500 beetles the scale insect rapidly declined in importance. It is not known exactly how many flies were released, although it is now known that these were also helpful in controlling the pest in some areas. The initial releases took place in 1888 and by the summer of 1889 control of the pest was just about complete. The whole cost of the programme has been calculated at *c.* $5000. Complete control of the pest was realised within 18 months. The beetle and the fly still control this scale insect today, over 100 years since the first introductions took place. The fly is the dominant biocontrol agent in coastal regions while the beetle is dominant in the interior. Following the success of this programme the beetle has been imported to control this scale insect in over 50 other countries.

More recently, another example of a successful classical biological control programme comprises that of the cassava mealybug. Cassava is an exotic crop that was introduced to Africa over 500 years ago. It is native to tropical Central and South America. It is now the staple food item of over 200 million people in an area known as the cassava belt, which extends from 15°N to 20°S. Until the 1970s cassava had remained relatively pest-free. In the 1970s, however, two pests of South American origin began to destroy the cassava crop. These were the cassava mealybug *Phenacoccus manihoti* and the cassava green mite *Mononychellus tanajoa.* Both of these pests spread rapidly throughout the cassava belt. In the case of the mealybug, it has been estimated that from its first detection in Zaire (DPR of Congo) in the early 1970s it spread at a rate of *c.* 300 km/year. By 1986 both pests had reached some 27 countries (*c.* 70% of the cassava belt) in Western and Central Africa. Since both these pests were exotic, a search was made by various research agencies for predators that could be used as biocontrol agents. In 1981 an encyrtid parasitoid (*Epidinocarsis lopezi*) was isolated from the mealybug and in 1988 a number of predatory mites were identified for green-mite control. The parasitoid was first released in Nigeria in 1981 and became established within a year. By 1993 the parasitoid had spread to at least 16 countries within the cassava belt and

Table 5.4. *Economics of successful classical biological control*

Pest	Country	Cost	Benefit/saving
Rufous scale	Peru	$1789	$226 916 p.a. in 1977 on chemical control
Sugarcane scale	Tanzania	$11 976	$67 832 p.a. in 1974 in increased yields
Coconut leafminer	Sri Lanka	$35 783	$12.4 million p.a. in 1977 in yield increases
Potato tuber moth	Zambia	$35 732	$547 560 p.a. in 1974 in yield increases
Mango mealybug	Togo	$175 000	$3.9 million p.a. in 1994

Source: Data collated from a range of sources. See further reading section for further details.

was successfully controlling mealybug populations. To control the green mite more than 6.1 million predatory phytoseiid mites of five species of Brazilian origin were released in 365 sites in 11 countries between 1989 and 1995. Three of these species (*Neoseiulus idaeus, Typhlodromalus manihoti* and *T. aripo*) are now established in Africa. The biggest success has occurred with *T. aripo* which is now established at more than 1000 locations in 11 countries. The control of these two pests is regarded as one of the most successful modern examples of classical biological control.

Since the first well-documented success with a biological control agent in 1888 to the present there have been more than 5000 different attempts at classical biological control. The success rate for these programmes seems to be around 30%. Table 5.4 gives sample data associated with the economics of establishing a classical biocontrol programme. The table shows quite clearly that, where control is achieved, the economic benefits are enormous. Most successes with classical biological control have been in agroecosystems that permit species to form stable predator–prey relationships and where the pest species is exotic. One of the reasons why classical biocontrol has not been a success with native pests is that introduced predators and parasitoids have to compete with preexisting natural enemy complexes. It is also a general rule that the best predators are those that have evolved with the pest to be controlled. Despite these problems, it has proven possible in some cases to control field-based native pests of annual crops using classical approaches.

These cases, although the exception rather than the rule, do demonstrate the wide applicability of the classical biological approach to pest control. However, because the aim of classical biological control is to establish a permanent association between pest and predator/parasitoid, such approaches have received little commercial investment. Although a classical biocontrol programme can provide substantial economic rewards, these programmes still rely on political support and on government funding. The next section briefly reviews the stages that are involved in setting up a classical biological control programme.

Procedure for setting up a classical biocontrol programme

Since the first successes with classical biological control, stepwise procedures have been developed that describe the processes involved in establishing such programmes. A schematic of a stepwise procedure is shown in Figure 5.13. The first stage in the procedure is to identify whether the pest species concerned is of native or foreign origin. This will require information connected with the pest's history, geographical distribution, ecology, taxonomy and, if possible, its parasitology. If the pest is exotic then the next stage is to collect information on the pest's status in its home location and to collect information on its native natural enemies. It has been shown on many occasions that the centre of origin of a pest species is where the best-developed natural enemy complex is likely to exist. This work, which often requires substantial government funding, will generally be carried out and/or supported by international agencies such as the European Biological Control Laboratory (EBCL), the Commonwealth Institute of Biological Control (CIBC)[4] or the International Organisation for Biological Control (IOBC). During the field collection process it will be important to collect species with as much genetic diversity as possible. This can usually be achieved by sampling over a large geographical area. Following collection, natural enemies are first tested in experiments in their native countries and then under quarantine conditions in importing countries. Quarantine is important to check for hyperparasitoids. Good natural enemies will generally have highly developed searching skills that permit them to control pest populations at relatively low densities. Following release from quarantine,

[4] The CIBC, which was formed in 1957, is now part of the organisation known as CABI Bioscience, an international research organisation that operates centres in Kenya, Malaysia, Trinidad, Pakistan, Switzerland and the UK.

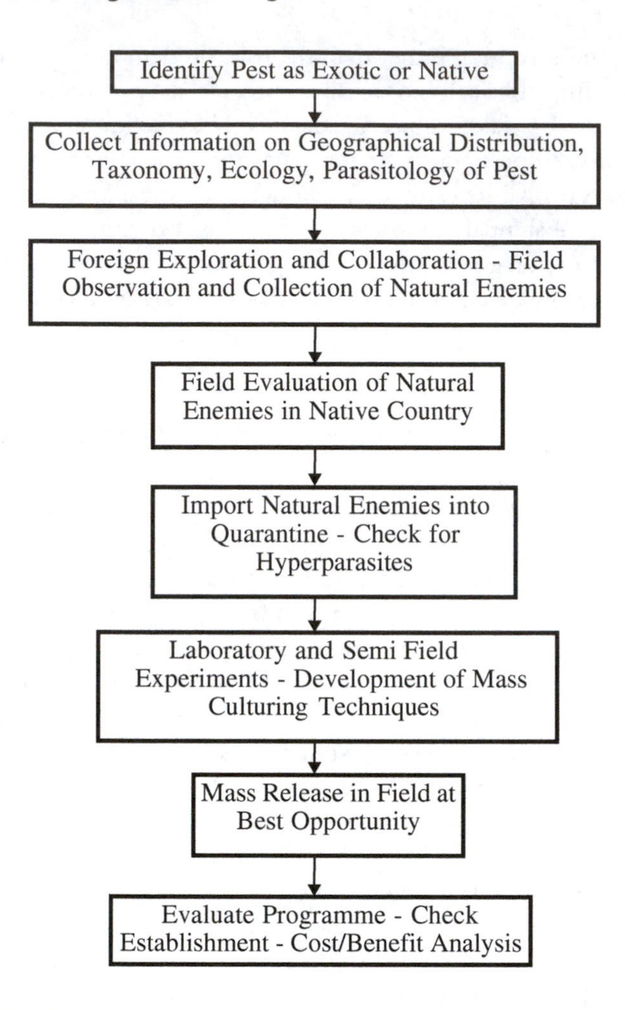

Fig. 5.13. Stepwise procedure for establishing a classical biological control programme.

further trials in laboratory and semifield (i.e. using field cages, glasshouses, etc.) conditions are then undertaken while methods for mass rearing are developed. Mass rearing or culturing is not always straightforward because natural enemies may need to be cultured in a tritrophic system in which plants are grown to support pest populations which then support natural enemy populations. In addition, it is also important to try to acclimatise natural enemies prior to their release.

Table 5.5. *Relationship between numbers released and biocontrol success in Canada*[a]

Successful introductions (%)	Number of individuals released
10%	<5000
40%	5000–31 200
78%	>31 200
Successful introductions (%)	**Number of releases**
10%	<10
70%	>20

Note:
[a] The table is intended to highlight the fact that both the number of individuals released on any one occasion and the number of separate releases can impact upon the likelihood of success. Data from Beirne, B.P. (1975). Biological control attempts by introductions against pest insects in the field in Canada. *Canadian Entomologist*, **107**, 225–36.

Following the above, natural enemies can then be mass-released. During this process care should be taken to ensure that a pest population exists for the natural enemies to develop upon. If the natural enemies, particularly adults, require alternative food, sources such as nectar should be available. The number of individuals released should also be considered. Table 5.5 shows quite clearly that there is an association between the percentage of successful introductions and the number of individuals released. The final stage in this process comprises the postrelease phase in which the aim is to monitor and determine whether releases have been effective or not. If the programme is not a success some attempt should be made to find out why. Unfortunately, it is this phase that is often the most neglected. If a biocontrol programme does not work then sponsors do not like to spend money on something that isn't working. Conversely, if the programme is working then sponsors will argue that further funding is no longer required, and this can hamper further development. Where postrelease studies have been undertaken, the data collected have proven invaluable. The time scale for all of these phases in a biocontrol programme is usually measured in years. In contrast to the classical approach which attracts little industrial sponsorship, using natural enemies as biological insecticides does represent an investment opportunity for industry since biological formulations will need to be purchased on an annual basis. These techniques are discussed is the next section.

Augmentation/inoculation techniques

The processes of augmentation or inoculation with natural enemies occur where natural enemies are absent or where they exist at levels that are ineffective for pest control. The aim is to establish a predatory population that is able to suppress pest numbers below economically damaging levels until harvest. Ecologically, the general equilibrium position of the pest population is not altered (as it is with classical biological control). Rather, the aim is to bring about a temporary dampening in peaks of pest activity so that economic crop losses are avoided. The rationale for this approach to biological control is that many effective natural enemies cannot survive on a year-round basis on a crop, either because of practical or geographical reasons. However, they are able to survive and effect control at certain times of the year, hence the need for annual releases. The critical difference between this approach and that of classical biological control therefore is that it is not long-term. Natural enemies are used as biological insecticides.

Some authors make an additional distinction between inundative and inoculative releases. In both cases the natural enemy is used as a short-term control measure. However, the former does not rely on reproduction of the natural enemy, while the latter does. This distinction is applicable where species reproduce rapidly within a season. For example, predatory mites are often used as inoculative releases on a seasonal basis in orchards. From an economic point of view inoculative releases are generally cheaper because natural enemies will reproduce once released. By contrast, inundative releases require culturing or rearing of natural enemies. This process adds to the cost of the control method. By far the greatest successes with augmentative/inoculative techniques have been with pathogenic organisms (discussed in Chapter 6); however some notable successes have also occurred with predatory invertebrates. Some of these successes are discussed below.

Because control is not usually long-term, using biocontrol agents as a temporary measure is often expensive, particularly in comparison to the costs associated with a conventional pesticide application. The expenses involved are incurred in rearing, distributing and applying the predators/parasitoids. For this reason, the use of biocontrol agents for augmentation/inoculation has been largely restricted to situations where the predators that are released can be confined. Such situations often exist with many high-value crops grown in greenhouses. In field situations where predators can easily disperse, the use of biocontrol agents is generally limited, although some successes have been achieved with the parasitoid *Trichogramma* spp. that parasitises the eggs of lepidopterous pests. This egg parasitoid is currently used as part of inundative control programmes in cotton crops in South America. The parasitoid is also marketed commercially in Europe for the control of lepidopterous pests of protected crops, while in China about 0.5 million hectares of field crops are treated annually with this insect. The two most widely quoted examples of successful inoculative releases involve the parasitoid *Encarsia formosa* against the greenhouse whitefly *Trialeuroides vaporariorum* and the predatory phytoseid mite *Phytoseilus persimilis* against the twospotted spider mite *Tetranychus urticae*. The use of *Encarsia* in greenhouses is now very well established throughout Europe on both a commercial scale and at the home-gardener level. The use of predatory mites for the control of phytophagous mites has also become well established in both greenhouses and in some orchard crops, again at both commercial and home-gardener levels.

Conserving natural enemies and habitat manipulation

In contrast to classical biological control and inoculative/augmentative releases, habitat manipulation does not involve collection and rearing of predators for release. Rather, this approach involves creating conditions under which predatory species are able to enhance their efficacy as biocontrol agents. For this reason, these techniques may also be regarded as examples of cultural techniques for pest control (discussed in more detail in Chapter 11).

A good starting point for many conservation techniques involves minimising pesticide use, using economic thresholds, using selective pesticides and applying pesticides when they are least likely to affect natural enemies adversely. As a technique, conservation can of course be used in addition to the other two techniques, but it can also be used on its own. Examples of habitat manipulation to attract and boost predator numbers include the use of alternative food sources other than prey, such as flowers, the use of behaviour-modifying attractant chemicals, the use of various agronomic practices and the creation of artificial predator-friendly habitats such as beetle banks. A couple of examples of these are given below.

Many predators and parasitoids require pollen and nectar to complete their life cycle and planting flowers around or under crops can boost their activity. For example, planting brightly coloured nasturtiums or phaselia flowers will attract beneficial species like parasitoids and hoverflies. Orchards undersown with wild flowers such as buckwheat, mustard and dill can similarly boost natural enemy numbers. Finally, planting wild blackberries around vineyards is known to increase the numbers of the parasitoid *Anagrus epos*. These wasps then play an important role in controlling leafhoppers (*Erythroneura elegantula*) on grapes.

Beetle banks were developed in the UK in the 1990s as artificial hedgerows that were established within fields with plants known to be favoured by overwintering carabid beetles. Providing these habitats can lead to an overall increase in the number of predatory beetles within the agricultural landscape and they also ensure that fields are rapidly colonised by predators following their overwintering period. In addition to specific effects upon predatory carabid beetles, it is now known that beetle banks will also provide habitats for other important predatory groups such as spiders. The artificial bank will also provide overwintering cover for insectivorous birds. Finally, beetle banks can also be used as habitats for flowers which can act as nectar sources for predators and parasitoids. In the UK at least, beetle banks are an integral part of what is now described as integrated crop management (ICM).

Vertebrate biocontrol agents

Fish, amphibians, reptiles, birds and mammals all consume a large number of insects, many of which are pest species, and it is known that in some circumstances they do make a significant contribution to pest control. Bark-foraging birds, for example, are known to feed extensively on wood borers and bark beetles. In fact, the very first known successful introduction of a natural enemy from one country to another occurred when mynah birds were sent from India to Mauritius in 1762 to control the locust *Nomadacris septemfasciata*.

Despite this, most vertebrates have received relatively little attention, primarily because they usually prey on a wide range of invertebrates and are unlikely to control any one species. There have also been mistakes. For example, the house sparrow was mistakenly introduced into North America from Europe in a failed attempt to control cankerworms. The attempt failed because house sparrows are primarily seed-eaters with a preference for living in close association with human habitation. One notable success with vertebrates concerns the use of the mosquito fish *Gambusia*, which has been introduced in both amenity areas and in agricultural crops such as rice. Other examples include the use of reptiles for the control of thrips and scale insects and the use of birds (*Alcippe brunea*) for the control of lepidopterous pests in greenhouses.

Table 5.6. *Biological control and habitat stability*

	Habitat	Partial control		Complete control	
		No.[a]	%	No.	%
r-end					
	Cereal crops	6	4	0	0
	Vegetable crops	21	14	0	0
	Pasture	19	12	2	8
	Tree crops	107	70	23	92
K-end					

Notes:
Table from Southwood, T.R.E. (1977). The relevance of population dynamic theory to pest status. In *Origins of Pest, Parasite, Disease and Weed Problems*, ed. J.M. Cherrett & C.R. Sagar, pp. 35–54. Table reproduced with permission from Blackwell Scientific Publications, Oxford.
[a] No. refers to the number of control programmes embarked upon; it does not refer to the number of releases or to the number of control species assayed.

Worldwide results from biocontrol programmes

Historical analysis of biocontrol programmes worldwide shows that the percentage of programmes in which complete control resulted is between 5 and 15%. In a significant number of cases (15–55%), introduced natural enemies have become established, but have not succeeded in controlling target pests. Overall, more than 400 pest species have been controlled worldwide using biological control. By far the greatest number of successes have been achieved in stable agroecosystems where pest and natural enemy species have time to form a permanent association. These data are shown in Table 5.6 which assesses cases of successful control against habitat stability.

Although biological control has been successful in a number of situations, these still represent a small proportion of all pest control solutions. It has been calculated that only about 4% of insect pests worldwide have been controlled by biological methods. Reasons for this somewhat small number are numerous but certainly include a lack of funding by both industry and government. The point has already been made that, from an industrial perspective, it does not make sense to develop a control programme that is permanent. However, from a governmental perspective it would make sense, and it has been suggested by many practitioners of biocontrol that a lot more could be done. More recently, it has been argued that researchers need to develop more integrated programmes and that the term 'integrated biological control' should be used to develop programmes that blend the products and techniques shown in Figure 5.12. Historically, this term has been used to describe releases of more than one natural enemy. It has been argued that an integrated approach would likely lead to a greater number of successes and hence greater government funding for such work.

Summary

Although biological control of pest species is often a cost-effective, environmentally friendly option, it only accounts for around 0.5% of the economic market for pest control solutions. At the close of the twentieth century only 12 pest species, seven of which are citrus pests, had been completely controlled by biocontrol agents in North America. The reasons for the relatively low level of success with biocontrol agents are numerous, but certainly include the following. First, biocontrol is not eradicative and it is often difficult to get risk-averse farmers to accept that a residual pest population will not be economically damaging. Second, biocontrol is a specific form of control and attracts little interest from companies that want to market products that can be used to control a wide range of pest species. Third, biocontrol agents take time to work and they do not have the immediate and dramatic effects that many chemicals do. Fourth, using biocontrol agents may require more labour and more specialised knowledge than is required with chemical pesticide applications. Fifth, because biocontrol agents are living organisms, they often have more problems associated with efficacy or product quality control than conventional chemicals. Finally, the introduction of biocontrol agents usually requires governmental support and organisation, a process that inevitably takes time.

The upshot of all this is that the option of using biocontrol agents is usually only explored when chemical solutions are not possible or practical, as is the case with country-wide outbreaks of pest species. It is not expected that this situation will change, although increasing familiarity with biocontrol agents may push farmers in the direction of regarding this as the first line of defence against crop pests in the future. In contrast to the use of predators and parasitoids, however, the one area of biocontrol that has had appreciable success in the pest-control market comprises the use of pathogenic species.

Pathogenic biocontrol agents (viruses, bacteria, fungi, nematodes and protozoa), which are almost always used as augmentative/inoculative releases (i.e. as microbial pesticides), are discussed in the next chapter.

Further reading

Barbosa, P. (1998). *Conservation Biological Control.* San Diego: Academic Press.

Beirne, B.P. (1975). Biological control attempts by introductions against pest insects in the field in Canada. *Canadian Entomologist*, **107**, 225–36.

Bottrell, D.G., Barbosa, P. & Gould, F. (1998). Manipulating natural enemies by plant variety selection and modification: a realistic strategy? *Annual Review of Entomology*, **43**, 347–67.

Caltagirone, L.E. (1981). Landmark examples in classical biological control. *Annual Review of Entomology*, **26**, 213–32.

Caltagirone, L.E. & Doutt, R.L. (1989). The history of the Vedalia beetle importation to California and its impact on the development of biological control. *Annual Review of Entomology*, **34**, 1–16.

Cherrett, J.M. & Sagar, C.R. (eds) (1977). *Origins of Pest, Parasite, Disease and Weed Problems.* Oxford: Blackwell Scientific Publications.

Copping, L.G. (ed.) (2001). *Biopesticide Manual.* Farnham, Surrey: British Crop Protection Council Publications.

Cortesero, A.M., Stapel, J.O. & Lewis, W.J. (2000). Understanding and manipulating plant attributes to enhance biological control. *Biological Control*, **17**, 35–49.

Cross, J.V., Easterbrook, M.A., Crook, A.M. *et al.* (2001). Review: natural enemies and biocontrol of pests of strawberry in northern and central Europe. *Biocontrol Science and Technology*, **11**, 165–216.

Damon, A. (2000). A review of the biology and control of the coffee berry borer, *Hypothenemus hampei* (Coleoptera: Scolytidae). *Bulletin of Entomological Research*, **90**, 453–65.

Debach, P. & Rosen, D. (1991) *Biological Control by Natural Enemies.* Cambridge: Cambridge University Press.

Gurr, G.M. & Wratten, S.D. (1999). Integrated biological control: a proposal for enhancing success in biological control. *International Journal of Pest Management*, **45**, 81–4.

Hokkanen, H.T.M. & Lynch, J.M. (eds.) (1995). *Biological Control: Benefits and Risks.* Cambridge: Cambridge University Press.

Landis, D.A., Wratten, S.D. & Gurr, G.M. (2000). Habitat management to conserve natural enemies of arthropod pests in agriculture. *Annual Review of Entomology*, **45**, 175–201.

Lomer, C.J., Bateman, R.P., Johnson, D.L., Langewald, J. & Thomas, M. (2001). Biological control of locusts and grasshoppers. *Annual Review of Entomology*, **46**, 667–702.

Mackauer, M., Ehler, L.E. & Roland, J. (eds.) (1990). *Critical Issues in Biological Control.* Andover: Intercept.

Neuenschwander, P. (2001). Biological control of the cassava mealybug in Africa: a review. *Biological Control*, **21**, 214–29.

Nordlund, D.A. (1996). Biological control, integrated pest management and conceptual models. *Biocontrol News and Information*, **17**, 35–44.

Obrycki, J.J. & Kring, T.J. (1998). Predaceous Coccinellidae in biological control. *Annual Review of Entomology*, **43**, 295–321.

Smith, S.M. (1996). Biological control with *Trichogramma*: advances, successes, and potential of their use. *Annual Review of Entomology*, **41**, 375–406.

Sunderland, K. & Samu, F. (2000). Effects of agricultural diversification on the abundance, distribution, and pest control potential of spiders: a review. *Entomologia Experimentalis et Applicata*, **95**, 1–13.

Van Driesche, R.G. & Bellows, T.S. (1996). *Biological Control.* New York: Chapman and Hall.

6

Microbial pest control

Introduction

The organisms that comprise the microbial pest control market are bacteria, fungi, nematodes, protozoa and viruses. Just one species though, the bacterium *Bacillus thuringiensis*, is responsible for 80–90% of all the sales in this market. In 1995, these sales were worth in the region of £75–100 million per year and, although this amount represents a tiny proportion of the total world market for agrochemicals, which currently comes in at £20–25 billion, it is a market that is growing far more rapidly. For example, for a large number of years the annual market for microbial pesticides held steady at around £20–25 million. However, in the past 10 years this has all changed, and while current growth in the chemical insecticide market is around 1–2% per year, growth in the microbial pest control market is now running at around 10% per year. Some estimates have even projected growth for microbial pesticides to be as high as 25% per year.[1]

The reasons for this increased interest in microbial products are many and certainly include the following: (1) the development of resistance to conventional synthetic insecticides; (2) a decline in the rate of discovery of novel insecticides and a comparatively high rate of discovery for novel microbial agents; (3) an increased public perception of some of the dangers associated with synthetic insecticides; (4) the host-specificity of microbial pesticides which, so far, has meant that their use has had virtually no harmful

[1] These are global data. Clearly there will be regional variations in growth in the use of microorganisms for pest control.

environmental impact; (5) improvements in the production and formulation technology of microbial pesticides; and (6) a relaxation of the regulations that govern the registration and use of many microbial pesticides (particularly in the USA[2]), which has made investment in research and development for these products a cheaper and more attractive option for industry.

In general, most microbial pesticides are applied as inundative releases (see Chapter 5) in which the microbe is analogous to the active ingredient in a conventional spray. Attempts at using microbes as either inoculative agents or as manipulable features of the environment have been made, but with far less success. Notable examples of successes with these latter two techniques include the control of rhinoceros beetle *Oryctes rhinoceros* in the South Pacific through the release of adults preinfected with a baculovirus (inoculative release) and the control of the lepidopterous pasture pest *Wiseana cervinata* in New Zealand, through a reduction in the frequency of ploughing (environmental manipulation) which permits a build-up of soil-borne viruses that can keep this pest under control. In the case of *Oryctes* it has been suggested that this may be a case of classical biological control since control of the target pest has been more or less permanent following initial releases.

This chapter is structured so that for each of the major groups of microbial pesticides I give an account of their mode of action, their perceived advantages and disadvantages and the areas in which future developments are likely to take place. Emphasis within the chapter is given to *B. thuringiensis,* simply because this is the organism which dominates the commercial microbial pest control market. Before the individual groups are considered, I will give a brief reminder of some of the problems involved in dealing with living organisms in pest control. Many of the points made below are also applicable to the use of invertebrate predators and parasitoids, as discussed in the preceding chapter.

Microbes and pest control – general considerations

All microbes, of course, are living organisms.[3] Because of this, the challenges that are involved in using these products for pest control differ in a number of ways from those that are considered important when using conventional

[2] In the USA the Environmental Protection Agency has stated that one of its goals is to fast-track the registration procedures for microbial pesticides in order to hasten their uptake. At the time of writing it was still not clear as to whether this goal had been realised.

[3] Viruses are a special case. All viruses are obligate intracellular parasites. Viruses cannot reproduce outside the cells of other living organisms.

synthetic insecticides. For example, living organisms will require culturing as opposed to synthesising and they will need to remain alive throughout the shelf-life of the product. They are likely to be susceptible to adverse environmental conditions when released into a crop, they can move within the environment and they are likely to kill their hosts slowly in order to maximise their own fecundity. Their survival may depend on the presence of a minimum host population density, they may not attack all stages of a pest population and individuals within a population may vary in their virulence towards target pests. Finally, they will need to come into direct contact with the target species in order to be effective. This last difference, which will be especially critical for the control of sessile pest species, starkly contrasts microbial pesticides with conventional pesticides where redistribution of the active ingredient may take place by systemic or vapour action. All of these factors mean that to achieve a high level of efficacy with microbial products requires a great deal of knowledge of pest species characteristics and the agroecosystem they are to be used in, as well as a detailed understanding of the biology of the microbe itself. Fortunately, much of this knowledge now exists, and it is one of the reasons for the upsurge in interest in microbial products. However, both growers and the general public will need to become far more informed of the constraints that are involved in dealing with living organisms, particularly since historically, many of these individuals will only be familiar with applying simple-to-use products on a routine basis.

Bacteria

More than 100 bacteria have been identified as arthropod pathogens. The only species to have achieved any significant commercial success for arthropod pest control is *Bacillus thuringiensis* (Table 6.1). In 1998, this bacterium was registered for use on a diversity of pests in over 100 different formulations throughout North America and Europe. Two other bacilli, *B. popilliae* and *B. sphaericus*, have been used in more specialised niches, and are likely to undergo further commercial development in the future. The only nonbacillus microbial insecticide to have been registered for use on an arthropod pest comprises the application of *Serratia entomophilia* to pastureland in New Zealand. This bacterium is used to control the pastureland grub *Costelytra zealandica* (Table 6.1). To date, bacterial pesticides have been used to control an enormous diversity of Coleoptera, Diptera and Lepidoptera in forestry and open-field crops, as well being used against various vectors of disease. Bacterial pesticides which are able to control hemipterous pests have yet to be developed.

Table 6.1. *Commercially available bacterial formulations for arthropod control*

Species	Target pests
Bacillus popilliae	Japanese beetles (*Popilla japonica*)
Bacillus thuringiensis subsp. *kurstaki*	Lepidopteran larvae
Bacillus thuringiensis subsp. *aizawai*	Lepidopteran larvae
Bacillus thuringiensis subsp. *israeliensis*	Dipteran larvae
Bacillus thuringiensis subsp. *tenebrionis*	Adults and larvae of Coleoptera
Bacillus thuringiensis subsp. *japonensis*	Soil beetles
Bacillus thuringiensis subsp. *aizawai*	Lepidopteran larvae (encapsulated delta-endotoxin)
Bacillus thuringiensis subsp. *kurstaki* (encapsulated deltaendotoxin)	Lepidopteran larvae and some beetles
Bacillus thuringiensis subsp. *morrisoni* (encapsulated deltaendotoxin)	Colorado potato beetle (*Leptinotarsa decemlineata*)
Bacillus sphaericus	Mosquito larvae
Serratia entomophila	The New Zealand grass grub *Costelytra zealandica*

Source: Data from Copping, L.G. (ed.) (2001). *The Biopesticide Manual,* 2nd edn. Farnham, Surrey: BCPC Publications.

Biology and mode of action of *B. thuringiensis*

The bacterium *B. thuringiensis* was first recognised in 1901 as the aetiological agent of sotto disease in the silkworm *Bombyx mori.* However, it was not until 1911 that it was given the name *Bacillus thuringiensis* by the German biologist Berliner after he recorded the bacteria in the pupae of the Mediterranean flour moth *Ephestia kuehniella* in grain stores in the German city of Thuringen. It is a Gram-positive, motile bacterium that is found in soils and insect-rich environments, such as grain stores and insect-rearing facilities, all over the world. Its growth cycle has two phases. Where nutrients are abundant the bacterium undergoes vegetative growth by binary fission. However, when environmental conditions deteriorate the bacterium sporulates, producing spores that are able to survive until conditions improve and vegetative reproduction can begin again. The critical feature for pest control of this life cycle is that concomitant with spore production is the production of a proteinaceous parasporal crystal

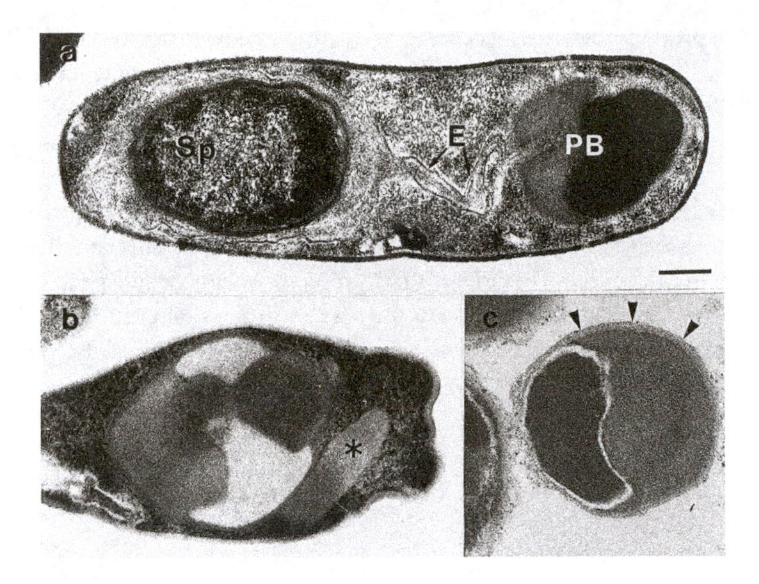

Fig. 6.1. Electron micrographs of the insecticidal parasporal body produced by *Bacillus thuringiensis* subsp. *israeliensis*. (*a*) Sporulating cell with developing parasporal body at the right. Sp, Spore; E, exosporium membrane; PB, parasporal body. (*b*) Completely formed parasporal body with cell just prior to lysis. The asterisk marks the insecticidal protein. (*c*) Parasporal body liberated from the cell. The arrows point to the fibrous envelope that holds the insecticidal proteins together. The bar in micrograph (*a*) is equivalent to 250 nm. Courtesy of B. A. Federici, Department of Entomology, University of California, Riverside.

that is toxic to a number of species of insect (Figure 6.1). When ingested, the crystal dissolves in the alkaline gut of the insect. Proteases within the insect gut digest a portion of the crystal, releasing a toxicologically active protease-resistant fragment, often referred to as the deltaendotoxin. The toxin diffuses through the peritrophic membrane of the gut wall and binds to receptors on the midgut epithelium. This binding leads to the formation of lesions in the gut wall which allow the gut contents to leak into the haemocoel. The insect then succumbs to a lethal septicaemia and, if bacterial spores are present, the abundance of nutrients stimulates germination and the onset of vegetative bacterial growth. The time scale and exact chronology of these events will depend on the insect species and toxin involved; however, it is clear that the time to mortality for susceptible species is measured in days rather than hours.

The genes that produce the proteinaceous crystal are located on large plasmids and, to date, more than 25 different crystals have been identified and sequenced. The genes code for proteins of 640–1200 amino acids which are classified into one of four classes known as CryI, CryII, CryIII and CryIV. Of these classes, CryI and CryII proteins are active against lepidopteran and/or dipteran species, CryIII are active against Coleoptera and CryIV are active against Diptera. Most *B. thuringiensis* strains are able to produce several different crystal proteins and the same crystal protein can be found in different *B. thuringiensis* strains or subspecies. The most important strains of *B. thuringiensis* for pest control, as determined by flagellar serotyping, comprise *B. thuringiensis aizawai*, *B. thuringiensis israeliensis*, *B. thuringiensis kurstaki* and *B. thuringiensis tenebrionis* (Table 6.1).

Advantages and disadvantages of *B. thuringiensis* as a pesticide

The advantages associated with the use of *B. thuringiensis*-based insecticidal products derive from its highly specific mode of action, from the ease of production of commercially available formulations with satisfactory shelf-lives (1–3 years), from the bacterium's comparatively low development costs when compared to conventional chemical compounds and from the current high rate of discovery of novel crystal proteins. At present, a novel crystal protein is discovered for every 1000 bacteria that are screened, while the rate of discovery of novel synthetic pesticides is in the region of one for every 30 000–50 000 compounds screened (see Chapter 1).

The specificity of the bacterial endotoxin, and its consequent safety to non-target species, derives from the fact that in order for the parasporal crystal to exert a toxic effect, specific binding of the toxic fragments, following protoxin enzymatic attack, to receptors on the insect midgut must occur. Protein fragments that are toxic have always been found to bind to midgut receptors and no binding between nontoxic fragments and midgut receptors has ever been recorded. To date, there is no evidence of any acute or chronic toxicity of any commercially available formulations to any mammalian species.

The production of commercial formulations of *B. thuringiensis* is carried out using batch deep liquid fermentation techniques, frequently followed by the production of freeze-dried spore–crystal mixtures. Initially spores are encouraged to germinate and change into vegetative cells in a strain-specific medum consisting of soya, fish protein, starch, yeast extract and trace minerals. These cells are transferred to a prefermenter and then a fermenter where growth continues until the nutrient supply is exhausted, cell growth ceases, and the production of spores and crystals occurs. Typical yields will be in the range 10^8–10^{10} spores/ml of fermentation broth. This broth is then concentrated in a centrifuge, dried and formulated for sale (Figure 6.2, colour plate).

The first field trials with *B. thuringiensis* preparations took place in the mid-1920s and the very first commercially available formulation was produced in France in 1938 and marketed under the trade name Sporeine. In 1995 30 formulations produced by 10 companies were available worldwide. At present, a great deal of research effort is focusing on using biotechnology to improve the efficacy of *B. thuringiensis*. This will be dealt with in detail in Chapter 13. One example is given by the product MVP that is produced by Mycogen. To produce this formulation the company transferred the genes producing the proteinaceous crystal into the genome of the bacterium *Pseudomonas fluorescens*. This bacterium does not produce spores but is able to produce the toxic crystal during its normal vegetative growth cycle. These cells can then be killed by chemical treatment and the remnants of the cell wall stick to the crystal, providing a biological coat that is claimed to improve field efficacy. These are the encapsulated deltaendotoxin formulations referred to in Table 6.1.

One final advantage that is frequently associated with many microbial pesticides, including *B. thuringiensis*, derives from the fact that because the bacterium is a living organism it can spread. Following product application to a crop and an initial cycle of primary infection the bacterium will eventually utilise all the nutrients within a pest species and then begin to sporulate, providing further toxic material for the control of other individuals that may have colonised the treated crop following the microbe's application. This process is known as secondary disease cycling.

The main limitations of *B. thuringiensis*-based formulations are associated with their lack of persistence in the field and their lack of systemic activity. For example, the deltaendotoxins are rapidly inactivated by environmental factors such as sunlight, rain and wind and will usually only remain infective for a few hours. This degradation leads to a rapid increase in the spore-to-crystal ratio, the significance of which will depend on whether both are required for effective pest control. The lack of systemic activity means that the bacterium cannot, at present, be used effectively against hemipterous pests such as aphids and whitefly. The other disadvantages with *B. thuringiensis* are concerned with its speed of action and with resistance development. Unlike synthetic chemical insecticides which often produce lethal effects within a few hours, *B. thuringiensis*-based formulations generally take from 1 to 7 days to cause target species mortality, although cessation of feeding of pest species will often occur within a few hours of ingestion of toxin.

At present, the development of pest species resistance to pesticides is primarily a problem associated with conventional synthetic pesticides. This material was covered in Chapters 1 and 4. Resistance development to *B. thuringiensis*-based sprays has begun to be observed. For example, in the laboratory, resistance to crystal proteins has been demonstrated with the Indian meal moth, the almond moth, the tobacco budworm, the diamondback moth and with the Colorado potato beetle. In field situations, up to 400-fold increases in resistance of diamondback moth populations have been recorded in populations that were heavily treated with *B. thuringiensis*-based sprays. The mechanism of this resistance development is generally believed to result from a reduced affinity of the midgut receptors for the toxic protein fragments. While *B. thuringiensis*-based insecticides may therefore be an improvement from an environmental point of view it would appear that they may not offer a technique that can be used on a sustainable basis. The importance of resistance development will be discussed in Chapter 13 in relation to transgenic plants.

Finally, although the specificity of *B. thuringiensis* is a bonus, it can also present a problem where a crop is attacked by a complex of pests. For example, in cotton and many fruit crops the range of arthropod pest species that need to be monitored and controlled can easily rise into double figures. Although commercial formulations of this bacterium are compatible with most synthetic insecticides, a complex of pests will mean that a grower's first choice of product will probably not be one that will only control one of the pests that is attacking the crop.

The bacteria *B. popilliae* and *B. sphaericus*

B. popilliae is the aetiological agent of milky diseases in beetles, especially the Scarabaeidae, and has been used for the control of larval Japanese beetle *Popillia japonica* in the USA for the past 60 years. *B. sphaericus* is toxic to the larvae of mosquitoes, particularly those in the genus *Culex,* and it has been used in their control for about the last 20 years. Unlike *B. thuringiensis*, neither of these bacteria produce free toxic crystals. In *B. sphaericus* the proteinaceous endotoxin, which is composed of several toxic proteins, remains located in the spore wall while *B. popilliae* produces no endotoxin at all. Both bacteria cause mortality in target species via a lethal septicaemia as a result of spore germination. The time to spore germination, and hence pest species death, with both these species is much slower than with *B. thuringiensis* and this is why neither of these bacteria has yet reached the stage of large-scale commercial production. For example, although a fully diseased Japanese beetle grub may contain 2–5 billion spores it may take months for an infected grub to die in the field. Some researchers have even questioned whether commercially available *B. popilliae* formulations work at all, particularly for localised beetle outbreaks in urban settings. In addition, *B. popilliae* also suffers because, as yet, it can only be produced *in vivo* and so its production requires a complicated tritrophic culturing system in which beetles are cultured on plants to produce food in which the bacterial culture can multiply. *B. sphaericus* can be produced by more conventional fermentation techniques and is regarded as a more promising candidate for further commercial development.

Future developments with bacterial pesticides

An expansion of the market for bacterial products in pest control in the future is likely to result due to advances in biotechnology and from advances in formulation technology. Examples of the former include the addition of toxic and nontoxic compounds to the bacterial suspension to enhance potency. Examples of the latter include the use of microencapsulation and other formulation techniques to improve stability in relation to pH, temperature and ultraviolet exposure. For example, studies on microencapsulation have been carried out with a range of bacterial species, including both *B. thuringiensis* and *B. sphaericus,* and have shown that encapsulation significantly enhances the stability of spores exposed to a variety of environmental conditions. It has also been demonstrated that the addition of spray adjuvant photostabilisers can enhance resistance to environmental degradation.

Candidate compounds for improving efficacy that have been evaluated so far include boric acid, chitinase, thuringiensin, emulsifying agents, protein solubilising agents and various nitrogenous compounds. The modes of action of these additives vary enormously, as do their effects on the potency of the bacterial formulation, while some are also toxic to vertebrates. For example, thuringiensin is a betaexotoxin that is produced by the vegetative cells of some *B. thuringiensis* isolates. It has a wide host spectrum, including both invertebrates and vertebrates, and works by interfering with the synthesis of DNA. Because of this, registration requirements for *B. thuringiensis*-based products currently require an absence of this compound.

Fungi

The very first microorganism to be recognised as an agent of disease in insects was a fungus. Towards the end of the nineteenth century Agostino Bassi

Table 6.2. *Commercially available fungal formulations for arthropod control*

Species	Target pests
Beauvaria bassiana	Different strains are available for control of corn borers, Homoptera, Thysanoptera and some coleopteran pests
Beauvaria brongniartii	Cockchafers
Metarhizium anisopliae[a]	Different strains are available for the control of various coleopteran and lepidopteran pests. Some strains also target thrips and termites
Metarhizium flavoviride	Cockchafers
Paecilomyces fumosoroseus	Whitefly
Verticillium lecanii	Whitefly and aphids

Note:
[a] A number of field trials have now also been undertaken with *Metarhizium anisopliae* for locust control. See, for example, Bateman R.P. and Alves, R.T. (2000). Delivery systems for mycoinsecticides using oil-based formulations. *Aspects of Applied Biology*, **57**, 163–70.

Source: Data collated from Copping, L.G. (ed.) (2001). *The Biopesticide Manual*, 2nd edn. Farnham, Surrey: BCPC Publications.

demonstrated that the fungus *Beauvaria bassiana* (Figure 6.3, colour plate) was the causal agent of the white muscardine disease of the silkworm *Bombyx mori*. Since then, over 800 species of entomopathogenic fungi have been identified. Of these, six species are commercially available (Table 6.2). Many more have been investigated and registered as possible pest control agents, e.g. *Hirsutella thompsonii* was registered for use against the citrus rust mite by the US Environmental Protection Agency as far back as 1981,[4] while the water fungus *Lagenidium giganteum* is still registered for mosquito control in North America. There are also two fungi in the genus *Metarhizium* that have been developed for pest control. *Metarhizium anisopliae* is currently registered in the USA for use in cockroach traps and for the control of termites. *M. flavoviride* is registered by the International Organisation for Biological Control (IOBC) for the control of grasshoppers and locusts in tropical and subtropical countries, while *M. anisopliae* is currently the subject of field trials for locust control. Finally, the

[4] Registration for this fungus was cancelled in 1988.

Figure 4.2a (above) Hydraulic pesticide application using a tractor-mounted system. Photograph © Micron Sprayers Ltd. Reproduced with permission.

Figure 4.2b Hydraulic pesticide application using a hose and lance in a greenhouse. Photograph © Micron Sprayers Ltd. Reproduced with permission.

Figure 4.3 Gaseous energy mist-blower applying pesticide to a cocoa crop in Brazil.

Figure 4.4 Centrifugal-energy spinning-disc sprayer applying pesticide to a tobacco crop. Photograph © Micron Sprayers Ltd. Reproduced with permission.

Figure 4.5 Rotating-head attachment fitted to a conventional mist-blower applying pesticide to blackcurrants. Photograph © Micron Sprayers Ltd. Reproduced with permission.

Figure 5.4 Braconid larvae emerging from a mature red admiral caterpillar. Photograph by Jim Kalisch. Reproduced with permission of the Department of Entomology, University of Nebraska.

Figure 5.5 The parasitic wasp *Peristenuc digoneutis* prepares to lay an egg in an aphid nymph. Photograph by Scott Bauer: reproduced with permission from USDA/ARS image gallery at:
http://www.ars.usda.gov/is/graphics/photos/mainmenu.htm

Figure 5.6 *Aleiodes indiscretus* wasp parasitising a gypsy moth caterpillar. Photograph by Scott Bauer: reproduced with permission from USDA/ARS image gallery at: http://www.ars.usda.gov/is/graphics/photos/mainmenu.htm

Figure 5.7 Egg parasitoids. Photograph by Jim Kalisch. Reproduced with permission of the Department of Entomology, University of Nebraska.

Figure 5.8 A 14-spot lady beetle eating a pea aphid. Photograph by Scott Bauer reproduced with permission from USDA/ARS image gallery at: http://www.ars.usda.gov/is/graphics/photos/mainmenu.htm

Figure 5.9 Predatory green lacewing larva feeding on whitefly. Photograph by Jack Dykinga reproduced with permission from USDA/ARS image gallery at: http://www.ars.usda.gov/is/graphics/photos/mainmenu.htm

Figure 5.10 Syrphid fly larva feeding on aphids.

Figure 5.11 Predatory hemipteran feeding on whiteflies. Photograph by Jack Dykinga reproduced with permission from USDA/ARS image gallery at: http://www.ars.usda.gov/is/graphics/photos/mainmenu.htm

Figure 6.2 Production of *Bacillus thuringiensis* formulations on a commercial scale. The photograph shows the large-scale fermenter used by Ciba Giegy in production laboratory no. 16 at its factory in Basel, Switzerland.

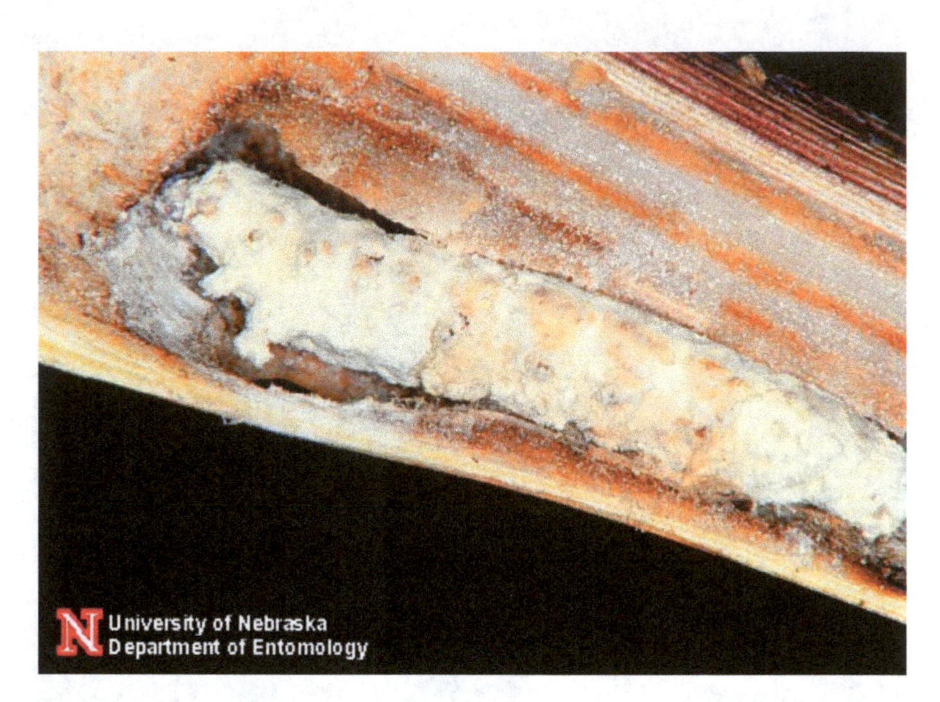

Figure 6.3 European corn borer infected with *Beauvaria bassiana*. Courtesy of Jim Kalisch. Reproduced with permission of the Department of Entomology, University of Nebraska.

Figure 6.4 Grasshoppers infected with *Nosema locustae*. Courtesy of Jim Kalisch. Reproduced with permission of the Department of Entomology, University of Nebraska.

fungus *Paecilomyces fumosoroseus* has recently been isolated and formulated for the control of whitefly, thrips and aphids. At the time of writing, this fungus had just been developed for commercial use in Europe. Overall, the most successful use of fungi in pest control has been in situations where it has been possible to maintain the high relative humidity needed for spore germination and pathogen development. These situations include the control of pests within soil and on protected crops.

Mode of action of fungal pathogens

The mode of action of the fungal pathogens that are used for pest control typically follows from the attachment of spores, which are the basis of all the commercially available preparations, to the target pest's cuticle. Following germination of the spore, and germ tube penetration of the body wall, the fungus then undergoes hyphal multiplication within the body cavity. Penetration of the arthropod cuticle is brought about by the combined action of mechanical pressure exerted by the germ tube and by the production of proteases and chitinases that are thought to be produced sequentially. For example, in laboratory culture, proteases are produced first, which suggests that degradation of the protein matrix of the cuticle is required before the fungus can degrade chitin. Hyphal multiplication within the pest and the associated physical damage to the pest's internal stucture are also often accompanied by the production of toxic metabolites which assist in causing death to the target species. The fungus *B. bassiana* produces the toxin beauvericin which has been shown to be toxic upon injection to insects, although *in vitro* experiments have also shown that the pathogenicity of strains is not always associated with their ability to produce this toxin. Despite this, the presence of toxins may be used to select strains for development since strains with toxins have still, so far, been found to kill target species far more effectively that strains without. Further mycelial growth within the dead organism then finally leads to the fungus sporulating and the production of further infective units, i.e. secondary disease cycling occurs. Like *B. thuringiensis*, therefore, commercial fungal preparations have the ability to spread from a site of primary infection to attack other target organisms. The time scale involved in the above events will vary with the ambient environmental conditions. However, even where conditions are optimal, the time to mortality is measured in days. For example, *B. bassiana* and *B. brongniartii* both take 3–5 days to produce mortality in their target pest species while *V. lecanii*, under optimal conditions, will generally kill whitefly and thrips 7–10 days after its application.

Advantages and disadvantages of using fungal pathogens in pest control

The advantages of using fungi for pest control are threefold. First, the species that are listed above have no known mammalian toxicity and only present a negligible risk to beneficial arthropod species, i.e. *V. lecanii* will only attack beneficial species under highly favourable environmental conditions (100% relative humidity, 20–25 °C). Second, they are simple to produce. Both *B. bassiana* and *B. brongniartii* can be cultured by solid-state fermentation on clay granules while *V. lecanii* is commercially produced on culture medium which is then concentrated and dried to harvest the spores. Typical commercial formulations will contain an array of spray additives that both stimulate spore germination and protect against rapid desiccation during periods of low humidity following their application to a crop. Third, fungi have a mode of action which makes it possible to use these microbes against pests that are hemipterous. In comparison to most other microbial organisms, which infect their hosts via the digestive tract, fungi can infect their hosts by direct penetration of the body wall. This means that hemipterous pests with sucking mouth parts can be targeted for control, as well as species with biting mouth parts, and that all the stages in a target species' life cycle can be attacked. In the case of *V. lecanii*, however, which primarily infects whitefly nymphs, adults are only killed when the relative humidity is greater than 90%, while eggs appear to be relatively immune to the fungus.

The primary problems associated with the use of fungal pathogens for pest control concern the environmental conditions, particularly a high relative humidity, that are required for optimal efficacy and the environmental conditions that are required for the storage of preparations. For example, infection of the pea aphid *Acyrthosiphon pisum* with *Entomophthora* species has been positively correlated with rainfall and it has also been possible to use rainfall and temperature to define ecological zones in the former Soviet Union where this aphid was consistently controlled by natural infestations of *Entomophthora*. Areas in which low rainfall was recorded were found consistently to support pea aphid populations that caused crop losses almost every year. Temperature is the other factor that has frequently been cited as limiting the use of entomopathogens in pest control. For example, most fungal species require a minimum temperature of 5 °C for spore germination; however, even at this temperature the development of the pathogen will not be optimal. The spores of *Metarhizum anisopliae*, a fungus that has been used to control spittlebugs on sugar cane and rhinoceros beetle on coconut palm, are rapidly inactivated by simulated sunlight with a half-life of 2–3 h, while the optimal temperature for the germination and growth of this pathogen varies between 20 and 30 °C.

Of the commercial formulations that are currently available in the UK shelf-lives are recorded as 1 year at 20 °C (*B. bassiana*), 1 year at 2 °C (*B. brongniartii*) and 6 months at 4 °C or 1 month at 20 °C (*V. lecanii*). In comparison to commercial pesticide formulations where a shelf-life of at least 18 months is regarded as desirable, these values do not hold up well. Of particular note is the fact that in many developing countries the specific conditions required for storage are unlikely to be available.

Future developments with fungal pesticides

Because of the problems encountered with field efficacy and storage, the prospects for pathogenic fungal-based pesticides do not look particularly good. Perhaps not surprisingly, the successes that have been realised with *V. lecanii* have primarily occurred in greenhouses where the ambient environmental conditions can be very accurately controlled. For the future it seems likely that developments in biotechnology may be used to manipulate desirable traits and to improve the field activity of many fungi. However, much of this work is still at a very seminal stage and a great deal of fundamental biochemical and molecular work with fungi still remains to be carried out.

Nematodes

By 1975, more than 3000 nematode–insect relationships had been described. However, only two families of nematode worm, the terrestrial Heterorhabditidae and Steinernematidae, have been extensively used to develop commercial formulations (Table 6.3). The first commercially available formulation comprised *Romanomermis culicivorax*, registered for use in the USA against mosquitoes in 1976. This registration was subsequently cancelled because of success with alternative products. Currently (2002), there are seven different nematode species that are available for arthropod pest control.

Attributes of these nematode-based formulations that make them amenable to production and use include ease of mass production, an efficacy comparable to synthetic pesticides, and a high level of safety to nontarget organisms. As with fungal pathogens, the most extensive use of nematode-based formulations has occurred in situations where it has been possible to maintain a high level of humidity. Table 6.3, which lists the major pests for

Table 6.3. *Commercially available nematode-based formulations for arthropod control*

Species	Primary target pests[a]
Heterorhabditis bacteriophora	Japanese beetles
Heterorhabditis megidis	Black vine weevils
Steinernema carpocapsae	Black vine weevil and cutworms
Steinernema feltiae	Fungus gnats
Steinernema glaseri	Larvae of scarab beetles
Steinernema riobravis	Mole crickets and citrus weevils
Steinernema scapterisci	Mole crickets

Note:
[a] Almost all of the pests controlled by nematode formulations are soil pests. This list represents major uses and is not complete.

Source: Data from Copping, L.G. (ed.) (2001). *The Biopesticide Manual*, 2nd edn. Farnham, Surrey: BCPC Publications.

commercially available formulations, shows that the most extensive use of these products has been in the control of soil-borne pests.

Mode of action of nematode-based formulations

Infective juveniles, in both the nematode families mentioned above, penetrate the host via the mouth, anus and spiracles while the Heterorhabditidae also have the ability to penetrate the cuticle directly. When compared with other microbial agents they kill their hosts relatively quickly. The life cycle of these nematodes involves four larval stages, the first of which develops inside the egg. It is the third-stage larvae that invade the host and release symbiotic gut bacteria (*Photorhabdus luminescens* in the Heterorhabditidae, *Xenorhabdus bovienii* in the Steinernematidae) which kill the host via their multiplication, and the hosts' subsequent and rapid (24–48 h) development of a lethal septicaemia. The fourth larval stage develops inside the target organism producing the reproductive stage. The nematodes then reproduce in the cadaver, creating infective juveniles that disperse and seek new hosts to attack, i.e. secondary disease cycling occurs. The whole life cycle takes about 6–14 days at 20–28 °C. For steinernematid worms this requires both sexes to be present in the host while heterorhabditid species are hermaphroditic and so only one nematode per host is sufficient for reproduction to occur.

Advantages and disadvantages of nematodes in pest control

In addition to the relatively rapid mode of action of nematode formulations mentioned above, the other advantages of using nematodes in pest control are their ability to seek out and locate hosts by responding to physical and chemical cues and that techniques for their commercial production are relatively simple. For example, the liquid culture of steinernematids in large-scale fermenters has produced yields as high as 150 000 infective juveniles per cm^3. In the USA, recent amendments by the Environmental Protection Agency have exempted nematodes from standard registration and regulation requirements, so simplifying their application, commercialisation and development, and they are now regarded as the second most promising group of organisms for further development after *B. thuringiensis*.

Unfortunately, like fungal formulations, most nematode-based products require a very high humidity to remain infective, in addition to being highly sensitive to drought, heat and ultraviolet radiation. At present, this has meant that their use in pest control has been largely limited to pests of protected crops and the soil (Table 6.3). The second difficulty with these formulations is associated with maintaining infectivity. All of the formulations listed in Table 6.3 have limited shelf-lives (Table 6.4). The development of water-dispersible granule formulations in which the nematodes enter into a partial anhydrobiotic state has recently been reported as a significant breakthrough, since it extends the survival and pathogenicity of formulations for up to 6 months at 4–25 °C. However, in comparison to conventional pesticides, this is still unsatisfactory, and it is likely to be a significant sticking point that will delay the commercial development of many nematode formulations.

Future developments with nematode-based formulations

In the future, improvements in formulation technology, an expansion of the species used for pest control and improvements in application technology are likely to occur with nematode-based products for pest control. At present, a great deal of effort has been directed at the control of pests of open-field crops. For example, it has recently been shown that the nematode *Steinernema riobravis* could be used to control the corn earworm *Helicoverpa zea* in fields of maize in the USA, although timing of the application and environmental conditions were both critical to success. Other studies have shown that nematodes may be useful for the control of cotton crop pests, for control of the

Table 6.4. *Storage stability of major nematode-based products*

Formulation	Storage stability[a]	
	20–25 °C	4–10 °C
Alginate gel (20×10^6 nematodes trapped into a gel matrix and coated on a mesh screen)	3–5 months	6 months
Clay (60×10^6 nematodes spread on 80 g clay)	0	3 months
Flowable gel (up to 1×10^9 nematodes suspended in a gel matrix enclosed in special film)	4–6 weeks	3 months
Water dispersible granules (100×10^6 nematodes in 350 g of granule material)	6 months	6 months

Note:
[a]Based on product labels.

Source: Table reproduced with permission from Hall, F.R. & Barry, J.W. (1995). *Biorational Pest Control Agents*. Washington, DC: American Chemical Society. Copyright © 1995 American Chemical Society.

sweet-potato weevil *Cyclas formicarius* and for control of the plum curculio *Conotrachelus nenuphar*. The other main research area which may impinge upon the success of nematodes within pest control comprises developments in genetic engineering. At present it is thought that it may be possible both to improve formulation stability, i.e. by inducing nematodes to enter a fully anhydrobiotic state, and to improve species virulence by manipulation of the nematode genome.

Protozoa

The protozoa are a large and diverse group of eukaryotic unicellular organisms that belong to the kingdom Protista. Approximately 60000 extant protists have been described and there are about the same number recorded from the fossil record. Within the kingdom Protista there are at least seven distinct phyla. These phyla are separated on the basis of their mode of locomotion and the structure of their locomotory organs. Of the seven phyla, four are known to parasitise invertebrates. These are the Apicomplexeta, Ciliophora, Microspora and Sarcomastigophora. Of these, however, only the phylum Microspora and the genus *Nosema* have received any attention in relation to arthropod pest control. The microsporidian *Nosema bombycis* causes the classic pebrine disease of the silkworm *Bombyx mori*.

At present, there is one commercially available protistan formulation. The protist *Nosema locustae* is marketed for the control of grasshoppers on rangeland (Figure 6.4, colour plate). Despite the availability of this formulation, its field effectiveness is known to vary. Unfortunately, protistan formulations are very expensive to manufacture and they tend to cause chronic rather than acute infections. At one time the recommendation with *N. locustae* was that it should only be applied if cost was not of primary concern.

Mode of action of microsporan protists

Most microsporans are transmitted orally by ingestion of spores. These spores develop inside a suitable host by the production of a hollow tube called the polar filament which places the infective agent called the sporoplasm inside or next to a gut epithelial cell. Repeated binary fission of the protozoan then occurs followed by the production of further spores. Unfortunately, for pest control, the time scale in these events, which may or may not lead to the death of the host, varies from days to weeks. For example, if first or second instar *Anopheles albimanus* larvae are infected with *N. algerae* then the larvae will usually die. However, if fourth instar larvae are infected then these individuals are able to support the protist while pupating and producing viable adults. With grasshoppers, infection of third instar nymphs, or earlier, is generally effective. However, if the grasshoppers are older then death will not occur, at least, not in the short term.

Advantages and disadvantages associated with protozoa

The main advantage associated with protozoa for pest control is their safety to nontarget species and, therefore, their compatibility with integrated control programmes. For example, *Nosema* spores can be combined with reduced-rate pesticide applications to produce mixtures that are highly toxic to grasshoppers.

Unfortunately for pest control, there are still a significant number of disadvantages associated with protozoan formulations. Paramount among these is the fact that they cause chronic rather than acute infections. This means that pest species do not die quickly, which is largely unacceptable to farmers. The production of protozoan formulations has also been a stumbling block in their development. All of the Microspora are obligate parasites that have to be cultured inside other living organisms. As yet, there is no way to produce *Nosema* on artificial media. One of the simple reasons why *N. locustae* has been so extensively evaluated is that this species has a high-enough spore production rate in culture. Current recommendations with this species are 2.5×10^9 spores per hectare for locust control and the cost of this comes in at a manageable $2–3 per hectare. There are other species (e.g. *N. cuneatum* and *N. acridophagous*) that kill grasshoppers, but their spore production rate in culture is not high enough to produce economically viable formulations. A similar situation exists with the production of *N. algerae* for mosquito control. At present, it is highly uneconomic.

Future developments with protozoan pesticides

Given the above, it is likely that most future research with protists will concentrate on improvements to production or culturing of these species for pest control. In addition, basic research is still needed on protozoan species that are toxic to insects. In the short term at least, it seems unlikely that this group will have much impact on arthropod pest control.

Viruses

The viruses that have been identified as causing diseases in arthropods come from seven different virus families, although all of the viruses that have been developed, so far, for commercial use in pest control belong to the Baculoviridae. The primary reason for this is that these viruses have only ever been isolated from invertebrates and so are thought to be completely safe to mammals and other higher life forms.

The first registration of a viral insecticide in the USA occurred in 1971. The formulations that are currently available for commercial use are shown in Table 6.5. These viral pesticides all control lepidopterous pests of agriculture and horticulture despite the fact that the most successful use of viruses to date has been against temperate forest pests where high damage thresholds and limited pest complexes exist. For example, both the pine sawfly *Neodiprion sertifer* and the Douglas fir tussock moth *Orgyia pseudotsugata* are successfully

Table 6.5. *Commercially available virus-based formulations for arthropod control*

Species (type)	Primary target pests
Adoxophyes orana (granulovirus)	Tortrix moths
Anagrapha falcifera (nucleopolyhedrosis)	Lepidopteran larvae
Anticarsia gemmatalis (nucleopolyhedrosis)	Velvet bean caterpillar and sugarcane borer
Autographa californica (nucleopolyhedrosis)	Lepidopteran larvae
Cydia pomonella (granulovirus)	Codling moth
Helicoverpa zea (nucleopolyhedrosis)	*Heliothis* spp. and *Helicoverpa* spp.
Lymantria dispar (nucleopolyhedrosis)	Gypsy moth
Mamestra brassicae (nucleopolyhedrosis)	Cabbage moth, armyworm and diamondback moth
Mamestra configurata (nucleopolyhedrosis)	Armyworm *Mamestra configurata*
Neodiprion sertifer (nucleopolyhedrosis)	Sawflies
Orgyia pseudotsugata (nucleopolyhedrosis)	Douglas fir tussock moth
Spodoptera exigua (nucleopolyhedrosis)	Beet armyworm
Synagrapha falcifera (nucleopolyhedrosis)	*Heliothis* spp. and *Helicoverpa* spp.

Source: Data collated from Copping, L.G. (ed.) (2001). *The Biopesticide Manual*, 2nd edn. Farnham, Surrey: BCPC Publications.

controlled by baculovirus pesticides. It is because these products are usually registered for use by government departments that they are not available to the general public. To date, no virus-based pesticides have been developed for the control of mosquitoes, flies or other dipterous pests. There are three important subgroups within the Baculoviridae that have been used to control other arthropod pests.

Baculoviridae

The Baculoviridae are viruses with double-stranded DNA in which three subgroups (A, B and C) have been recognised. Subgroups A and B comprise occluded viruses in which the virus particles are enclosed by occlusion bodies while subgroup C comprises nonoccluded viruses that are transmitted as free particles, as occurs with many virus diseases of plants and higher animals. All

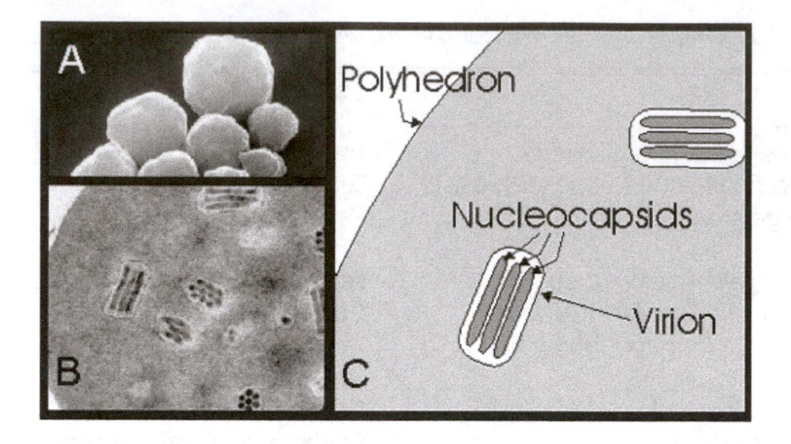

Fig. 6.5. A) Baculovirus particles, or polyhedra; B) Cross-section of a polyhedron; C) Diagram of polyhedron cross-section. Electron micrographs (A&B) by Jean Adams, graphic by V. D'Amico. Copied with permission from website at: http://www.nysaes.cornell.edu/ent/biocontrol/pathogens/baculoviruses.html#suppliers. Cornell University, USA.

of the subgroups have a similar nucleocapsid (a protein capsid which surrounds the double-stranded DNA core), whether occluded or not. The virion is then completed by an outer lipoprotein envelope which surrounds the nucleocapsid (Figure 6.5). With occluded viruses the virions (or virion) are then surrounded by a paracrystalline protein matrix. In subgroup A Baculoviridae both the number of occluded virions and the number of enveloped nucleocapsids can vary, whereas in subgroups B and C the lipoprotein envelope will typically only contain a single nucleocapsid. The products that are commercially available for pest control represent viruses that belong to subgroups A and B. These subgroups are known as the nuclear polyhedrosis viruses (NPVs) and the granulosis viruses (GVs), respectively. Nonoccluded viruses (subgroup C) have been developed for use in pest control. However, because they will not survive well once released into the environment and therefore cannot be sprayed on to crops, their potential is highly constrained to specific situations that involve the release of preinfected individuals into a crop. Research with nonoccluded viruses has been primarily undertaken by national and international government research organisations.

The NPV-based formulations that are commercially available for arthropod pest control are shown in Table 6.4. Because viruses cannot replicate outside living cells, all of these formulations are produced in a complicated tritrophic culture system in which insects are cultured on plants in order to support a culture of the virus within the pest species. The occlusion bodies of NPVs, which contain many viral envelopes, are composed of a protein called polyhedrin and, as their name suggests, these appear to be polyhedral in shape in scanning electron micrographs. Following their ingestion, the occlusion body dissolves in the insect gut, releasing the viral particles. These particles invade cells lining the gut, replicate and then spread throughout the insect as nonoccluded particles. As the host dies the virus then becomes occluded, releasing further infective units, i.e. secondary disease cycling occurs. A single caterpillar at its death may contain over 10^9 occlusion bodies from an initial dose of 1000. The speed with which this occurs is determined in part by environmental conditions. Under optimal conditions target pests will be killed in 3–7 days. Where conditions are not ideal death may take from 3–4 weeks.

The first NPV to be commercially developed was for control of the cotton bollworm *Helicoverpa zea*. At present, one of the most extensive uses of NPVs occurs in Brazil where the soybean looper or velvetbean caterpillar *Anticarsia gemmetalis* is controlled on over 1 million hectares annually. GVs differ from NPVs in that their protein coat is composed of granulin and that their occlusion bodies contain only one viral envelope. These viruses also replicate inside the nuclei of cells and appear to be ovoid or elliptical in shape in scanning electron micrographs. Their mode of action is thought to be essentially the same as NPVs.

Nonoccluded viruses do not survive well outside cells. Their most extensive use to date has been in the control of the rhinoceros beetle *Oryctes rhinoceros* on palm trees, particularly in the South Pacific. The virus, known as *Rhabdionvirus oryctes*, is used as an inoculative agent (in contrast to the inundative use of NPVs and GVs) by preinfecting adults with the virus. The virus will replicate inside adults but is not particularly toxic to them. In contrast, the virus is lethal to larvae. The adults thus disseminate the virus throughout palm plantations through their excreta. The virus was first used in Malaysia in 1967 and since then has been used in over 40 other countries worldwide. More recently, the virus was successfully used in the Middle Eastern country of Oman to control rhinoceros beetle on date palm. In all of the above cases the beetle population has been reduced to a point where it no longer represents an economic threat to farmers (see also comments on p. 136).

Advantages and disadvantages of viruses in pest control

The main advantages associated with using baculoviruses are their high specificity, safety to nontarget species and ease of application. Like other microbial

Table 6.6. *Summary of the desirable characteristics required by microbial pesticides. For microbes that have the desired character, a score of 1 is given; for microbes without, a score of 0 is given. Scores in parentheses*[a]

	Bacteria	Fungi	Nematodes	Protozoa	Viruses
Time to kill	good (1)	poor (0)	good (1)	poor (0)	poor (0)
Easy to apply	yes (1)	yes (1)	yes (1)	yes (1)	yes (1)
Storage characteristics	good (1)	poor (0)	poor (0)	poor (0)	good (1)
Environmental stability	poor (0)	poor (0)	poor (0)	poor (0)	poor (0)
Safe to nontarget species	yes (1)	yes (1)	yes (1)	yes (1)	yes (1)
Easy to produce	yes (1)	yes (1)	yes (1)	poor (0)	poor (0)
Total scores	5	3	4	2	3

Source: [a] This table is illustrative and not scientific. For example, there are degrees of environmental stability and safety. The table is intended to highlight areas that require further research for each of the microbial groups listed.

formulations, they can be applied using conventional pesticide application equipment. Despite these advantages, there are problems which still need to be overcome before they will be more extensively used. One of the main problems to have constrained their commercial development is their high host-specificity. Both farmers and companies are much more interested in using (or producing) products with a wide spectrum of activity. The fact that they work relatively slowly (days) is also a problem since farmers are used to using products that work quickly (hours). They are also sensitive to environmental degradation. Both NPVs and GVs are inactivated by exposure to sunlight. Finally, they are expensive to produce because of the need for tritrophic culturing systems.

Future developments with viral formulations

The primary areas for development at present concern environmental stability and production. For example, optical brighteners and various dyes have been shown to reduce viral sensitivity to ultraviolet radiation. Improvements to production are likely to derive from cell culture systems. At present, arthropod cell lines exist which can be used to culture these viruses. Unfortunately, the yield from these cell lines tends to be low and the viral particles also appear to lose infectivity with cell cycles. However, it is highly likely that these problems will be overcome in the near future. Finally, it is likely that developments in biotechnology will lead to the production of transgenic viruses with enhanced toxicity. For example, it is already possible to get viruses to express the *Bacillus thuringiensis* deltaendotoxin. These supermicrobes will definitely have an impact on pest control in the next 20 years.

Summary

Table 6.6 provides a summary of the major features of microbes that are currently used for arthropod pest control. By using a simple scoring system the table identifies areas in which improvements have been made and those in which improvements are still to be realised. The main areas for improvement comprise production and environmental stability. Overall, the future for microbial products in pest control looks good. All of these products have a high level of safety which will make them increasingly attractive to farmers who want to produce food that is uncontaminated with pesticides. The relative safety of microbes also means they will integrate well with other control methods. Finally, the fact that most microbial formulations can be applied using standard pesticide application equipment is a significant bonus. The next chapter describes another of the techniques that can be used in integrated programmes – the use of pheromones for arthropod pest control.

Further reading

Bateman, R.P. (1997). Methods of application of microbial pesticide formulations for the control of grasshoppers and locusts. *Memoirs of the Entomological Society of Canada*, **171**, 67–79.

Bateman, R.P. & Alves, R.T. (2000). Delivery systems for mycoinsecticides using oil-based formulations. *Aspects of Applied Biology*, **57**, 163–70.

Burnell, A.M. & Stock, S.P. (2000). *Heterorhabditis*, *Steinernema* and their bacterial symbionts – lethal pathogens of insects. *Nematology*, **2**, 31–42.

Butt, T.M., Jackson, C.W. & Magan, N. (eds) (2001). *Fungi as Biocontrol Agents*. Wallingford: CABI Publishing.

Copping, L.G. (ed.) (2001). *The Biopesticide Manual*, 2nd edn. Farnham, Surrey: BCPC Publications.

Ehlers, R.U. (2001). Mass production of entomopathogenic nematodes for plant protection. *Applied Microbiology and Biotechnology*, **56**, 623–33.

Entwhistle, P.F., Cory, J.S., Bailey, M.J. & Higgs, S. (1993). Bacillus thuringiensis, *an Environmental Biopesticide: Theory and Practice*. New York: John Wiley.

Federici, B.A. (1995). The future of microbial insecticides as vector control agents. *Journal of the American Mosquito Control Association*, **11**, 260–8.

Gill, S.S., Cowles, E.A. & Pietrantonio, P.V. (1992). The mode of action of *Bacillus thuringiensis* endotoxins. *Annual Review of Entomology*, **37**, 615–36.

Hall, F.R. & Barry, J.W. (eds) (1995). *Biorational Pest Control Agents – Formulation and Delivery*. American Chemical Society Symposium Series 595. Washington, DC: American Chemical Society.

Kaya, H.K. & Gaugler, R. (1993). Entomopathogenic nematodes. *Annual Review of Entomology*, **38**, 181–206.

Lambert, B. & Peferoen, M. (1992). Insecticidal promise of *Bacillus thuringiensis*. *Bioscience*, **42**, 112–22.

Milner, R.J. (1997). Prospects for biopesticides for aphid control. *Entomophaga*, **42**, 227–39.

Moscardi, F. (1999). Assessment of the application of baculoviruses for control of Lepidoptera. *Annual Review of Entomology*, **44**, 257–89.

Navon, A. (2000). *Bacillus thuringiensis* insecticides in crop protection – reality and prospects. *Crop Protection*, **19**, 669–76.

Starnes, R.L., Liu, C.L. & Marrone, P.G. (1993). History, use, and future of microbial insecticides. *American Entomologist*, **39**, 83–91.

Strasser, H., Vey, A. & Butt, T.M. (2000). Are there any risks in using entomopathogenic fungi for pest control, with particular reference to the bioactive metabolites of *Metarhizium*, *Tolypocladium* and *Beauvaria* species? *Biocontrol Science and Technology*, **10**, 717–35.

7

Pheromones and pest control

Introduction

Pheromones are chemicals that mediate intraspecific interactions within species. They belong to a group of chemicals known as semiochemicals or communication chemicals. This larger grouping also includes chemicals that mediate interactions between species called allelochemicals. Allelochemicals are briefly discussed at the end of this chapter. Pheromones have been defined as 'externally released chemicals that cause a specific reaction in a receiving organism of the same species'. The first observations on chemical communication between species occurred in the late nineteenth century. However, it was not until 1959, when the first pheromone was identified, that the word pheromone (from the Greek words 'to carry' and 'to excite') was first coined.

The very first pheromone to be identified was the sex pheromone of the silkworm moth *Bombyx mori* (Figure 7.1). This identification was undertaken in the late 1950s and involved the dissection of *c.* 500000 female moths. The yield from these moths was 12 mg of pheromone. The first trials that involved assaying pheromones for arthropod pest control occurred in the mid-1960s. Since then, hundreds of other pheromones have been identified and there are now more than 50 (in over 300 different formulations) that can be used in arthropod pest control programmes (Table 7.1). Fortunately, the technology for pheromone identification has progressed and it is no longer necessary to dissect large numbers of individuals. For example, identification of the sex pheromone of the artichoke moth in the late 1970s required only 20–30 adult females. In relation to arthropod pest control, pheromones have been

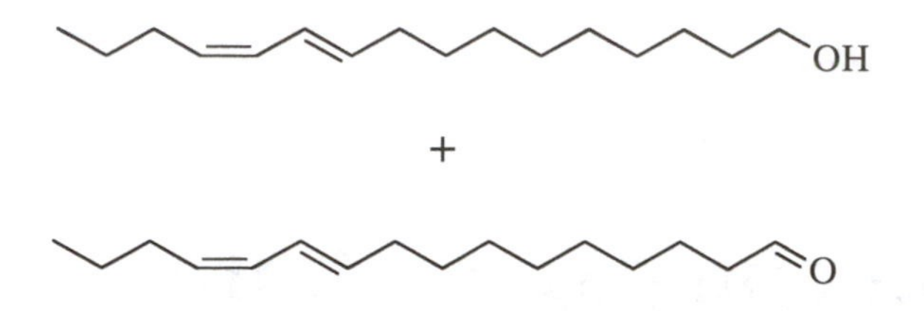

Fig. 7.1. Sex pheromone of the silkworm moth *Bombyx mori*. The pheromone comprises two compounds: *trans*-10, *cis*-12-hexadecadienol and *trans*-10, *cis*-12-hexadecadienal.

developed that can be used in the control of Coleoptera, Diptera, Lepidoptera and Acarina. Pheromones from other important pest groups, e.g. Hemiptera, have been identified, but they have not been developed commercially.

Types of pheromone

Pheromones are used by insects to locate mates, for dispersion, for trail formation, to signal alarm and for maturation. Of these, it is pheromones that are used for mate location that have probably received the most attention in relation to pest control. The pheromones that fall into this category comprise both sex pheromones and aggregation pheromones. The best-studied sex pheromones are those produced by Lepidoptera. In most cases it is females who release the pheromone in order to attract males. The pheromone disperses down-wind and receptive males then respond via a process known as chemoanemotaxis, i.e. chemically mediated, wind-mediated movement. The distances over which pheromones are able to elicit a response are situation- and species-specific. In some cases pheromones may act over many kilometres; in other instances it may just be a few hundred metres. The concentration of pheromone required to elicit a response will also vary with circumstances and species. Laboratory studies have shown that just one molecule of pheromone may be enough to elicit a neurological response in a receiving species. Field experiments with bombycol (sex pheromone of *B. mori*) have shown that 200 molecules/ml air are enough to elicit a behavioural response in a down-wind, receptive male. Pheromones are therefore active at very low concentrations. When males approach females both chemical and visual cues then serve to bring the individuals together. In pest control, these pheromones have been used extensively to monitor pest species in order to improve the timing of other control measures such as pesticide applications.

Table 7.1. *Commercially available pheromone formulations*

Target species	Common name	Pheromone	Mode of action
Adoxophyces orana	Leaf roller	Sex pheromone	Mating disruption of male moths
Anarsia lineatella	Peach tree borer	Sex pheromone	Mating disruption of male moths
Anthonomus grandis	Boll weevil	Sex pheromone	Monitoring and mass trapping of both males and females
Bactrocera oleae	Olive fly	Sex pheromone	Monitoring and lure and kill of male flies
Carposina niponensis	Peach fruit moth	Sex pheromone	Mating disruption of male moths
Ceratitis capitata	Mediterranean fruit fly[a]	Synthetic attractant	Lure and kill of male fruit flies
Chilo suppressalis	Rice stem borer	Sex pheromone	Mating disruption of male moths
Cosmopolites sordidus	Banana corm weevil	Aggregation pheromone	Mass trapping of males and females
Cydia funebrana	Plum fruit moth	Sex pheromone	Monitoring, mating disruption and lure and kill of male moths
Cydia nigricana	Pea moth	Sex pheromone	Monitoring male populations
Cydia pomonella	Codling moth	Sex pheromone	Monitoring, mating disruption, and lure and kill of male moths
Dacus cucurbitae	Melon fly	Sex pheromone	Lure and kill of female flies
Dendroctonus frontalis	Southern pine beetle	Aggregation pheromone	Mass trapping of males and females
Dendroctonus ponderosae	Mountain pine beetle	Aggregation pheromone	Mass trapping of males and females
Dendroctonus pseudotsuga	Douglas fir beetle	Aggregation pheromone	Mass trapping of males and females
Dendroctonus rufipennis	Spruce beetle	Aggregation pheromone	Mass trapping of males and females

Table 7.1. (*cont.*)

Target species	Common name	Pheromone	Mode of action
Dyocoetes confusus	Balsam bark beetle	Aggregation pheromone	Mass trapping of males and females
Eucosma gloriola	Pine shoot borer	Sex pheromone	Mating disruption of male moths
Eupoecilia ambiguella	Grape berry moth	Sex pheromone	Mating disruption of male moths
Grapholitha molesta	Oriental fruit moth	Sex pheromone	Mating disruption of male moths
Heliothis virescens	Tobacco budworm	Sex pheromone	Mating disruption of male moths
Homona magnanima	Tea tortrix moth	Sex pheromone	Mating disruption of male moths
Ips sexdantatus	Six-spined ips beetle	Aggregation pheromone	Mass trapping of male and females
Ips typographus	Spruce bark beetle	Aggregation pheromone	Mass trapping of male and females
Keiferia lycopersicella	Tomato pinworm moth	Sex pheromone	Mating disruption of male moths
Lobesia botrana	Grapevine moth	Sex pheromone	Mating disruption of male moths
Lymantria dispar	Gypsy moth	Sex pheromone	Mating disruption of male moths
Musca domestica	Housefly	Sex pheromone	Mating disruption of male flies
Oryctes rhinoceros	Rhinoceros beetle	Aggregation pheromone	Mass trapping of males and females
Ostrinia nubilalis	Corn stem borer	Sex pheromone	Monitoring adult males
Pectinophora gossypiella	Pink bollworm	Sex pheromone	Mating disruption of male moths
Pityogenes chalcographus	Spruce bark beetle	Aggregation pheromone	Mass trapping of males and females
Platynota idaeusalis	Tufted apple moth	Sex pheromone	Mating disruption and lure and kill of male moths
Platyptilia carduidactyla	Artichoke plume moth	Sex pheromone	Mating disruption of male moths
Plutella xylostella	Diamond moth	Sex pheromone	Mating disruption of male moths
Popilla japonica	Japanese beetle	Sex pheromone	Lure and kill of both males and females

Prays citri	Citrus flower moth	Sex pheromone	Mating disruption of male moths
Prays oleae	Olive moth	Sex pheromone	Mating disruption of male moths
Rhyacionia buoliana	Pine shoot moth	Sex pheromone	Mating disruption of male moths
Rhynchophorus ferrugineus	Red palm weevil	Aggregation pheromone	Mass trapping of males and females
Rhynchophorus palmarum	American palm weevil	Aggregation pheromone	Mass trapping of males and females
Sanninoidea exitiosa	Peach twig borer	Sex pheromone	Mating disruption of male moths
Scolytus multistriatus	European elm bark beetle	Sex pheromone	Monitoring and lure and kill of male beetles
Spodoptera exigua	Beet armyworm	Sex pheromone	Mating disruption of male moths
Spodoptera frugiperda	Fall armyworm	Sex pheromone	Mating disruption of male moths
Synanthedon hector	Cherry tree borer	Sex pheromone	Mating disruption of male moths
Synanthedon myopaeformis	Apple clearwing moth	Sex pheromone	Mating disruption of male moths
Synanthedon pictipes	Lesser peach tree borer	Sex pheromone	Mating disruption of male moths
Synanthedon tipuliformis	Currant clearwing moth	Sex pheromone	Mating disruption of male moths
Tetranychus urticae	Twospotted spider mite	Alarm pheromone	Improved pesticide activity
Thaumetopoea pityocampa	Pine processionary moth	Sex pheromone	Monitoring and mass trapping of male moths
Trypodendron lineatum	Ambrosia beetle	Aggregation pheromone	Mass trapping of males and females
Zeuzera pyrina	Leopard moth	Sex pheromone	Mating disruption of male moths

Note:
[a] Not strictly a pheromone, the compound trimedlure is actually an example of a kairomone that is used for pest control.

Source: Data collated from Copping, L.G. (ed.) (2001). *Biopesticide Manual*. Farnham, Surrey: BCPC Publications

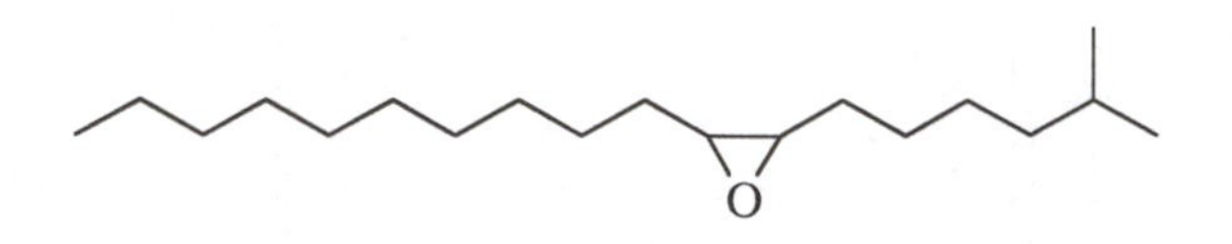

Fig. 7.2. Sex pheromone of the gypsy moth *Lymantria dispar* (+*cis*-7, 8-epoxy-2-methyl octadecane).

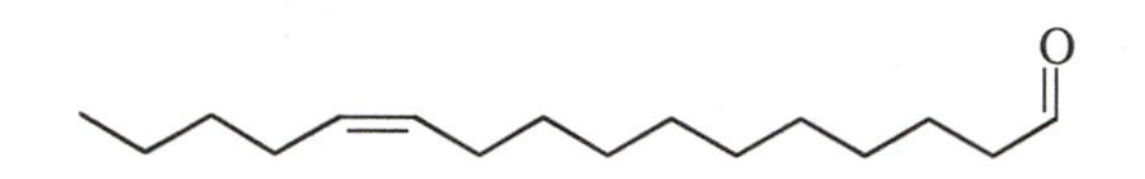

Fig. 7.3. Sex pheromone of the artichoke moth *Platyptilia carduidactyla* (*cis*-11-hexadecanal).

Aggregation pheromones also bring individuals together to mate. However, they may also serve to overcome host defences. They have been most extensively used in pest control with various species of bark beetle. Alarm pheromones are produced in response to attack by predators. They therefore serve as a warning and generally elicit an escape response in receiving and responsive individuals.

Chemistry of pheromones

Chemically, most pheromones are comprised of single or multiple low-molecular-weight volatile molecules. This makes sense because pheromones function to elicit specific responses that are time-limited. To date, most of the pheromones that have been chemically characterised are produced by lepidopterous pests. Typically, these pheromones are C_{10}–C_{21} unsaturated aliphatic alcohol, acetate, aldehyde or ketone derivatives (e.g. Figures 7.2 and 7.3). Most insects synthesise pheromones via specific biochemical pathways in which geometrical and optical isomerism are strictly controlled. The result of this is that most insects only respond to specific isomers. This increases production costs for synthetic pheromones. However, techniques have been developed for the commercial production of a large number of synthetic pheromones. In addition to their specific responses to particular isomers, many insects also require specific ratios of different compounds within a pheromone formulation. For example, the moths *Clepsis spectrana* and

Adoxophyces orana have an identical sex pheromone that is composed of *cis*-9 and *cis*-11-tetradecenyl acetates. However, it is the precise ratio of these two components that determines which species will respond.

Formulation of pheromones

Because pheromones are volatile compounds, specific formulations have been developed for their release. Collectively, these are called controlled-release formulations. The aim is to control release rate *per se*, the uniformity of release rate and the period of time over which the pheromone will be active. Where pheromones are blends of more than one compound, it will also be important that the pheromone dispenser releases each compound in the correct proportions. The main formulations that have been developed over the last 40 years are plastic fibres, laminate flakes, microcapsules and pheromone-impregnated rubber septa. If these specific formulations were not used then a pheromone that was applied in a crop would rapidly evaporate.

Hollow fibres comprise short sections of plastic tubing, sealed at one end and filled with pheromone. The plastic is impermeable to the pheromone and release occurs by evaporation from the liquid–air interface followed by diffusion through the air column to the open end of the fibre (Figure 7.4). With fibres the release rate will decline as the tube empties. The release rate itself will be determined by the diameter of the fibre as well as by the chemical composition of the pheromone. Most fibre formulations are dispensed by hand within crops.

Laminate flakes comprise a central porous layer containing the pheromone sandwiched between two permeable vinyl layers (Figure 7.5). As with fibres, most of these formulations are dispensed by hand. Pheromone is released from the flake following diffusion through the vinyl layers. Release rate can therefore be partially controlled by altering the thickness of these layers.

Microencapsulated pheromone formulations have the advantage over flakes and fibres in that they can be applied using conventional spray application equipment. Microcaps have diameters of 5–20 μm and comprise pheromone enclosed by a plastic polymeric layer that is 0.1–2 μm thick.[1] Pheromone release rate can be controlled by altering the chemical composition of the microcap, by altering the thickness of the microcap and by altering its size.

[1] Microcapsules can also be used with insecticide formulations. The aim with these formulations is also to get controlled release of the formulation. They were discussed in Chapter 4.

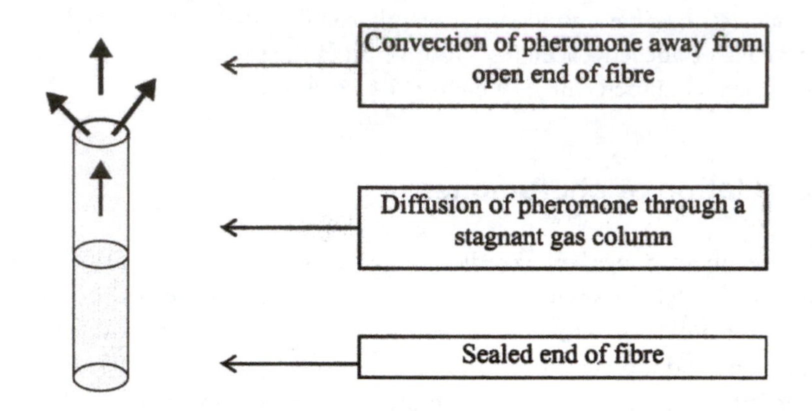

Fig. 7.4. Hollow fibre for pheromone release. Fibres are typically 100–500 μm in diameter. Diameter is the main determinant of release rate. Adapted from Birch, M.C. & Haynes, K.F. (1982). *Insect Pheromones*. London: Edward Arnold.

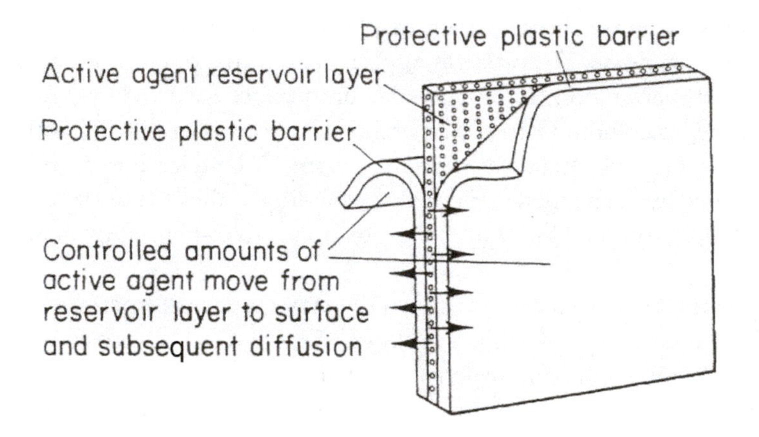

Fig. 7.5. Schematic representation of a laminate flake dispenser. Adapted from Birch, M.C. & Haynes, K.F. (1982). *Insect Pheromones*. London: Edward Arnold.

Various adjuvants can also be incorporated within a microcap formulation to modify the pheromone release rate.

Rubber septa are the most widely used pheromone dispensers and they are used as a standard in tests with new controlled-release systems. A solvent (usually dichloromethane) is used to aid absorption of pheromone into the rubber septa. The solvent evaporates and the impregnated septa are then stored in sealed vials or in hermetically sealed foil envelopes. Research in the late 1970s showed that aldehydic pheromones would interact adversely with chemicals already present in the septa. Pretreatment of septa with dianilino-ethane solved these problems. However, as more complex pheromones are identified it is important to be aware that such problems may occur.

In addition to the above, there are also a variety of miscellaneous controlled-release systems. These include the use of solid polyvinyl rods, cigarette filters, polyethylene vials, dental roll, rice seeds, leather and electronically driven controlled-release devices. The aim with all of these systems is to get controlled release of the pheromone. Whichever system of release is used, the pheromone will usually need to be associated with some type of insect-trapping device. Even where pheromones are sprayed as microcaps it is likely that pheromone traps will be used to monitor the efficacy of the treatment. Pheromone traps are discussed in the next section.

Pheromone traps

Pheromone traps serve to catch insects that are attracted by the pheromone that is released. Perhaps the most obvious advantage associated with these traps is that they should be highly species-specific, and so simple to sort. Expert knowledge may not be required to count insects that are caught in a pheromone trap.[2] Because insects use visual cues as well as chemical and behavioural cues over very short distances, trap design is often crucial. Pheromones will attract insects to a particular area but these insects still need to be caught. Crucial features of a trap include its colour, size, shape and height above the ground. Not surprisingly, a number of different trap designs have been used over the past 40 years. Some of these are shown in Figure 7.6.

We can divide traps into three categories: those that trap insects on a sticky surface, those that kill insects in a liquid and those that kill insects by exposure to pesticide-impregnated plastic strips. Where liquid is used it is usually ethylene glycol (antifreeze). Each of these trap types has advantages and disadvantages. For example, the sticky traps that are widely used to monitor Lepidoptera are cheap, easy to transport and store and simple to use.

[2] Although this is generally true, some pheromone traps (e.g. sticky traps) may collect a diversity of species. This, and the fact that some pest species are polymorphic, may mean that sorting trap catches may not always be a straightforward process.

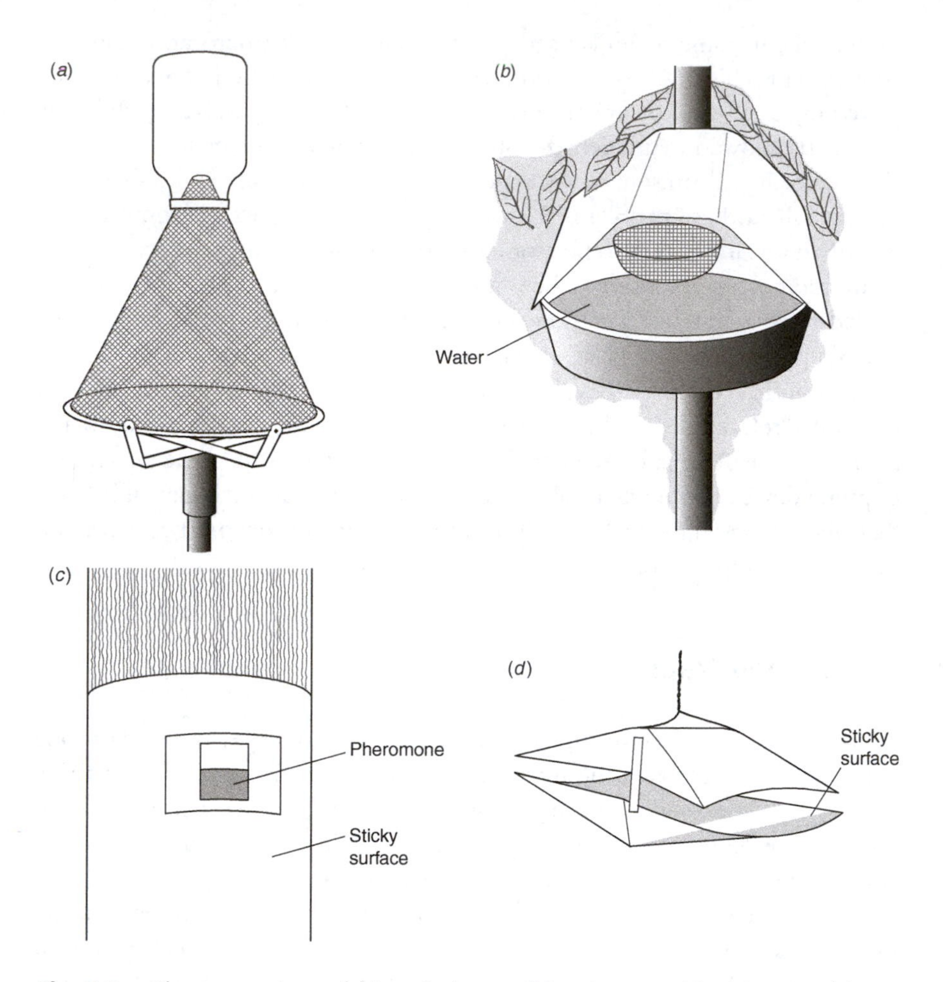

Fig. 7.6. Pheromone traps. (*a*) inverted cone, (*b*) water trap, (*c*) sticky trap, (*d*) sticky trap. Adapted from Birch, M.C. & Haynes, K.F. (1982). *Insect Pheromones*. London: Edward Arnold.

However, their sticky surface will wear down over time and will become overloaded with insects if there are sufficient present. Liquid and insecticidal traps may trap more insects than sticky traps but they are also more difficult to handle. With insecticidal traps there may be additional safety issues associated with handling and possible exposure to the biocide.

The number of insects that a pheromone trap will catch will depend on a series of factors. These factors are either associated with the trap or with the pest. Important trap factors include those mentioned above, i.e. trap colour, size, shape and placement. For example, gypsy moth pheromone traps are more effective when placed adjacent to tree trunks. Important pest population factors include the size of pest population present, the proportion of individuals attracted to the trap, the proportion entering the trap and the proportion that are retained by the trap. Of course, all of these parameters can also be impacted upon by the prevailing environmental conditions, especially the weather. Unfortunately, most studies on trap design concentrate on the total trap catch and do not attempt to measure these factors individually. Further research on important trap characteristics will undoubtedly help if pheromones are to be used with maximum efficacy.

Using pheromones

Pheromones can be used in four different ways in pest control. First, they can be used to monitor pest species numbers. Second, they can be used to mass-trap pest species. Third, they can be used to disrupt mating. Fourth, they can be used to attract species to poisoned bait. This approach is known as 'lure and kill'. Of these, by far the most successful application of pheromones has been in monitoring.

In monitoring systems the number of individuals trapped (usually males) is used as an indicator of pest presence. Mass trapping involves population reduction by collection of a large proportion of the pest population. Because sex pheromones generally only attract males, this technique has been exploited most efficiently with aggregation pheromones that attract both sexes. Mating disruption involves swamping an area with pheromone (a sex pheromone) so that males are unable to locate females because of sensory adaptation (habituation) or competition. The 'lure and kill' approach involves attracting individuals to a source that is treated with a poison, usually a pesticide. The pest picks up a lethal dose and subsequently dies. Each of these techniques will be discussed in more detail in the following sections.

Monitoring

Many pest species are now routinely monitored using pheromone traps. Such monitoring is generally carried out in order to target other control measures or to detect pest presence, particularly at ports and airports. In addition, monitoring of pest populations may provide more fundamental data on pest population trends, especially if the monitoring is undertaken on a long-term

Table 7.2. *Application of pheromone monitoring*

Application category	Subcategory	Specific example
Pest detection	Early warning	Cotton leafworm in Egypt
	Distribution	Gypsy moth in forestry
	Quarantine	Japanese beetle in the USA
Thresholds for control measures	Timing	Codling moth in orchards
	Risk assessment	Bark beetles in forestry
Pest population density	Population trends	Armyworms in East Africa
	Dispersion	Boll weevil in Texas
	Control measure	Pink bollworm in cotton efficacy

Source: Data collated from Jutsum, A.R. & Gordon, R.F.S. (eds) (1989). *Insect Pheromones in Plant Protection*. New York: John Wiley.

basis. A summary of the applications of monitoring systems is provided in Table 7.2. The table gives a breakdown concerning categories of information that derive from monitoring programmes. These categories are not mutually exclusive and pheromone trap catches may provide information for more than one category. For example, many farmers in the USA use pheromone traps over extended periods to monitor pest populations in orchards (pest population trend data) on a year-round basis. Exactly the same data are also used to make decisions about when to use pesticides (timing) and they can also be used to determine if control is effective (treatment efficacy). In the case of the codling moth *Cydia pomonella*, such monitoring has led to 50–75% reductions in pesticide applications. A threshold of 1–5 moths per trap per week has been established as the point at which pesticides need to be applied depending upon location.

The most widely used pheromones for monitoring are the sex pheromones produced by Lepidoptera. One major drawback of this is that only males are caught in traps. The size of the egg-laying population (females) is thus inferred from the male trap catch. This is not ideal as it has been shown on many occasions that trap efficiency will decline with an increase in population density. The reason for this is thought to be enhanced competition (by pheromone traps) with wild females as the pest population increases.[3] Pheromone traps may therefore be ideal for monitoring at low pest population densities,

[3] As the number of females increases in a wild pest population, the ratio of pheromone traps to females declines. From a male point of view this will mean an increase in the likelihood of following a 'real' trail rather than one produced by a synthetic pheromone, all other factors being equal.

but may be less appropriate where pest densities are high. From a pest control point of view, this will only be relevant if the pest species has a high economic threshold. Species with relatively low economic thresholds will have already been exposed to control measures anyway. In summary, there are now a large number of species that are monitored using pheromones. These pheromone traps are cheap, simple to use and are highly effective at low population densities. This makes them ideal for pest detection and for use with economic thresholds for pesticide application.

Mass trapping

The aim of mass trapping is simple – to reduce the pest population below economically damaging levels. Mass trapping, unlike monitoring, is therefore a pest control measure in its own right. Table 7.1 shows that it is a technique that has been most extensively applied to the control of bark beetles in forestry. In these situations pheromones are used to concentrate the beetles on particular trees, which are then felled before the brood can emerge. Mass trapping, as a technique, is ideal in situations where the pest population is highly dispersed, where conventional control using pesticides is inappropriate and where the crop exists as an 'island' with a reduced probability of short-term reinvasion by a pest. Ideally, mass trapping should also reduce both male and female populations. However, this is not essential. For example, the citrus flower moth *Prays citri*, a pest of lemons in Israel, has been successfully controlled using the female sex pheromone. Sticky traps, baited with pheromone, are used to mass-trap male moths. The result is that females remain unfertilised and pesticides are no longer required to control this pest. This is an exception. The most extensive use of mass trapping has been to control bark beetles in forestry.

Many bark beetles have evolved chemically mediated behaviour patterns in which both sexes cooperate to attack trees. Beetles are often attracted to trees by both host odours and by pheromones given off by pioneer attackers. The combination of host odours and attractant pheromones serves to attract the large numbers of beetles that are required to overcome the host trees' defences, i.e. the production of resin. Many beetles also act as a vector for the pathogenic fungus *Ceratocystis minor* which invades the sapwood, further stressing the tree and reducing its resistance to attack by beetles. Large-scale outbreaks of beetles are usually the result of unusual weather conditions. Depending upon weather conditions there may be 3–4 generations a year. The results of a mass attack by bark beetles are often dramatic. For example, in the early 1990s *c.* 15% of the Torrey pine *Pinus torreyana* were destroyed in California by the action of the bark beetle *Ips paraconfusus*. This amounted to some 900 trees from an intact stand of the rarest pine in North America.

The aggregation behaviour of bark beetles has been known for at least the last 200 years and trap trees had been used in Europe to control these beetles long before their pheromones were identified. Using trap trees is a relatively straightforward process – live trees are felled and left in a forest to attract beetles. Following colonisation by bark beetles the trees are then removed and destroyed before brood emergence. The discovery of beetle pheromones in the late 1960s led to the use of these as an alternative to trap trees. The first bark beetle pheromones to be identified were from the western pine beetle *Dendroctonus brevicomis*. With this species females are the pioneers and attack trees first. Females are attracted to pine trees by myrcene, the host terpenoid. Once they arrive, females then release a pheromone called *exo*-brevicomin. This pheromone, in conjunction with the host terpene, serves to attract large numbers of both males and females to the tree. Responding males release a further pheromone called frontalin that further synergises the attack. Once males and females begin to mate, further pheromones are produced (verbenone and *trans*-verbenol) that serve to deter further beetles from attacking the tree. These chemicals thus put an upper limit on the pest population that is supported by the host tree. The discovery of these pheromones and the development of pheromone traps has led to their use in the management of a large number of bark beetle species (Table 7.1). In almost all these cases, the strategy is similar: mass-trap the beetles on to trees, which are then destroyed. In some mass-trapping programmes, antiaggregation pheromones (e.g. verbenone) have also been used to deter attack.

Unfortunately, many researchers are not convinced about the efficacy of the mass-trapping approach. For example, large-scale experiments in the USA and Scandinavia in which millions of beetles have been trapped appear to have had very little effect upon overall tree mortality. Control areas are generally not used (because of the potential economic losses) and so genuine conclusions regarding efficacy are hard to make. Even in the case of the Torrey Pine, where bark beetle populations have been reduced following a pheromone-based mass-trapping strategy, it cannot be proven scientifically that the reduction was attributable to the control programme. However, that the reduction occurred, and that significant numbers of beetles were trapped, do suggest that this technique may have some part to play, particularly in the management of forestry pests. A schematic of the control programme used for beetles in the Torrey Pine nature reserve is shown in Figure 7.7.

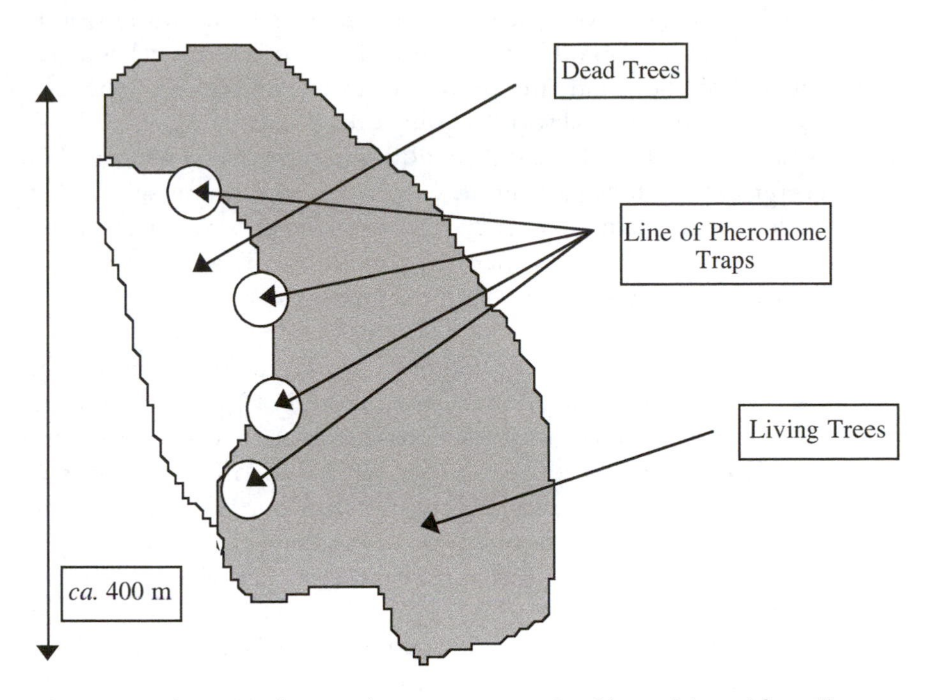

Fig. 7.7. Schematic of Torrey pine state reserve, San Diego. Adapted from Shea, P.J. (1995). Use of insect pheromones to manage forest insects: present and future. In *Biorational Pest Control Agents*, ed. F.R. Hall & J.W. Barry, pp. 272–83. Washington, DC: American Chemical Society. Between 1989 and 1991 *c.* 850 trees were killed by beetles. After the trapping programme was completed no more trees died. Over 300 000 beetles were caught between 1991 and 1993.

Mating disruption

Mating disruption, like mass trapping, is a control technique in its own right. The aim of the technique is to disrupt mating by swamping an area with sex pheromone so that males are unable to locate females. This may occur because the central nervous systems of responding males may habituate to the pheromone and so no longer respond to it or because real pheromone trails are hidden and males simply cannot locate females. Males end up spending almost all their time following false trails created by pheromone traps. Because sex pheromones are the basis of the technique it has been most widely applied to lepidopterous pests (Table 7.1). The most widely quoted example of successful mating disruption comprises control of the pink bollworm *Pectinophora gossypiella*. This example is discussed in more detail below.

The pink bollworm is one the most important pests of cotton worldwide. It is difficult to control because newly hatched larvae quickly penetrate the cotton boll and are protected from insecticides. The sex pheromone of the pink bollworm is a mixture of *cis-cis* and *cis-trans* 7,11-hexadecadienyl acetate and was first identified in the 1960s. The first trials with this pheromone took place in the early 1970s when it was shown that traps could reduce mating frequency, leading to lower larval incidence in bolls and to a reduction in resident adult populations. This was important because farmers were concerned that the traps might attract moths, which would then lay more eggs in their crops. During the late 1970s synthetic versions of the pheromone were registered for use and *c.* 20000 hectares of cotton in the USA were treated with the pheromone in 1978. Since then, the use of the pheromone has increased and insecticide use in cotton crops has decreased. However, it has not been a total success and many farmers still prefer to use insecticides to control this pest. It has been suggested that one of the reasons for this is that the pheromone must be applied by hand. This makes application very expensive in comparison to a pesticide application. In contrast, much more success with this pheromone has been achieved in developing countries in South America, Africa and Asia, particularly Peru, Egypt and Pakistan. Many experimental trials have now shown that the pheromone can be used to control the pest successfully, that it is economic to use the pheromone, that control is comparable to or better than that achieved with pesticides and that the technique preserves natural enemies. Overall, it would seem that this technique has a great deal to offer in terms of pest control. More recently, studies have shown that the technique may be even more effective if high concentrations of pheromone are used. As long as this is economic, then we should expect more from mating disruption in the future.

Lure and kill

The simple aim of 'lure and kill' strategies is to attract an insect pest to a trap and then to kill that pest. Since this is clearly what happens with all pheromone traps, it is questionable whether this is a separate category, i.e. when a pheromone attracts a pest to a sticky trap, the insect is lured and killed. In the context of arthropod pest control it has become common to use the term 'lure and kill' to refer to situations where species are attracted to traps that are baited with an insecticide. The insect spends long enough at the trap to pick up a lethal dose and so is lured and then killed. Clearly, one big advantage of this is judicious use of pesticide, i.e. the pest comes to the pesticide rather than the alternative of trying to target the pest with the pesticide, a process which is often both harmful and wasteful (see Chapter 4).

It is a technique that has been used with a large number of pest species (Table 7.1). Some of the more notable successes comprise control of the cotton boll weevil and the redbanded leafroller in the USA. Trials have also been carried out to assay the utility of this approach for the control of lepidopterous *Spodoptera* spp. and dipterous *Culex* spp. In the case of *Culex*, research has shown that the pheromone can be combined with a growth regulator (these are discussed in detail in Chapter 8). For example, trials in Kenya demonstrated that spraying pools of water with a combination of oviposition pheromone and growth regulator could result in 100% pupal mortality in the target mosquito population.

Other applications of pheromones

Alarm pheromones and oviposition-deterring pheromones

Many pest species mark where they have laid eggs with a pheromone to deter other females of that species from laying eggs nearby. In theory, these pheromones could be used to reduce egg-laying in a crop or to redirect egg-laying elsewhere. Although these pheromones have not been developed commercially, experimental work has shown that they may be useful in the future. For example, field trials with the oviposition-deterring pheromones of both the cherry fruit fly *Rhagoletis cerasi* and the apple maggot fly *R. pomonella* indicated that these chemicals were effective in reducing egg-laying on treated fruit. Studies with a variety of other pest species have produced similar results. These data suggest that these pheromones may be commercially developed within the near future.

Alarm pheromones cause pest species to move. They are produced when pests become aware of predators. At present, there is one commercially available formulation, the spider mite *Tetranychus urticae* alarm pheromone (Table 7.1). These pheromones are intended to be used as mixtures with pesticides. The aim is to get the pest species to increase its movement and so enhance its exposure to the pesticide. Increased movement is clearly of most relevance where contact pesticides are used; however, alarm pheromones can also increase the toxicity of systemic products, permitting reduced doses of pesticide to be applied. Trials with the aphid alarm pheromone farnesene have shown that combinations of reduced-rate insecticide and pheromone are effective at controlling cereal aphids. This pheromone is not yet commercially available.

Development of pheromones for pest control

Thirty years ago it was thought that pheromones offered great opportunities for arthropod pest control. Pheromones are highly specific, they have negligible toxicity to nontarget species, only very small amounts are required and they can increase the efficacy of conventional control methods. Moreover, they are simple to use and do not usually require specialised knowledge. However, despite the fact that over 50 different pheromones are commercially available for pest control, many more have been identified and synthesised but have not yet been used. Figure 7.8 shows that more than 500 different pheromonal compounds have been identified from the Lepidoptera alone. Of these, only a small proportion have been developed for use in pest control. The use of pheromones has not expanded as many people thought. The reasons for this lack of development are many, but certainly include the following:

- lack of basic knowledge concerning insect communication;
- lack of investment in developing these products;
- lack of detailed field evaluation – identifying a pheromone is not enough;
- obtaining enough material to undertake preliminary experiments;
- lack of detailed studies on trap designs;
- lack of studies on blends and release rates;
- lack of studies on the importance of isomerism.

Of the above, a lack of investment in research and development stands out, particularly since it has a bearing on all the other categories. Although pheromones are highly specific, this also presents problems for their commercial development. Problems exist at both the user end and at the manufacturer's end. At the user end, the difficulty is that many crops are often attacked by a pest species complex. For a grower this may mean that a pheromone trap is not a replacement control measure but rather an extra control measure. At the manufacturer end, the difficulty is that pheromones are active in extremely small amounts. This may make the production of effective pheromone delivery systems expensive and it may also make it difficult to take advantage of economies of scale in the production process. It is not surprising that the most widespread applications of pheromones are to be found in key pests of major worldwide crops, e.g. pink bollworm in cotton.

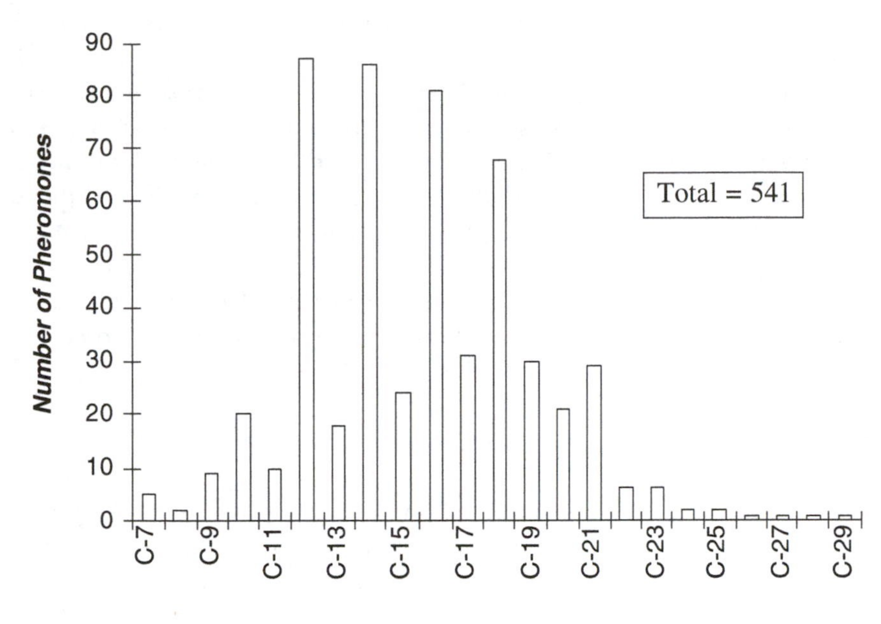

Fig. 7.8. The number of pheromone compounds that had been identified in Lepidoptera by 2000. The figure presents the number against carbon chain length.

Allelochemicals

Chemicals that mediate interactions between different species are called allelochemicals. These interactions may benefit both species, the emitting organism, or the receiving organism. Technically, they are referred to as synomones, allomones and kairomones. Among these, it is kairomones that have received the most attention in relation to arthropod pest control. These are chemicals that insects use as attractants (benefit to receiver). It has been suggested that these chemicals could be used to disrupt the location of food crops by pests. However, by far their greatest use has been in trapping programmes. A simple example of the use of attractants comprises the domestic use of beer to attract slugs in the home garden. Slugs, of course, are not arthropod pests. It has also been suggested that attractants could be used to improve the efficacy of biological control programmes. For example, it is known that many predators and parasitoids are attracted to host pests by pheromones and by volatiles released from damaged plants. However, despite numerous research efforts, no large-scale programmes to exploit such behaviour have so far been developed.

At the agricultural level probably the single most extensive use of an attractant allelochemical comprises the use of trimedlure to attract the Mediterranean fruit fly *Ceratitis capitata*. This chemical compound (chloromethylcyclohexane carbonic acid butyl ester) is not a pheromone, but is a synthetic attractant which can be used to lure insects either to sticky traps or to pesticide-treated traps. Baited traps were first developed in the 1970s and they are still in use in many countries today. The lure has proven particularly useful for monitoring pest presence within export/import crops at airports and seaports.

Summary

Although a significant number of pheromones now make a contribution to arthropod pest control, more could be achieved. Given that so many pheromones have already been identified, further research is now needed on how best to deploy these compounds within pest control programmes. It has been estimated that only 5–10% of all pheromones that have been discovered have been evaluated for commercial development. One of the differences between 40 years ago, when pheromones were first discovered, and now is that pheromones were once viewed as a tool to be used in isolation. This view unfortunately set pheromones up for problems. Today, pheromones are usually regarded as tools that are part of integrated control programmes. This shift in perception should ensure that pheromone use in pest control expands during the next 20 years. The development of integrated pest management, and the use of pheromones within this paradigm, is discussed in detail in Chapter 12. The next chapter considers another approach towards pest control that is based on manipulating species behaviour. This approach is based upon manipulating behaviour by altering the chemical control of growth and development in pest species.

Further reading

Birch, M.C. & Haynes, K.F. (1982). *Insect Pheromones*. London: Edward Arnold.

Carde, R.T. & Bell, W.J. (eds.) (1995). *Chemical Ecology of Insects*. Chapman and Hall, New York.

Carde, R.T. & Minks, A.K. (eds) (1997). *Insect Pheromone Research: New Directions*. New York: Chapman and Hall.

Carde, R.T. & Minks, A.K. (1995). Control of moth pests by mating disruption: successes and constraints. *Annual Review of Entomology*, **40**, 559–85.

Copping, L.G. (ed.) (1998). *Biopesticide Manual.* Farnham, Surrey: British Crop Protection Council Publications.

Gut, L.J. & Brunner, J.F. (1998). Pheromone-based management of codling moth (Lepidoptera: Tortricidae) in Washington apple orchards. *Journal of Agricultural Entomology*, **15**, 387–406.

Hall, F.R. & Barry, J.W. (eds) (1995). *Biorational Pest Control Agents.* Washington, DC: American Chemical Society.

Howse, P., Stevens, I. & Jones, O. (1998). *Insect Pheromones and their use in Pest Management.* New York: Chapman and Hall.

Isaacs, R., Ulczynski, M., Wright, B., Gut, L.J. & Miller, J.R. (1999). Performance of the microsprayer, with application for pheromone-mediated control of insect pests. *Journal of Economic Entomology*, **92**, 1157–64.

Jutsum, A.R. & Gordon, R.F.S. (eds) (1989). *Insect Pheromones in Plant Protection.* New York: John Wiley.

Millar, J.G. (2000). Polyene hydrocarbons and epoxides: a second major class of lepidopteran sex attractant pheromones. *Annual Review of Entomology*, **45**, 575–604.

Norin, T. (2001). Pheromones and kairomones for control of pest insects. Some current results from a Swedish research program. *Pure and Applied Chemistry*, **73**, 607–12.

Ogawa, K. (2000). Pest control by pheromone mating disruption and the role of natural enemies. *Journal of Pesticide Science*, **25**, 456–61.

Pickett, J.A., Wadhams, L.J. & Woodcock, C.M. (1995). *Exploiting Chemical Ecology for Sustainable Pest Control.* British Crop Protection Council Monograph 63. Farnham, Surrey: British Crop Protection Council.

Renou, M. & Guerrero, A. (2000). Insect parapheromones in olfaction research and semiochemical-based pest control strategies. *Annual Review of Entomology*, **45**, 605–30.

Seybold, S.J., Bohlmann, J. & Raffa, K.F. (2000). Biosynthesis of coniferophagous bark beetle pheromones and conifer isoprenoids: evolutionary perspective and synthesis. *Canadian Entomologist*, **132**, 697–753.

Suckling, D.M. (2000). Issues affecting the use of pheromones and other semiochemicals in orchards. *Crop Protection*, **19**, 677–83.

Trimble, R.M., Pree, D.J. & Carter, N.J. (2001). Integrated control of oriental fruit moth (Lepidoptera: Tortricidae) in peach orchards using insecticide and mating disruption. *Journal of Economic Entomology*, **94**, 476–85.

8

Insect growth regulators

Introduction

Chemicals that affect the growth and development of insects are called insect growth regulators. Insecticidal versions of these compounds are often included within the group of chemicals that are known as biorational pest control formulations (a grouping which also includes pheromones and microbial formulations – chapters 6 and 7). Biorational compounds are so named because of their high level of environmental safety.

Insect growth regulators that have been developed for arthropod pest control can usually be applied using the same equipment as conventional insecticides. However, since they affect growth and development, they are primarily effective against immature insect stages. With the exception of the order Thysanura, adult insects do not moult. The upshot of this is that timing of an application is often critical to ensure that maximum product efficacy is achieved. The growth regulators that are of interest for pest control include moulting hormone, juvenile hormone and chitin synthesis inhibitors. Before these chemicals are discussed in detail, a brief review of normal arthropod development is provided.

Arthropod growth and development

All arthropods possess tough exoskeletons that support internal organs and muscles in addition to presenting a barrier to water loss. For an arthropod to grow, these exoskeletons must be shed periodically. The moulting process is controlled by a combination of hormones and requires the successful production of a new exoskeleton. The most important hormones involved comprise juvenile hormone, moulting hormone, eclosion[1] hormone, and ecdysis[2]-triggering hormone. In general, the level of moulting hormone within an insect's circulatory system determines whether a moult will occur, while the level of juvenile hormone determines whether the organism remains at an immature stage or not. Juvenile hormone levels decline as arthropods go through successive moults to become adults. Moulting hormone is produced in response to the release of prothoracicotropic hormone from brain cells. This occurs in response to signals from mechanical stretch receptors that indicate a new moult is required to sustain further growth. Following the production of a new cuticle, eclosion hormone and/or ecdysis-triggering hormone are produced. Insects then shed the old cuticle or exuvium.

The notion of using an insect's hormones to control pest species is not new. The concept was first described almost 40 years ago in the mid-1960s. For example, a juvenile hormone analogue was reported to have successfully controlled the stored product pest *Tribolium castaneum* in 1968. The first major review of insect growth regulators for pest control was published in the 1975 *Annual Review of Entomology*.

Juvenile hormone mimics

Early attempts to interfere with moulting and insect development concentrated on juvenile and moulting hormones. For example, removal of the corpora allata (the gland which secretes juvenile hormone) from insects causes the premature production of adults. Similarly, exposure of pupae to active corpora allata causes the production of adult monsters. Further, more detailed studies revealed that juvenile hormone would penetrate insect cuticles and could be topically applied to test species.

Studies on plants further enhanced this work when it was shown that some plants (especially firs and ferns) can produce chemicals that have juvenile hormone-like activity. This research led to the identification of juvabione in balsam fir trees. Juvabione, and other chemicals that have subsequently been identified, are now referred to as juvenile hormone mimics or analogues and

[1] Eclosion is the process of adult emergence from a pupa or, in species with partial metamorphosis, adult emergence from the last nymphal instar.

[2] Ecdysis means to cast off or shed. The same word is used whether with reference to a reptile shedding its skin or with reference to an arthropod shedding its cuticle. In human terms we could legitimately refer to a stripper as an ecdysiast!

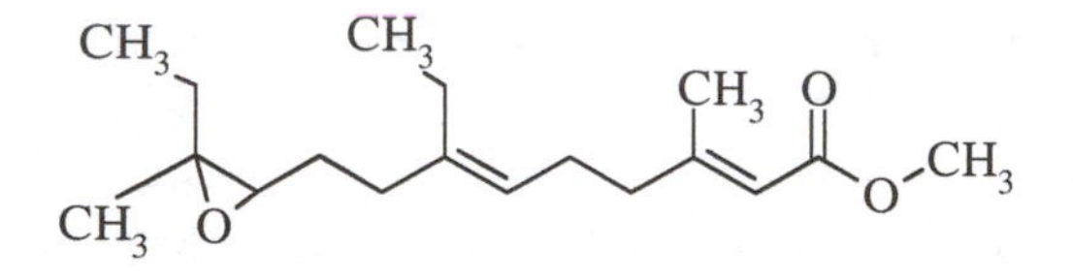

Fig. 8.1. Chemical structure of juvenile hormone.

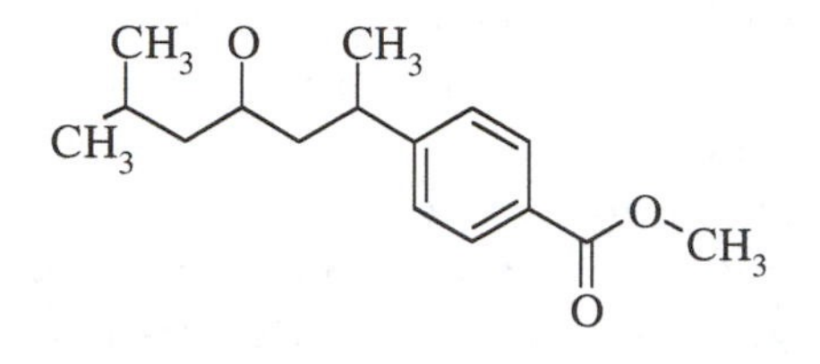

Fig. 8.2. Chemical structure of juvabione.

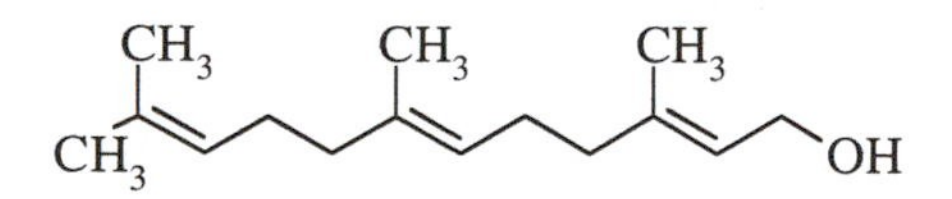

Fig. 8.3. Chemical structure of farnesol.

they seem to be part of a plant's defence against insect attack. It was this research that led to the production of the first synthetic hormone (farnesol) for use in pest control. The chemical structures of juvenile hormone, naturally occurring juvabione and farnesol are shown in Figures 8.1–8.3. Despite the production of farnesol as a juvenile hormone mimic, its use in pest control has been limited. Unfortunately, farnesol does not penetrate insect cuticles well, it maintains insects in the juvenile (and possibly injurous) stage and it does not have a great deal of residual activity following application.

However, some successful analogues of juvenile hormone have been developed commercially. These analogues include methoprene, hydroprene, kinoprene, fenoxycarb and pyriproxyfen (Table 8.1). Methoprene is marketed for the control of mosquito larvae, horn flies, pests of stored tobacco (*Lasioderma serricorne* and *Ephestia elutella*), pests of stored grain (*Rhyzopertha*

Table 8.1. *Juvenile hormone formulations available in 2000*

Common name	Trade names	Primary target pests
Kinoprene	Enstar II	Variety of Homoptera
Fenoxycarb	Insegar	Variety of Homoptera, Lepidoptera, Diptera, Dictyoptera and Coleoptera
Hydroprene	GenTrol, Mator, Gencor	Variety of Homoptera, Lepidoptera, Dictyoptera and Coleoptera
Methoprene	Kabat, Apex, Precor, Pharorid, Minex	Variety of Homoptera, Lepidoptera, Diptera, Dictyoptera, Coleoptera and Siphonaptera
Pyriproxyfen	Sumilarv	Variety of Homoptera, Diptera and Coleoptera

Source: Data collated from Tomlin, C. (ed.) (2000). *The Pesticide Manual*, 12th edn. Farnham, Surrey: British Crop Protection Council.

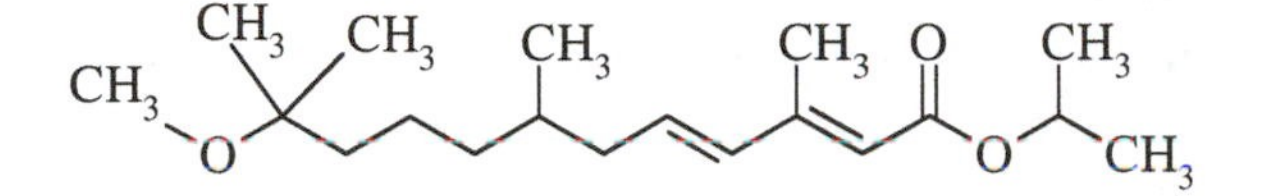

Fig. 8.4. Chemical structure of methoprene.

dominica, *Oryzaephilus surinamensis* and *Tribolium castaneum*), fungus gnats and pharaoh ants. Methoprene has also been used in combination with the organophosphate insecticide methacrifos to control the grain pests *Sitophilus oryzae* and *S. granarius*. Hydroprene is marketed for the control of cockroaches and fire ants. The structure of methoprene is shown in Figure 8.4. Kinoprene is a juvenile hormone analogue that is specific to Homoptera. Unfortunately, it does not compete economically with conventional insecticides and so, although effective, it has largely been withdrawn from use. Fenoxycarb is a a carbamate juvenile hormone mimic and pyriproxyfen is a phenyl ether juvenile hormone mimic. Fenoxycarb is marketed for the control of a diversity of pest species while pyriproxyfen is marketed for the control of public health pests. For example, pyriproxyfen can be used in cat collars for flea control. The chemical structures of fenoxycarb and pyriproxyfen are shown in Figures 8.5 and 8.6.

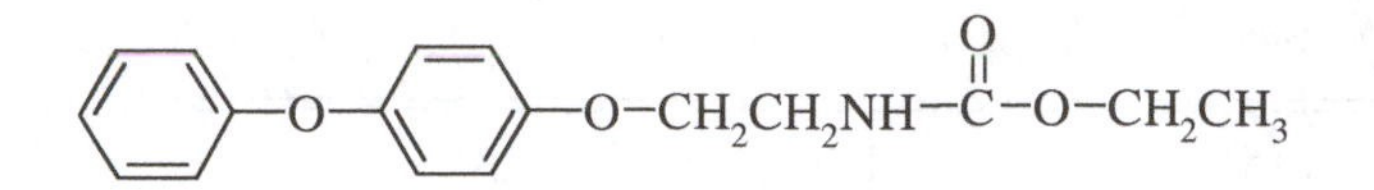

Fig. 8.5. Chemical structure of fenoxycarb.

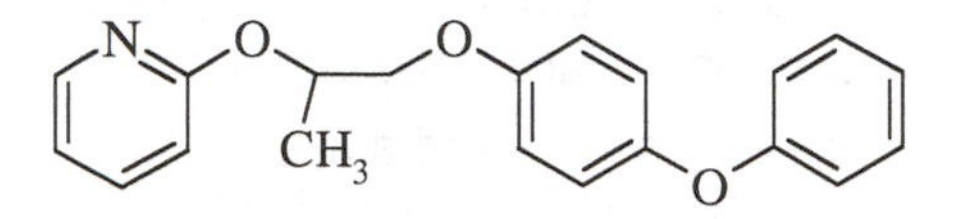

Fig. 8.6. Chemical structure of pyriproxyfen.

The main advantage associated with using juvenile hormone mimics is their high specificity. However, juvenile hormone mimics have not been without their problems, including the following. First, many insects that are resistant to conventional insecticides appear to be resistant to juvenile hormone mimics, possibly because routes of enzymatic molecular degradation are similar. Second, hormone mimics do not disrupt the normal development of larvae and so they can only really be used against pests in which it is the adult that is the noxious organism. Third and last, to be effective these mimics must be applied at the correct developmental stage of the target organism and it is also desirable that they should have a fair degree of environmental persistence. Unfortunately, most of the juvenile hormone mimics break down fairly rapidly on exposure to light and air. Because of these problems, juvenile hormone mimics have really only been successfully used against dipterous pests of public and animal health.

The alternative approach to disrupting levels of juvenile hormone within pest species is to try to decrease levels that circulate with the bodies of pests. Clearly it is not practical to remove the glands that secrete juvenile hormone. However the knowledge that some plants produce juvenile hormone mimics has stimulated the search for antagonists of juvenile hormones within plants. Inhibiting juvenile hormone activity may be a much more useful route for pest control because it would disrupt successful larval development. One such antagonist has been found in the bedding plant *Ageratum*. This compound was found to cause precocious development to take place in a number of pest species by causing premature metamorphosis. The active ingredients, once

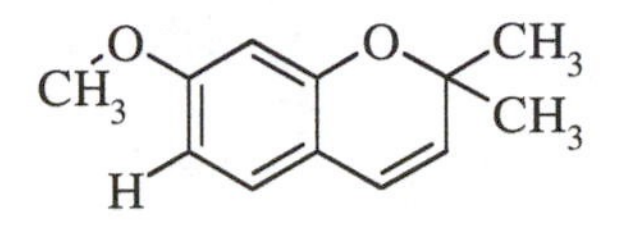

Fig. 8.7. Chemical structure of precocene I.

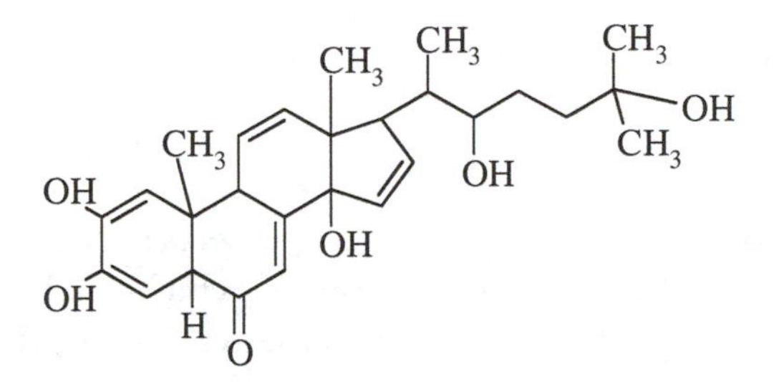

Fig. 8.8. Chemical structure of ecdysone.

identified, were therefore called precocenes. The structure of precocene I is shown in Figure 8.7. Despite the discovery of these compounds, none has proven sufficiently active for use in pest control.

Moulting hormone mimics

While the level of juvenile hormone within an insect determines whether the organism remains at an immature stage or not when a moult occurs, the hormone that determines whether a moult will take place is called moulting hormone. This hormone is secreted by the insect's prothoracic gland and is called ecdysone (Figure 8.8). This compound is a complex steroidal molecule, and while some plants (conifers and ferns) seem to produce similar compounds, these appear to have no effect upon moulting.

Moulting hormone mimics represent the newest group of insect growth regulators to be developed for commercial release. In contrast to juvenile hormone mimics, ecdysone agonists have the significant advantage that they can be used to control arthropod pests during larval/nymphal stages. The best-known representatives of this group comprise the chemical compounds tebufenozide, methoxyfenozide and halofenozide. These compounds appear

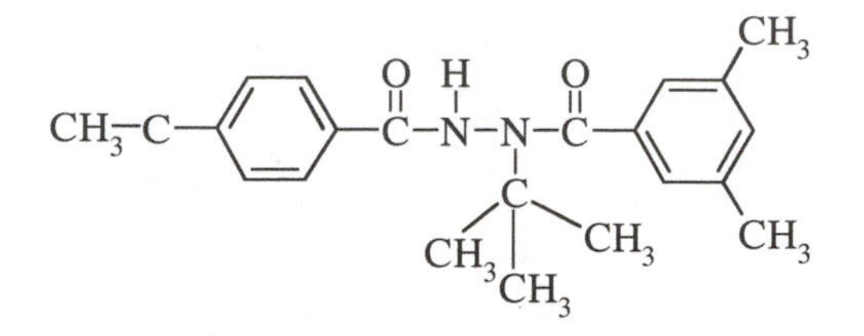

Fig. 8.9. Chemical structure of tebufenozide.

to work by binding to the ecdysone receptor protein, causing a precocious and unsuccessful moult to take place. The chemical structure of tebufenozide, which is marketed for the control of a range of lepidopterous crop pests, is shown in Figure 8.9. Because tebufenozide is not toxic to predatory mites or beneficial Coleoptera it can be used in integrated control programmes. Research is currently under way to evaluate ecdysone agonists for the control of stored product pests.

The only other compound to be developed on a commercial basis that would appear to have moulting hormone-like activity comprises the extract from the neem tree (azadirachtin). This chemical appears to function at least partially in the disruption of moulting and there are now a number of commercial formulations of neem that are available. This chemical was discussed in detail in Chapter 2.

Chitin synthesis inhibitors

Although no great success for pest control has so far been achieved with moulting hormone mimics, a great deal of success has been had with compounds that interfere with the breakdown of the old cuticle and the production of a new cuticle during a moult. These compounds interfere with the production of chitin, the structural polysaccharide found in arthropod cuticles. Exposure to a chitin synthesis inhibitor causes a lack of chitin in a new exoskeleton following a moult. Insects then die from desiccation, starvation or predation. There are now at least 10 different active ingredients that are collectively referred to as benzoylphenylureas (Table 8.2). The chemical structure of one of the best-known of these, diflubenzuron, is shown in Figure 8.10.

Table 8.2. *Benzoylphenylurea (chitin synthesis inhibitor) formulations available in 2000*

Common name	Trade names	Primary target pests
Buprofezin[a]	Applaud	Various Homoptera
Chlorfluazuron	Atabron, Helix	Various Lepidoptera and Homoptera
Diflubenzuron	Dimilin	Various Lepidoptera, Diptera, Homoptera and Orthoptera
Fluazuron	Acatak	Cattle ticks
Flucycloxuron	Andalin	Spider mites and some Lepidoptera and Diptera
Flufenoxuron	Cascade	Phytophagous mites
Hexaflumuron	Consult	Various Lepidoptera, Diptera, Homoptera and Coleoptera
Lufenuron	Match	Various Lepidoptera, Diptera, Homoptera and Coleoptera
Novaluron	Rimon	Various Lepidoptera, Diptera and Homoptera
Teflubenzuron	Nomolt	Various Lepidoptera, Diptera, Homoptera, Hemiptera, Dictyoptera and Coleoptera
Triflumuron	Alsystin, Baycidal	Various Lepidoptera, Diptera, Homoptera, Hemiptera, Dictyoptera and Coleoptera

Note:
[a] Listed as a probable chitin synthesis inhibitor.
Source: Data collated from Tomlin, C. (ed.) (2000). *The Pesticide Manual*, 12th edn. Farnham, Surrey: British Crop Protection Council.

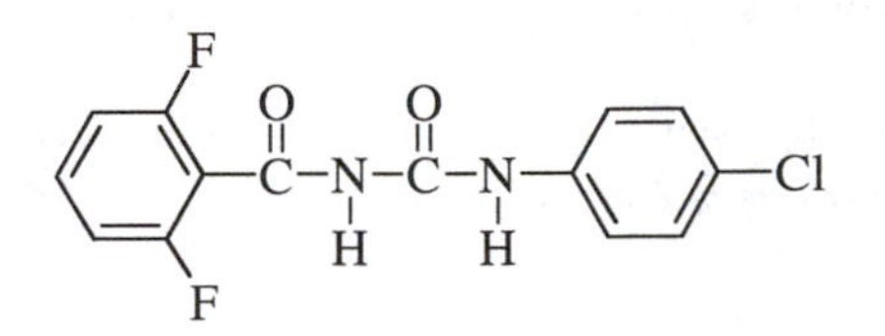

Fig. 8.10. Chemical structure of diflubenzuron.

Diflubenzuron, marketed under the trade name Dimilin, is toxic to a wide range of pests, has a low (negligible) mammalian toxicity, a low toxicity to most beneficial species[3] and a high degree of environmental persistence. It is thought to work by disrupting the enzyme chitin synthetase, particularly in the final stages of chitin synthesis, namely the polymerisation of acetylglucosamine. However, the precise mode of action of this compound is still not known. Because of its low toxicity to nontarget species, diflubenzuron can be used as part of integrated pest management programmes. Overall, benzoylphenylureas are marketed for control of a number of pest species (Table 8.2) and they have been especially successful in Africa in locust control programmes.

Summary

In summary, most success with insect growth regulators has been achieved with chitin synthesis inhibitors. These compounds are targeted at larvae as they are ineffective against adults. There have been some additional successes with the control of public and veterinary pests, particularly ants, with the juvenile hormone mimics methoprene and hydroprene. At present, insect growth regulators play a relatively minor role in overall arthropod pest control. However, it is highly likely that as our knowledge of insect growth and development expands, more opportunities for developing growth regulators for use in pest control will occur. In the next chapter we will consider pest control techniques that are also based on disrupting species development. However, these techniques are concerned with disrupting the mating process. Collectively, they have been called genetic control techniques.

Further reading

Cohen, E. (1993). Chitin synthesis and degradation as targets for pesticide action. *Archives of Insect Biochemistry and Physiology*, **22**, 245–61.

Cohen, E. (2001). Chitin synthesis and inhibition: a revisit. *Pest Management Science*, **57**, 946–50.

Dhadialla, T.S., Carlson, G.R. & Le, D.P. (1998). New insecticides with ecdysteroidal and juvenile hormone activity. *Annual Review of Entomology*, **43**, 545–69.

[3] Exposure of streams and rivers to this compound should be avoided because of its known adverse effect on crustacean populations.

Dobson, P., Tinembart, O., Fisch, R.D. & Junquera, P. (2000). Efficacy of nitenpyram as a systemic flea adulticide in dogs and cats. *Veterinary Record*, **147**, 709–13.

Gernier, S. & Grenier, A.M. (1993). Fenoxycarb, a fairly new insect growth regulator: a review of its effects on insects. *Annals of Applied Biology*, **122**, 369–403.

Horodyski, F.M. (1996). Neuroendocrine control of insect ecdysis by eclosion hormone, *Journal of Insect Physiology*, **42**, 917–24.

Horowitz, A.R., Mendelson, Z., Cahill, M., Denholm, I. & Ishaaya, I. (1999). Managing resistance to the insect growth regulator, pyriproxyfen, in *Bemisia tabaci*. *Pesticide Science*, **55**, 272–6.

Ishaaya, I. & Horowitz, A.R. (1995). Pyriproxyfen, a novel insect growth regulator for controlling whiteflies: mechanisms and resistance management. *Pesticide Science*, **43**, 227–32.

Kramer, K.J. & Muthukrishnan, S. (1997). Insect chitinases: molecular biology and potential use as biopesticides. *Insect Biochemistry and Molecular Biology*, **27**, 887–900.

Nijout, F.H. (1994). *Insect Hormones*. Princeton: Princeton University Press.

Nomura, M. & Miyata, T. (2000). Effects of pyriproxyfen, insect growth regulator on reproduction of common cutworm, *Spodoptera litura* (Fabricius) (Lepidoptera: Noctuidae). *Japanese Journal of Applied Entomology*, **44**, 81–8.

Pons, S., Riedl, H. & Avilla, J. (1999). Toxicity of the ecdysone agonist tebufenozide to codling moth (Lepidoptera Tortricidae). *Journal of Economic Entomology*, **92**, 1344–51.

Staal, G.B. (1975). Insect growth regulators with juvenile hormone activity. *Annual Review of Entomology*, **20**, 417–60.

Tomlin, C. (ed.) (2000). *The Pesticide Manual*, 12th edn. Farnham, Surrey: British Crop Protection Council Publications.

Waldstein, D.E. & Reissig, W.H. (2001). Apple damage, pest phenology, and factors influencing the efficacy of tebufenozide for control of oblique banded leafroller (Lepidoptera: Tortricidae). *Journal of Economic Entomology*, **94**, 673–9.

9

Genetic manipulation of pest species

Introduction

The process of genetic manipulation for pest control purposes involves deleterious interference in the functioning of genes. Such an interference is usually brought about by the introduction of a population of laboratory-reared and genetically impaired pest species into a wild pest population. The aim is to bring about a reduction in the size of the wild population by disseminating harmful or defective genetic material. This is typically achieved when wild individuals mate with the laboratory-reared, impaired individuals.

Genetic manipulation of arthropod pests can be achieved by using at least two different approaches. Both of these approaches involve rearing and release of pest species, and both involve the exposure of pest species to mates that will not produce viable offspring. Differences occur in the technology that is used to produce these deleterious mates. One approach uses preexisting incompatibility and/or induces it by sterilisation. The other approach uses recombinant DNA technology to produce mates that are harmful to the target pest population. This latter approach is discussed in the final chapter of this book (Chapter 13). This chapter is solely concerned with the former approach. There have also been attempts to use chemosterilants to control pests directly. For example, trials have shown that it is feasible to reduce populations of the housefly *Musca domestica* by using poisoned baits. However, since chemosterilants are toxic to a wide spectrum of life, they have not found widespread application. Direct control by using chemosterilants is also more akin to using a chemical pesticide, and is therefore not a true example of genetic manipulation, and so is not discussed further in this chapter.

The rationale for producing and/or releasing laboratory pests that are deleterious to wild populations is that this can lead to autocidal ('self-killing') control of pest species. It is therefore a highly specific form of pest control, much like the use of pheromones, discussed in Chapter 7. The biggest obstacle to this approach has not been the creation and/or use of deleterious genetic traits but rather their incorporation into wild populations. Clearly, there is no selective advantage in a population driving itself to extinction. This, however, is exactly what genetic manipulation seeks to achieve. The techniques that have been used to produce species that will bring about autocidal control of a pest population are fourfold. They are referred to as: (1) the sterile insect release technique (SIRT); (2) the use of hybrid sterility; (3) the use of cytoplasmic incompatibility; and (4) the use of chromosomal translocations. Of these, by far the most successful has been the first. However, before SIRT is discussed in detail, a brief description is provided for each of these techniques.

Methods of genetic manipulation

SIRT (also known as the sterile insect technique or the sterile male technique) involves the release of male insects that have been mass-reared and sterilised in a laboratory. When these individuals mate with wild females, no viable offspring are produced and so the pest population declines. This technique has been most effectively exploited with the American screwworm *Cochliomyia hominivorax*, with the Mediterranean fruit fly *Ceratitis capitata* and with the tsetse fly *Glossina* spp. The technique is discussed in more detail later in this chapter. The other techniques that have been used for autocidal control have been assayed in field trials, but none has been developed on the scale of SIRT.

It has been known for a long time that when mating occurs between strains or subspecies in species complexes that the male (heterogametic) offspring are usually sterile. This is known as Haldane's rule. Theoretically, it should be possible to use these sterile hybrids to bring about autocidal control of a pest population. This approach was used in 1942 in the very first field trial of a genetic control method. Field-collected specimens of the tsetse fly *Glossina mortisans centralis* were released into a population of *G. swynnertoni* that had caused an outbreak of human trypanosomiasis in Tanzania. It was already known that if these two species mated the female offspring were semisterile and male offspring were completely sterile. The result of the trial was that the *G. swynnertoni* population was replaced by *G. mortisans centralis*. This species did not survive well and so, locally at least, the area became tsetse-free.

Since the 1940s other trials have been undertaken with mosquitoes and with the tobacco budworm *Heliothis virescens*. For example, in the 1970s attempts were made to control the vector of filariasis *Anopheles gambiae* by releasing sterile male hybrids derived from crosses between male *A. arabiensis* and female *A. melas*. Unfortunately, despite the release of *c.* 300 000 hybrid pupae, the sterile males failed to mate with wild females, possibly because of behavioural premating barriers. More success with this technique has been achieved in laboratory experiments with tobacco budworm sterile hybrids. These are produced from crosses between male *Heliothis virescens* and female *H. subflexa*. However, this technique has not yet been successfully used in a large-scale field trial. Although the use of hybrid sterility has not yet reached large-scale development, it still remains an attractive option for species that exist as complexes, such as the tsetse fly, and it is a technique that definitely requires further investigation.

Cytoplasmic incompatibility occurs where two individuals of apparently the same species mate, but do not produce offspring. It is thought that, although the spermatozoa of the male enter the egg, fusion between the nucleus of the sperm and the egg does not occur. Biologically, it has been suggested that this is because of a cytoplasmic factor – a rickettsia (an obligate, intracelluar bacterial parasite) that lives in sperm cytoplasm and initiates sperm inactivation, so preventing successful fertilisation. Like hybrid sterility, it is a process that has been best studied with species complexes. The best-documented example of this technique comprises control of the southern house mosquito *Culex quinquefasciatus* in Burma. This insect was eradicated locally following the release of cytoplasmically incompatible males. Further research with the filarial vector *Aedes scutellaris* in Papua New Guinea also produced some promising results, but neither of these two studies were followed up in any detail. It is thought that, like hybrid sterility, this is a research area that deserves further attention.

It has been known for a long time that the normal arrangement of chromosomes can be disrupted by exposure to chemicals or ionising radiation. Such rearrangements are called chromosome translocations, and while adults carrying the translocation may remain fertile, the offspring that result from crosses with normal individuals are usually fully sterile. It has been suggested that such translocations could be used to introduce useful genes into pest populations, i.e. genes for insecticide resistance, reduced vectoring capacity, temperature sensitivity and nondiapausing. At present, although translocation strains have been produced in laboratories, field application of these strains appears to be a long way off. The use of translocation strains produced using recombinant techniques is discussed in more detail in Chapter 13. There is one

Table 9.1. *Theoretical population decline in each generation following constant release of sterilised males into a wild population of 1 million virgin females and 1 million virgin males*[a]

Gen.	No. of virgin females	No. of sterile males released	Ratio of sterile to wild males	Percentage sterile matings	No. of females in next generation
F_1	1000000	2000000	2 : 1	66.66%	333333
F_2	333333	2000000	6 : 1	85.71%	47619
F_3	47619	2000000	42 : 1	97.70%	1107
F_4	1107	2000000	1807 : 1	99.95%	Less than 1

Note:

[a] These data are taken from Knipling, E.F. (1955). Possibilities of insect control or eradication through the use of sexually sterile males. *Journal of Economic Entomology*, **48**, 459–69. The data assume that reproduction is not successful among fertile wild individuals and that wild males die after they have mated once. These assumptions are discussed in the text.

example, however, where the use of radiation to alter the genetic make-up of a species has proven useful. This example comprises the use of radiation to sterilise pest species. This technique is discussed in detail below.

Sterile insect release technique

In 1916 it was shown that X-rays could be used to sterilise the cigarette beetle *Lasioderma serricorne*. However, it was not until 1937 that the suggestion was made to use this procedure in the control of arthropod pests. The idea is simple. Mass-rear a pest, sterilise it and then release the pest into a population of wild individuals. Mating between sterile and wild individuals will not produce viable offspring and this may then lead to autocidal control of the pest. This is SIRT (Table 9.1). Clearly it will be advantageous if females only mate once and if the species is limited in its dispersal. A ratio of between 10 and 100 sterile-to-wild males is generally thought necessary (i.e. 10:1 up to 100:1), although the exact ratio will depend on the mating habits of the pest. The technique is 100% species-specific and is unlikely to have any negative impact on the environment. In the case of the American screwworm *Cochliomyia hominivorax* (the best-documented use of SIRT to date) there is some evidence to suggest that deer outbreaks have occurred following control of the fly. However, these outbreaks should not be regarded as huge environmental problems, especially in light of the pest control that has been achieved.

Clearly, species for release must not act as vectors for disease or be able to feed on humans or other species, since such a technique would be unacceptable to the general public. In general, for SIRT to be effective requires that target pests have certain characteristics. Target pests should be reasonably well known ecologically and they should also be making a substantial economic impact. In addition, control of the pest and management of the SIRT programme will need to be carried out by governments since the process typically involves the release of living organisms in huge numbers over very large areas (i.e. over numerous jurisdictions). It is for the above reasons that SIRT is not an attractive option for business. All the cases of SIRT that have been carried out to date have been government-sponsored. Finally, for success to be realised there must be techniques for rearing, sterilising and releasing insects. Since the technique requires sterilisation of males, its success will depend on factors such as the competitiveness of the released individuals, the ratio of sterile to wild individuals and the number of matings that individual insects attempt. Mass rearing, culturing and release can all have a bearing on success. Each of these factors is discussed in more detail below.

Mass rearing

The aim of mass rearing is to produce millions or billions of sterile males that can successfully compete with fertile individuals in the wild population. Because of this, it is essential that culturing is easy. This means that the species should ideally have simple dietary requirements; more specifically, it should be possible to culture target species on an artificial diet. This diet should have no effect upon the species' field behaviour following release. The target species should also have a high fecundity and not suffer a high level of larval mortality. The alternatives of field collecting or using live diets are usually not economically viable. During the culturing process *per se* it is also critical that there is no loss of fitness. It has been known for a long time that culturing insects often leads to laboratory strains that are no longer as fit as wild individuals. This was one of the reasons for a breakdown in the screwworm control programme in the mid-1970s (Figure 9.1). The process will therefore usually require a frequent input of wild genetic material. It is also important that the climatic conditions during culturing match those likely to be met in the field. At the end of the culturing process it is also important that a stage of the

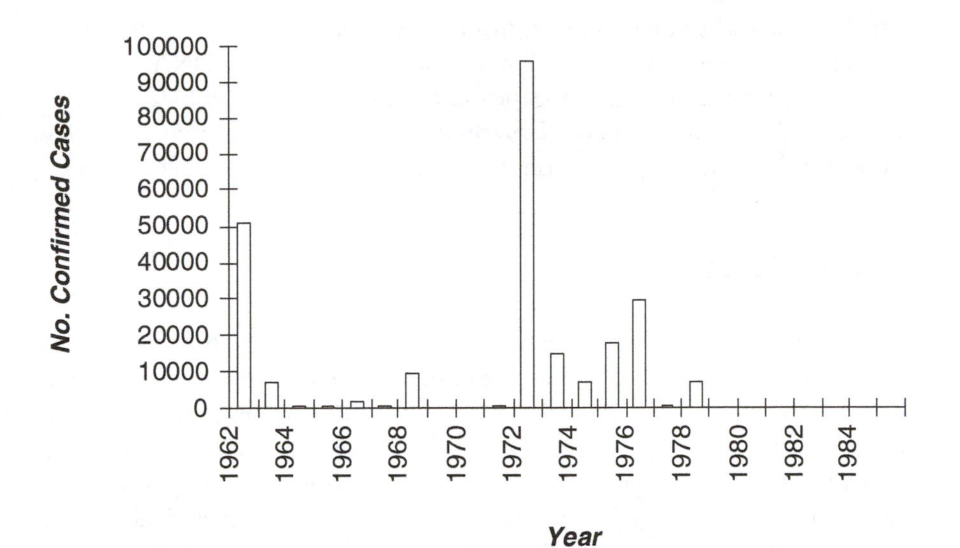

Fig. 9.1. Confirmed cases of screwworm in the south-western USA during the sterile release programme that began in 1962. Prior to the programme there were over 1 000 000 cases of screwworm *per annum*. Notice that control broke down during the 1970s. Data collated from Krasfur, E.S. (1999). Sterile insect technique for suppressing and eradicating insect population: 55 years and counting. *Journal of Agricultural Entomology*, **15**, 303–13.

insect can be harvested and packaged for field release. Finally, in any cultures there is also the possibility that disease will be a problem. This means that culturing facilities must be well maintained.[1]

Sterilisation

The most commonly used techniques for sterilisation involve irradiation or chemosterilisation. These techniques induce dominant lethal mutations, usually in sperm. Chemosterilants have been assayed historically but are inherently more dangerous to use. Irradiative techniques involve the use of X-rays or gamma-rays. Obviously the key with sterilisation is to avoid any somatic damage that could cause behavioural changes. The sterilised individuals must act completely normally. It has been found that a loss of competitiveness

[1] Many of the issues associated with culturing also apply to predators and parasitoids that are cultured for biological control programmes (Chapter 5).

usually occurs when organisms are made completely sterile so the aim of sterilisation is usually to cause a lethal mutation in sperm or egg DNA. This can clearly only happen after gametogenesis, so the stage that is sterilised has to be carefully selected. In a number of successful SIRT cases it has been found that pupae are the easiest stage to sterilise, since they are relatively simple to handle.

Insect release

Successful release of sterilised individuals requires that a technique exists to do this and that the ratio of released to wild individuals is sufficient to engender control. This will require knowledge of the current wild population density, and it will also require knowledge of the wild population's developmental stage. Knowledge of target species dispersal is also important. For most situations it is thought that the ratio of sterile to wild individuals should be about 40:1 (i.e. between 10:1 and 100:1; see earlier comments). Because one of the aims of the release is to mix up the wild and sterile populations it is usually necessary to disperse individuals over a very wide area. Repeated release will also be likely until control is achieved. Finally, the issue of whether to release one or both sexes will also need to be considered. This will depend on whether it is possible, or economically feasible, to separate sexes during culturing and the harm that they might do. For example, it would not be a good idea to release millions of female mosquitoes (even if they were sterile) for the simple reason that they will feed on the human population.

Mating

Following release of sterilised insects, these individuals then need to mate with wild females to bring about a population decline. Important factors that can influence the success of the strategy comprise the ratio of sterile to wild individuals, whether females mate just once and whether important density-dependent factors exist that serve to regulate the pest population as a whole. For example, if the pest population responds to a decline in numbers by increasing reproductive output then this can modulate the success of the programme. The outcome will then be determined by the relationship between the selective force driving the population downwards (via sterile matings) and the density-dependent force driving the population upwards (via increased 'fertile' reproductive effort). Even where species do not respond to a population decline by boosting reproductive effort, success will be partially deter-

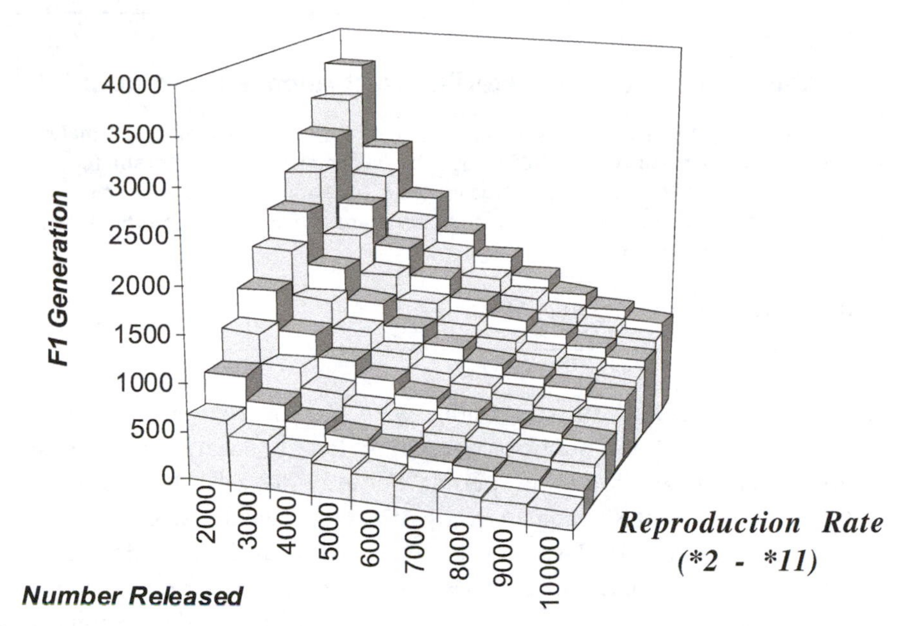

Fig. 9.2. Schematic of the effects of release number and reproduction rate on the change in a population of 1000. Data assume no immigration/emigration and that released males are equally competitive with wild males.

mined by the natural rate of increase of the population. For example, if a population tends to increase fivefold each generation, then a decline brought about by sterile releases must more than compensate for this. Mathematical models have been proposed that can account for this (Figure 9.2 and Box 9.1). Overall, these models indicate that the initial ratio of sterile to wild individuals must be greater than the pest's net increase potential. As indicated above, typical ratios that are quoted for success appear to vary from 10:1 to 100:1. In some cases insecticides have been used partially to suppress pest populations, prior to release. These applications can improve the sterile-to-wild ratio before onset of sterile male releases. Clearly, the ambient environmental conditions will also have an impact on the rate of development of a pest population, and so upon the success of an individual SIRT programme.

Control of the American screwworm fly

The American screwworm *Cochliomyia hominivorax* (Figure 9.3, colour plate – also sometimes called the primary screwworm) is a parasite of mammals,

Box 9.1
Mathematical model for sterile insect release technique:

W = egg-laying females (first W), males (all other Ws); R = number of males released; C = competitiveness (1.00 = equal to wild males); M = immigrant fertile females; I = effective rate of population increase. Equation reprinted by permission from Pedigo, L. (1999). *Entomology and Pest Management*. Upper Saddle River, New Jersey: Prentice Hall.

$$W_{n+1} = [(W_n \times (W_n/W_n + (R \times C))) + M] \times I$$

including humans. Its distribution limits it to the Americas where it exists throughout tropical and subtropical regions. It also penetrates temperate regions when climatic conditions permit. For example, it has been recorded in Canada, although it is unable to overwinter there. Females of the fly lay up to 500 eggs in open animal wounds that have been caused by accident, fly or tick bites, or by management practices such as castration or dehorning. In total, females may lay up to 3000 eggs. However, this is under ideal conditions. The larvae that hatch feed on flesh for 4–9 days before falling off and pupating in the soil. Infected animals tend to attract further flies and in some cases the secondary screwworm *C. macellaria* has also been found to feed opportunistically on infected animals. This is a species which is normally restricted to dead animals. Pupation takes approximately 8 days, adults will mate 2–5 days after emergence and oviposition will take place 3 days after mating. Under ideal conditions, the whole life cycle can be completed in 3–4 weeks. If the weather becomes unfavourable, pupae can enter a period of diapause for up to 2 months.

The larvae are screw-shaped to prevent animals dislodging them. The damage they cause is mainly to animal hides; however, heavily infested wounds can also function as foci for disease entry. If animals are left untreated at this stage, it is highly likely that they will die. There are no effective methods for controlling the fly on open-range grazing systems, particularly since the absolute pest population density of the fly is relatively low and the area over which the host population extends is typically huge. Before the onset of control measures for the fly, it was estimated that this pest was causing losses of $50–100 million per year to farmers in the USA. In 1935 alone, a large outbreak year, there were 1.2 million recorded cases of screwworm, including 55 cases of human exposure, and *c.* 180 000 cattle died.

In the early 1950s laboratory experiments first demonstrated that screwworms could be sterilised by exposure to ionising radiation. Following the development of techniques for mass rearing the screwworm, trials with SIRT were initiated in the 1950s. The screwworm was reared on an artificial diet of ground beef and it was easy to separate male and female pupae as they were different sizes. Sterilisation occurred by exposing pupae to gamma-rays produced by a cobalt (C^{60}) source. The very first field trial involving the release of sterilised flies took place on Sanibel Island, near Tampa, Florida. This was a relatively small-scale trial and involved a clearly defined area. The first large-scale trials, again involving a clearly defined area, involved releases on the Caribbean island of Curaçao. These trials were carried out in 1954. Sterile flies were released at a rate of three flies per hectare per week. The island is 444 km^2 or 444000 hectares, i.e. approximately 1.2 million flies were released per week. The result was eradication in four generations or about 6 months. Following this success a programme was initiated in Florida in 1958. In this programme four flies per hectare per week were released over an 18-month period. Approximately 50 million sterile flies were released each week and, in total, more than 2 billion flies were released. The screwworm problem was eradicated from Florida in just over a year at a cost of $7–10 million dollars.

The next attempts to control the screwworm concentrated on the southern states in the USA with the aim of creating a sterility barrier at the US–Mexico border. This programme involved weekly releases of 180 million sterilised flies along the south-western border of the USA. The programme started in 1962 using flies that had been reared at a factory in Texas, and was so successful that by the mid-1960s the USA was declared pest-free (Figure 9.4). The programme did break down in the 1970s but this was due to poor-quality flies (Figure 9.1). Overall, the number of recorded cases of screwworm dropped from 50000 in 1962 to fewer than 100 in 1970.

The next stage in the eradication programme was to extend the sterility barrier southwards. By the mid-1980s the barrier had been pushed back to the southern Mexican border. By 1994, Belize, El Salvador and Guatemala were reported to be pest-free. By the late 1990s the programme had extended through Nicaragua and Honduras and by 2000 it had been pushed back to the isthmus of Panama (Figure 9.4). This barrier, which is maintained today, is only 225 km long and so is much easier (and cheaper) to sustain. The barrier is also maintained by the use of poison baits that are treated with insecticide. Further experiments are currently under way in Cuba and Jamaica. Overall cost–benefit analyses of the whole control programme have shown that $14 have been saved for every $1 spent. The New World screwworm control programme represents one of the most successful uses of SIRT known.

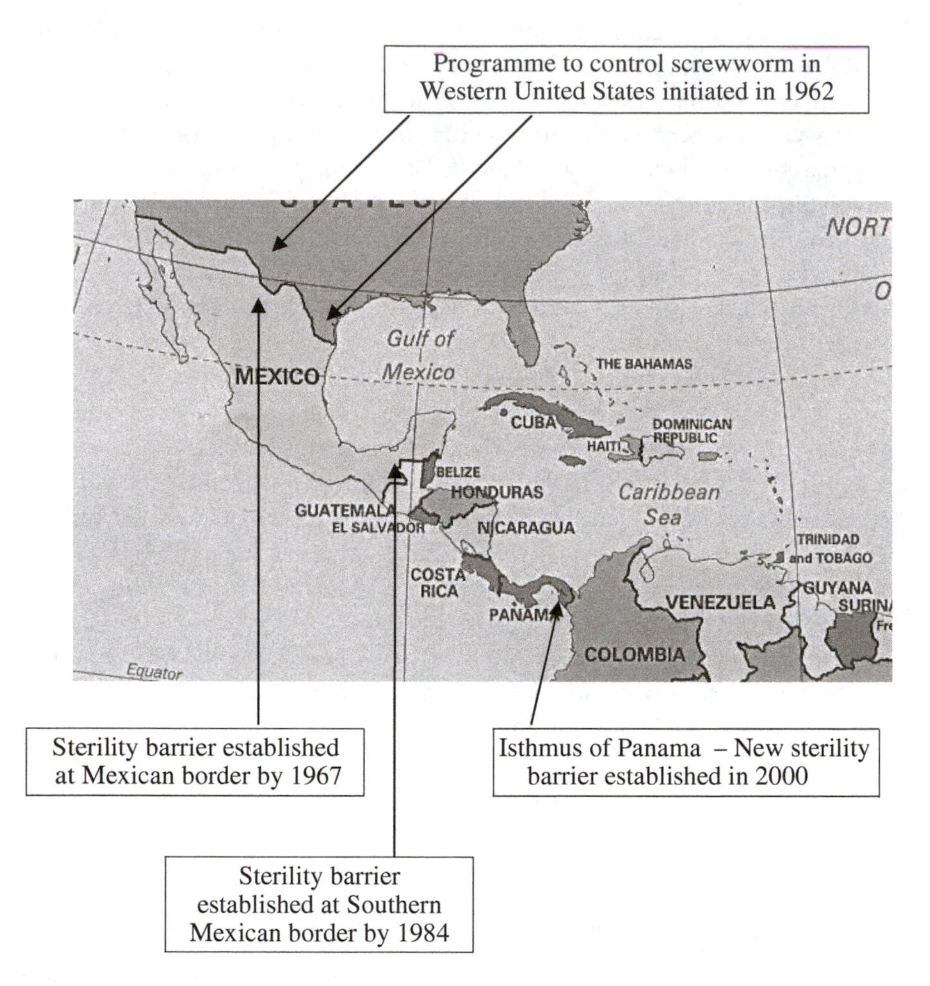

Fig. 9.4. Eradication of the screwworm *Cochliomyia hominovorax*. At the start of 2000 a sterility barrier was established in Panama. All of the countries north of the barrier are screwworm-free. Data collated from a variety of sources. The USDA maintains a website on this programme at http://www.aphis.usda.gov/oa/pubs/fsscworm.html

Control of the screwworm in Libya

In 1988 the screwworm was detected for the first time in Libya. Pests were detected in several thousand sheep that had been imported from Uruguay. In order to prevent the pest from spreading, a major control programme was rapidly initiated by the Libyan government in collaboration with the Food and

Agricultural Organisation (FAO) of the United Nations and with the International Atomic Energy Agency (IAEA). The primary concern at the time was that, if the pest spread, it could devastate the livestock industry in Africa, Asia and Europe and that endemic African wildlife, with no previous exposure to this pest, could be especially at risk. The initial programme focused on quarantine and treatment of infected animals with insecticides. However, by 1990 a sterile release programme was initiated in which 3.5 million flies were released per week. This increased to 40 million flies per week by the middle of 1991 once a large-scale rearing and sterilisation facility was established. These releases stopped towards the end of 1991. Surveys continued throughout 1992 and by 1993 the screwworm was officially declared eradicated from Libya. By the end of the twentieth century, there were no further outbreaks of this pest in the Old World.

Control of the Mediterranean fruit fly

The Mediterranean fruit fly *Ceratitis capitata* (Figure 9.5, colour plate) is a fruit pest of major global importance. The fly has been found to feed on more than 100 different species of fruit. Female flies lay 2–10 eggs under the skin of fruit and the larvae that hatch then feed on the fruit pulp. Females have been recorded as being able to lay up to 600 eggs in total. Generation time is 3 weeks to 3 months, depending on climatic conditions. During the late 1970s a huge eradication programme, involving SIRT, was launched for the control of this pest in Central America. This programme was called 'Programa Moscamed'. The fly (of European origin) was first detected in Costa Rica in 1955 and by the 1970s it had spread thoughout Central America and was also moving north through Mexico to the USA. The eradication programme took an integrated approach that involved quarantine, mass trapping using bait sprays and sterile insect releases. A rearing facility was set up in Metapa, Mexico. By 1981, this facility was producing *c.* 500 million flies each week. These flies were released at a rate of 1000 sterile flies per hectare per week and by the end of 1982 it was claimed that the fly had been eradicated within Mexico. Since then, eradication based on sterile releases has continued in Central America, and promising results have also been achieved in the Canary Islands and Hawaii.

In California, sterile fly releases are still used to suppress local outbreaks of the fly that are thought to originate in illegally imported food. More recently, it has been proposed that SIRT be used to control the Mediterranean fruit fly in the Near East. Economic studies have shown that the fly causes over $250 million worth of damage to crops in Jordan, Israel, Palestine and the Lebanon.

Preliminary studies by the FAO/IAEA have shown that it would be economically feasible to undertake SIRT programmes within this region. To this end, it was proposed that these projects should start in 1999–2000. No data on the success of these trials are as yet available.

Control of the tsetse fly

The tsetse fly *Glossina* spp. (Figure 9.6, colour plate) is a pest of major economic and human significance. It is found throughout sub-Saharan Africa and is important as a vector of trypanosomes. These are flagellate protozoans that cause a disease known as sleeping sickness in humans and as nagana in cattle. Approximately 300000 cases of human trypanosomiasis are recorded each year, of which 2–3% are fatal. The exact number of cattle fatalities is unknown but is likely to be much higher. Both male and female tsetse flies can transmit the disease.

The first attempts to control tsetse fly were made in the 1940s, and were discussed under hybrid sterility. However, attempts were also made in the 1980s to use SIRT as a control technique for this pest. These trials were carried out at a number of locations in Africa. It has been reported that *G. palpalis palpalis* was totally suppressed in central Nigeria in 1985 by using an integrated approach that involved SIRT and insecticide-impregnated screens. In the mid-1990s the IAEA and the Ethiopian government initiated trials to control *G. palpalis* in southern Ethiopia. The results of this research are still not final. However, in 1996 *G. austeni* was eradicated from Zanzibar (island of Unguja), using a mixture of SIRT, insecticide-impregnated screens and insecticide-treated cattle in another IAEA-supported research programme. In this 2-year programme over 8.5 million sterile males were released. The males that were released were raised on diets that incorporated trypanocidal drugs to ensure that they would not add to the indigenous vector population. Prior to the onset of this control programme, annual losses attributable to the tsetse fly were estimated at $2 million. The control programme took *c.* 2 years to complete and the economic benefits, following eradication, have been enormous. Overall, the conclusions from the programme were that an integrated approach was needed in order to enhance the effectiveness of SIRT. The effective threshold ratio for success was greater than 10:1 sterile-to-wild individuals. At ratios lower than this, tsetse populations did not decline.

It seems clear therefore that SIRT has a lot to offer in relation to tsetse fly control. It is highly likely that SIRT will be used far more widely in countries where tsetse flies are endemic. It has been argued that most of the barriers to implementing the technique on a large scale appear to be political, as well as economic. If these barriers can be overcome, then it seems that this technique could prove very useful for the control of this, as well as other, pest species.

Other species controlled using SIRT

In addition to the three species discussed above, SIRT has been attempted with a number of other pests. These include a number of fruit flies, bollworms and other lepidopterous pests, coleopterous pests and hemipterous pests. A list of the more notable trials that have been carried out to date is given in Table 9.2. Despite these trials, most have not found practical and widespread application. For example, after 30 years' research on the use of SIRT for control of the codling moth, the primary conclusion was that, while control of this pest was feasible, it was not yet economic because chemical control costs about half as much. Since this situation may change, we might expect the use of SIRT to become more widespread in the future.

Insect orders not controlled by SIRT

In addition to species that have been controlled (whether economically or not) using SIRT, there are also those for which the technique seems inapplicable. In particular, it seems that SIRT cannot be used against some of the most destructive and sporadic pests that have wide host ranges and low economic thresholds. This includes many hemipterous pests, as well as periodic and highly damaging orthopterous pest (e.g. locust) outbreaks. In the case of many hemipterous pests, which often reproduce parthenogenetically, control by interference with mating is clearly not appropriate. As indicated above, the use of SIRT is also highly dependent upon having established satisfactory procedures for rearing, sterilisation and release.

Summary

The reasons why SIRT has been so successful, especially with control of the American screwworm, can be summarised as follows.

- Billions of flies can be produced on a regular basis.
- Control of this pest by conventional means is difficult.

Table 9.2. *Summary of major field trials in which species have been assayed for control using the sterile insect release technique (SIRT)*[a]

Common name	Latin name	Trial locations
Dipterous pests		
American screwworm	*Cochliomyia hominivorax*	North America, Central America, Libya
Mediterranean fruit fly	*Ceratitis capitata*	Central America, Hawaii, Canary Islands
Mexican fruit fly	*Anastrepha ludens*	Mexico
Queensland fruit fly	*Dacus tryoni*	Australia
Cherry fruit fly	*Rhagoletis cerasi*	Switzerland
Oriental fruit fly	*Dacus dorsalis*	Guam
Melon fly	*Dacus cucurbitae*	Various Japanese islands
Onion fly	*Delia antiqua*	Netherlands
Olive fly	*Dacus oleae*	Various countries in Europe
House mosquito	*Culex quinquefasciatus*	Florida, USA
Malarial mosquito	*Anopheles ludens*	El Salvador
Stable fly	*Stomoxys calcitrans*	St Croix (Virgin Islands)
Tsetse fly	*Glossina palpalis*	Nigeria, Upper Volta
Lepidopterous pests		
Grape moth	*Eupoecilia ambiguella*	Various countries in Europe
Codling moth	*Cydia pomonella*	USA, Canada, Switzerland
Coleopterous pests		
European cockchafer	*Melolontha melolontha*	Switzerland
Pink bollworm	*Pectinophora gossypiella*	California, USA
Boll weevil	*Anthonomus grandis*	Louisiana, USA

Notes:

[a] In all of these trials control has been successful. Some of the trials involved geographically restricted local pest populations. Some of the trials used SIRT as one part of an integrated programme. Data collected from a range of sources.

- The distribution of the screwworm is largely known.
- The public supported the programme.
- The government supported the programme.

It is likely that the use of SIRT will expand in the future. However, it is also likely to remain a government-supported control option. The clear advantage of SIRT is that is a highly specific technique. It is also a technique that improves as it works, i.e. as the pest population declines, the ratio of sterile to wild individuals increases. This is in complete contrast to almost all other arthropod pest species control techniques where control typically becomes less effective as the pest population falls. Finally, it is also permanent, barring reintroductions of the pest controlled, and is completely free from harmful environmental side-effects. As a technique, it may be restricted to control of a few pests with particular ecological characteristics, but where it does work, control can be dramatic. It is highly likely that other pest species will be controlled with this technique in the future, particularly if the technique is integrated with other control measures, as is the case with control of the Mediterranean fruit fly in North America.

Further reading

Allsopp, R. (2001). Options for vector control against trypanosomiasis in Africa. *Trends in Parasitology*, **17**, 15–19.

Calvitti, M., Remotti, P.C., Pasquali, A. & Cirio, U. (1998). First results in the use of the sterile insect technique against *Trialeurodes vaporariorum* (Homoptera: Aleyroididae) in greenhouses. *Annals of the Entomological Society of America*, **91**, 813–17.

Dominiak, B.C., McLeod, L.J., Landon, R. & Nicol, H.I. (2000). Development of a low-cost pupal release strategy for sterile insect technique (SIT) with Queensland fruit fly and assessment of climatic constraints for SIT in rural New South Wales. *Australian Journal of Experimental Agriculture*, **40**, 1021–32.

Kakinohana, H. (1998). Eradication of the melon fly by means of the sterile insect technique. *Journal of the Atomic Energy Society of Japan*, **40**, 91–100.

Katsoyannos, B.I., Papadopoulos, N.T., Kouloussis, N.A., Heath, R. & Hendrichs, J. (1999). Method of assessing the fertility of wild *Ceratitis capitata* (Diptera: Tephritidae) females for use in sterile insect technique programs. *Journal of Economic Entomology*, **92**, 590–7.

Knipling, E.F. (1955). Possibilities of insect control or eradication through the use of sexually sterile males. *Journal of Economic Entomology*, **48**, 459–69.

Krasfur, E.S. (1999). Sterile insect technique for suppressing and eradicating insect population: 55 years and counting. *Journal of Agricultural Entomology*, **15**, 303–13.

Lindquist, D.A., Abusowa, M. & Hall, M.J.R. (1992). The New World screwworm in Libya: a review of its introduction and eradication. *Medical and Veterinary Entomology*, **6**, 2–8.

Mayer, D.G., Atzeni, M.G., Stuart, M.A., Anaman, K.A. & Butler, D.G. (1998). Mating competitiveness of irradiated flies for screwworm fly eradication campaigns. *Preventive Veterinary Medicine*, **36**, 1–9.

Pettigrew, M.M. & O'Neill, S.L. (1997). Control of vector-borne disease by genetic manipulation of insect populations: technological requirements and research priorities. *Australian Journal of Entomology*, **36**, 309–17.

Reichard, R. (1999). Case studies of emergency management of screwworm. *Revue Scientifique et Technique de l'Office International des Epizooties*, **18**, 145–63.

Rossler, Y., Ravins, E. & Gomes, P.J. (2000). Sterile insect technique (SIT) in the near east – a transboundary bridge for development and peace. *Crop Protection*, **19**, 733–8.

Schliekelman, P. & Gould, F. (2000). Pest control by the introduction of a conditional lethal trait on multiple loci: potential, limitations, and optimal strategies. *Journal of Economic Entomology*, **93**, 1543–65.

Vreysen, M.J.B., Saleh, K.M., Ali, M.Y. *et al.* (2000). *Glossina austeni* (Diptera: Glossinidae) eradicated on the island of Unguja, Zanzibar, using the sterile insect technique. *Journal of Economic Entomology*, **93**, 123–135.

10

Host-plant resistance

Introduction

Using plants that are resistant to attack by arthropod pests is clearly an attractive option, particularly if the resistance is complete in the sense that the attacking organism is no longer able to cause economic damage. To use resistant plants will usually require no more knowledge than a farmer already possesses. Most resistant plants should be environmentally benign (in fact, they may even promote environmental well-being). Resistant plants are likely to be highly compatible with other forms of pest control (chemicals, natural enemies, etc.). Finally, resistant plants may afford plants complete protection from pests for the duration of the growing season. A number of definitions for resistance have been proposed (Box 10.1), all of which refer in some way to the genotype of the host plant. Clearly, the term resistance, like the term pest, is a relative one, i.e. resistance can only be defined in relation to other varieties of a species. Resistance will also vary with ambient abiotic and biotic conditions.

Historical observations

The first observations concerning host-plant resistance were made in the third century BC when Theophrastus recorded differences in the susceptibility of plants to fungal pathogens. The first documented record concerning resistance to insects is found in 1782 when the wheat cultivar 'Underhill' was reported as resistant to the Hessian fly *Mayetiola destructor*.

Box 10.1
Some definitions of host-plant resistance[a]

The collective heritable characteristics by which a plant may reduce the probability of its utilisation as a host by a pest.
The inherent ability of a crop plant to restrict, retard or overcome a pest infestation.
Any inherited characteristics that lessen the effects of attack by crop pests.
Genotype-based characteristics of a plant that are able to reduce or tolerate the feeding impacts of pests.

Note:
[a] Definitions collated from a range of sources.

However, although the use of resistant plants became established as a pest control technique in the nineteenth century, following the successful control of the grape phylloxera, it was not until the twentieth century that breeding for arthropod resistance became established as a scientific discipline. The development of scientific breeding programmes at the start of the twentieth century was made possible because of discoveries in genetics that took place at this time. The story concerning the grape phylloxera is described in more detail later in this chapter. Table 10.1 provides a simple historical account of some of the milestones in the development of host-plant resistance.

Since the start of modern breeding programmes, many thousands of cultivars have been developed that have either complete or partial resistance to a range of pest species. Modern developments in biotechnology and genetic engineering (see Chapter 13) have helped this process, although it has been suggested that progress with this technology may not be as dramatic as some people may think. For example, the first transgenic plants with insect resistance were produced in the early 1980s. However, by the year 2000 only a small number (*c.* 20) of insect-resistant transgenic plants had been developed for commercial release and almost all of these were based on the *Bacillus thuringiensis* endotoxin (see Chapters 6 and 13).

This chapter is organised so that crop plant development and secondary plant substances (key to much host-plant resistance) are considered first. A summary of the types of resistance that can be developed is then provided, together with a review of the techniques that are available for producing resistant plants. Specific examples of resistant plants and their utility in pest

Table 10.1. *Some milestones in the development of host-plant resistance*

Date	Development
Third century BC	Theophrastus records differences in disease-susceptibility among crops
1782	Wheat variety 'Underhill' reported resistant to Hessian fly *Mayetiola destructor*
1817	Sorghum *Sorghum vulgare* reported resistant to grasshoppers *Melanoplus* spp.
1831	Apple variety winter Majetin reported resistant to woolly aphid *Eriosma lanigerum*
1885	North American grape rootstocks (*Vitis* spp.) reported resistant to the grape phylloxera *Phylloxera vitifoliae*
1935	Cotton with resistance to leafhoppers *Empoasca* spp. reported
1984	Transgenic tobacco expressing the *Bacillus thuringiensis* endotoxin produced
1994	First field releases (USA) of transgenic plants (corn, cotton, potatoes) resistant to insect attack

control programmes are then given. Finally, the chapter concludes by considering some of the problems that are associated with using resistant plants to control arthropod pests.

Crop plant development

Some 10 000–15 000 years ago, human beings began to domesticate plants and animals, i.e. to farm (see Chapter 1). The selection of plant species for domestication would have been driven by numerous factors, including resistance to attack from pest species. These first domesticates have been called landraces or field varieties. In parallel with plant landrace development would have been the development of wild weed races.[1] These weeds have exchanged and continue to exchange genetic material with cultivated varieties today. Landraces

[1] Some wild weed races are the ancestral forms of landraces, i.e. some of the first plants to be cultivated were selected from species we would now regard as weeds.

of plants form the basis of subsistence agriculture and constitute the material from which most modern-day cultivars have been developed.

Modern cultivars began to be produced at the end of the nineteenth and the beginning of the twentieth century (see above). Many of these breeding programmes incorporated resistance genes from wild relatives of crop plants. It has been suggested that there are many examples of situations where only a few genes from wild relatives stand between humanity and starvation, if not economic ruin. These genes promote resistance to pests in a number of different ways. One of the most widespread defences in plants against attack by herbivores comprises the use of chemicals. These chemicals, which are often referred to as secondary plant substances, are described in the next section.

Secondary plant substances

Secondary plant substances are so named because they were originally thought to play no part in primary biochemical reactions concerned with growth and reproduction. Their main function was thought to be as a defence against herbivorous attack by acting as repellents, inhibitors or toxins (see Chapter 2). More recently, this classification has been questioned since there is evidence that many secondary substances have been critical in sustaining plants throughout their evolution. For example, many secondary substances can act as stores for carbon and nitrogen. It has also been argued that many primary substances can act as defences. For example, it is now known that amino acid composition can affect the relative success of many pest species that feed on plants.

The diversity of secondary plant substances involved in host-plant resistance is shown in Table 10.2. The effects of these compounds vary considerably. They deter feeding, disrupt development, provide barriers to attack, assist with wound healing, disrupt digestion and many are neurotoxic to herbivorous pests. They can be stored inside or outside cells (intra- and extracytoplasmically); some are stored for relatively long periods of time while others undergo rapid turnover within plants. They vary in expression levels within a plant, both in space and in time. For example, it is known that expression levels of secondary plant substances will vary not only with age and with season, but also diurnally. Finally, expression levels of these substances can be modified by ambient environmental conditions.

In addition to chemical defences, plants also use a range of morphological features to defend against insect attack. For example, plant hair (trichome)

Table 10.2. *Secondary plant metabolites involved in host-plant resistance to insects*

Substance	Putative functions
Alkanes, aldehydes, ketones, long-chain waxes	Barrier to insect attack, involved in wound healing
Lignins and tannins	Barrier to insect attack, involved in wound healing, affect insect digestion
Monoterpenoids	Neurotoxic (pyrethroids), deterrents (citronellol, iridoids)
Sesquiterpenoids	Feeding deterrents (polygodial), toxic (lactones, gossypol), hormonal interference (juvenile hormones, precocenes)
Diterpenoids	Feeding inhibition (resins in conifers)
Triterpenoids	Feeding deterrents (cucurbitacins), disrupt development (azadirachtin), toxins and deterrents (saponins)
Steroids	Developmental interference (ecdysterols)
Phenolic compounds	Toxic (chlorogenic acid, coumarins)
Flavonoids	Feeding deterrents (flavonols), toxic (flavonoids)
Quinones	Feeding deterrents and toxic (quinones)
Alkaloids	Neurotoxic (nicotine, physostigmine), antifeedants (pyrrolidines)
Nonprotein amino acids	Antimetabolic and toxic action (dihydroxyphenylalanine, canavanine)
Cyanogenic glycosides	Toxic and feeding deterrents (cyanide)
Glucosinolates	Toxic and feeding deterrents (mustard oils)
Lectins	Toxic (agglutination of cells and gut disruption)
Chitin-binding proteins	Toxic
Amylase inhibitors	Disruption of gut enzymes

Source: Table modified from Panda, N. & Khush, G.S. (1995). *Host Plant Resistance to Insects*. Wallingford: CAB International. Most of the above substances are found in a range of plant species. The exact mode of action(s) for many of the above chemicals has yet to be determined.

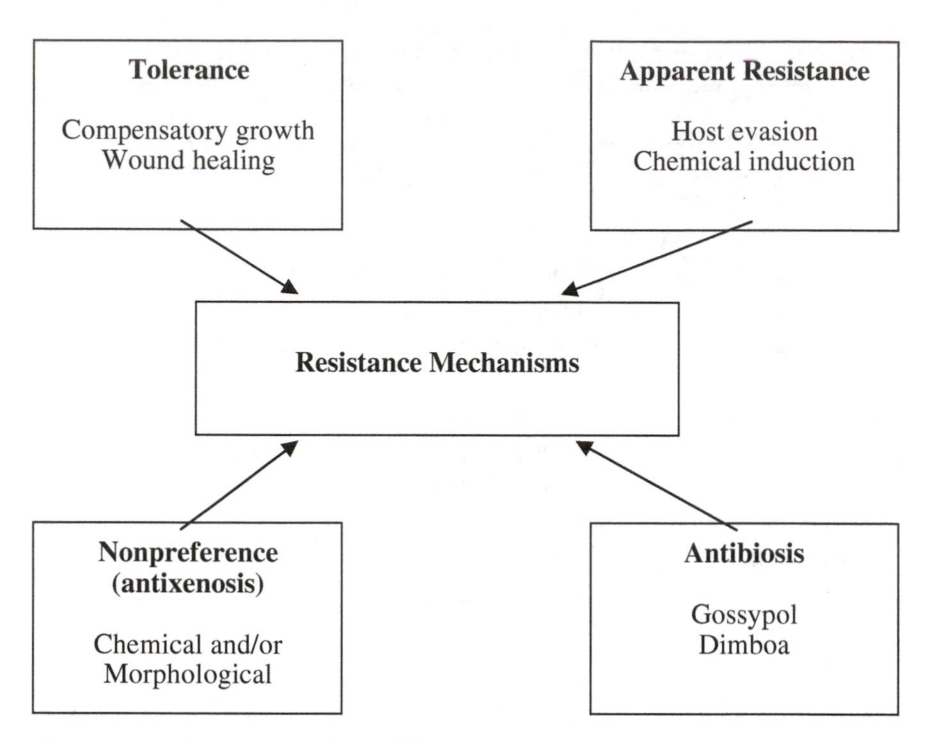

Fig. 10.1. Schematic describing different resistance mechanisms. Within each box, examples are provided.

density can alter susceptibility to attack, as can leaf waxiness. Plants can also avoid attack by evading hosts (apparent resistance). Finally, the environment can have a bearing on the degree of resistance of a plant to attack, both directly (through plant health) and indirectly (by modifying the efficacy of known genetic resistance). Mechanisms of resistance to attack by pests are discussed below.

Types of resistance

Most textbooks that describe host-plant resistance follow a classification that was first proposed in the early 1950s. This classification identifies three categories: antixenosis (nonpreference), antibiosis and tolerance. In addition to these primary categories some authors have also identified a fourth category called apparent resistance. A schematic of these categories is shown in Figure

10.1. It is important to realise that individual plants may utilise more than one resistance mechanism. For example, resistance in the alfalfa variety Cody is attributable to a mixture of antixenosis, antibiosis and tolerance. The environment may also impact significantly on the efficacy of a resistance mechanism. For example, it is well known that in plants that are stressed, i.e. lacking water and/or nutrients, the proportional effects of pest attack are greater. This can be equated to the human condition where stressed individuals are known to be more susceptible to infection with pathogenic organisms.

Nonpreference or antixenosis

Nonpreference or antixenosis occurs where pests tend to avoid or reduce their feeding rate on a particular genotype, usually when presented with a choice. For example, the green spruce aphid *Elatobium abietinum* is known to prefer American to European spruce species, despite being able to feed on both. The distinction between the above terms is that nonpreference is a characteristic of the pest while antixenosis is a characteristic of the plant. It has been suggested that antixenosis should only include the phenomenon of nonacceptance in a no-choice situation. Some scientists subscribe to this distinction, while others do not.

The mechanisms of antixenosis may be chemical or morphological. For example, many terpenoids (Table 10.2; see also glossary for Chapter 7) act as feeding deterrents to pest species. There are also numerous studies that have documented the importance of amino acid content in plants. For example, the development of pea aphids *Acyrthosiphon pisum* on peas, sorghum aphids *Melanophis sacchari* on sorghum, and peach-potato aphids *Myzus persicae* on potatoes are all significantly correlated with the free amino acid content of the host plant. Morphological features of plants that can cause antixenosis include hairiness (pubescence), waxiness and colour. For example, many sap-feeding species (hemipterous pests) find it difficult to feed on hairy leaves.

From a practical point of view, the use of nonpreference as a control measure will depend on its efficacy. For example, with crops that are planted as monocultures, weak nonpreference breaks down if pests are present. A pest species may not prefer the crop, but will use it as a resource if nothing else exists. That said, the rate of development of a pest population can be lower and consequent economic damage lower on a less preferred cultivar. It has also been shown on numerous occasions that natural enemies are often more effective in controlling pests on plants that are partly resistant to pest

Fig. 10.2. Synergism between host-plant resistance and a predatory mirid bug *Cytorhinus lividipennis* in controlling the development of green leafhopper populations on susceptible and resistant rice strains. Figure modified from Panda, N. & Khush, G.S. (1995). *Host Plant Resistance to Insects*. Wallingford: CAB International.

species. Figure 10.2 shows the results of cage studies with the green leafhopper *Nephotettix virescens* on rice. On resistant rice varieties, predation by a mirid bug is much more effective. In this example therefore, there is synergism between the predator and the antixenotic host-plant.

Antibiosis

Antibiosis can be distinguished from antixenosis because this resistance mechanism involves an impairment of a pest's metabolic processes. The survival, growth, and fecundity of a pest may be directly affected. In some situations the feeding insect may die. This is different from simply not liking the look, touch or taste of a plant. Antibiosis affects insect performance rather than insect preference. It is secondary plant substances that are most often associated with antibiosis (Table 10.2). The best-studied compounds include hydroxamic acids in corn and wheat, and gossypol in cotton.

Many plants contain compounds that are directly toxic to insects. Many of these compounds were discussed in Chapter 2. Well-known neurotoxins include pyrethrum, nicotine and rotenone. Secondary plant substances, like gossypol, by contrast, are usually referred to as growth inhibitors. Gossypol slows the development of a number of lepidopterous pests and has now been incorporated into a number of partly resistant cotton cultivars. Other compounds with antibiotic activity include those that affect insect development, e.g. azadirachtin (see Chapter 2). Overall, antibiosis has been a target for a number of research programmes that are concerned with developing resistant plants. This is most clearly seen historically with the production of cotton plants expressing gossypol. However, this form of resistance is also in evidence in the new transgenic plants that express the *Bacillus thuringiensis* endotoxin (see Chapter 13).

Tolerance

Plants that can grow, develop and produce sufficient yield despite being attacked by pests are said to be tolerant to attack. This category of resistance therefore represents resistance to damage rather than resistance to the insect. Tolerance is also often referred to as compensation or compensatory growth. Typically, a tolerant plant will be defined in comparison to a susceptible plant whose yield is far lower under the same level of insect attack. Some researchers do not consider this to be a true example of resistance.

The factors that are involved in tolerance are not always clear but they may involve compensatory growth, wound healing and changes in photosynthesate partitioning. From a breeding point of view, tolerance is attractive because there is no pressure on an insect pest to attempt to overcome the resistance. Tolerant varieties will also have higher economic thresholds for pesticide applications and may integrate well with integrated pest management programmes (see Chapter 12). A good example of tolerance concerns corn plants that are resistant to the corn rootworm *Diabrotica virgifera.* The pest feeds on the roots, and in varieties that are tolerant, plants compensate by developing larger root systems. Another example of tolerance concerns various cereal plants that, when attacked early in the season, respond by producing larger plants with more ear-bearing shoots. Finally, in many developing countries cotton plants with resistance to aphids are used. The resistance mechanism with these cotton plants is believed to involve compensatory growth.

Apparent resistance

Apparent resistance is also known as ecological resistance or pseudoresistance. It is a form of resistance that is primarily dependent upon environmental conditions. Plants are susceptible to attack, but resistance is achieved by avoiding potential pests, by maximising plant health or by damage-driven production of chemical compounds (induced resistance). Since this last category clearly has a genetic basis, it is questionable whether this is not actually a type of antixenosis.

Examples of apparent resistance include the use of fast-growing, short-season varieties that are able to escape or avoid attack by pest species. For example, early-maturing pigeon pea plants are able to avoid attack by the pod fly *Melanagromyza obtusa*. This strategy has also been used successfully with cotton plants. Keeping plants healthy (well-watered and well-fertilised) is a simple way to try to minimise the impacts of pests (see earlier). The use of inducible chemical defences is an approach that has been recognised, but not well-understood, for a number of years. For example, exposing grapevines to the Williamette mite *Eotetranychus willametti* in California seems to make vines resistant to the much more damaging Pacific spider mite *Tetranychus urticae*.

In summary, most plants will use a mixture of resistance mechanisms to defend against attack. The goal of many breeding programmes is to identify the genes responsible for these resistance mechanisms and to transfer them to plants with desirable agronomic qualities. In the next sections the sources of resistant genes and the methods used to produce resistant plants are described.

Germplasm collections

Source material for plant-breeding programmes today is largely held in specialised germplasm collections. The Russian geneticist Vavilov, working in the 1920s, was the first person to describe the need for such collections. Over a 20-year period that began in 1916 he established the world's first germplasm collection in St Petersburg with over 250000 accessions. As a result of this work, Vavilov proposed that there were eight centres of varietal wealth (Figure 10.3) and that these centres represented the centres of origin of most cultivated plants. Although it has now been established that many crops did not originate in a Vavilovian centre, his ideas have driven many successful exploration programmes.

Today, thousands of germplasm collections are held worldwide. In 1973 the International Board for Plant Genetic Resources (IBPGR) was established to coordinate the collection and distribution of material to international research institutes. This organisation is now known as the International Plant Genetic Resources Institute (IPGRI) and is based in

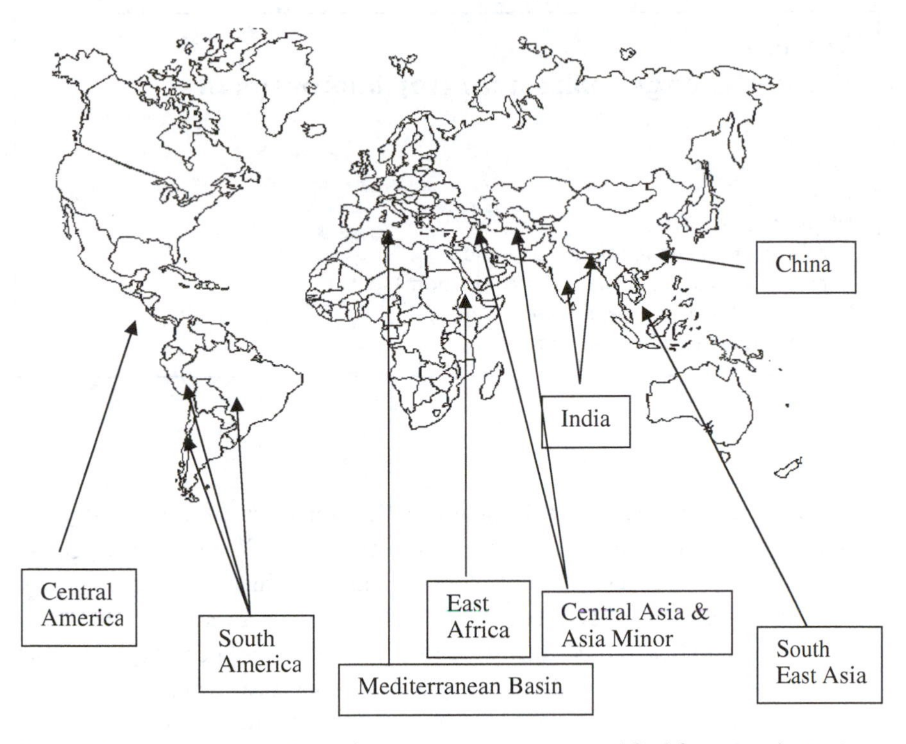

Fig. 10.3. Vavilovian centres of crop diversity. Figure modified from Panda, N. & Khush, G.S. (1995). *Host Plant Resistance to Insects*. Wallingford: CAB International.

Rome, Italy. The organisation is international and nongovernmental, and it has produced databases that list sources for germplasm on a global basis. The major centres for germplasm collections, and associated breeding programmes, include the following: the International Centre for Maize and Wheat Improvement (CIMMYT), Mexico, the International Rice Research Institute (IRRI), Philippines, the International Potato Centre (CIP), Peru, the International Institute for Tropical Agriculture (IITA), Nigeria, the International Crop Research Institute for Semi-Arid Tropics (ICRISAT), India, and the International Centre for Agricultural Research in Dry Areas (ICARDA), Syria.

The sources of material for most germplasm collections vary (Box 10.2). It is common now for source material to be collected during major expeditions. One of the reasons for establishing an intergovernmental organisation was the scale of the need to collect and because of a concern that global development would lead to the loss of potentially useful, local varieties. With

Box 10.2
Categories of germplasm for crop improvement

Wild species
Weed races
Landraces
Unimproved cultivars
Purified cultivars no longer in cultivation
Improved modern cultivars under cultivation
Breeding cultivars not yet released
Mutants developed by breeding

thousands of germplasm banks worldwide, enormous progress has been made in the conservation of plant genetic resources. However, since germplasm collections represent *ex situ* conservation in which evolutionary processes are essentially frozen, some researchers have also argued that more resources should be devoted to *in situ* conservation. It has been suggested that *in situ* conservation would allow germplasm to coevolve with pests, farming practices and climatic changes and so represents a better method for germplasm conservation. The problem with *in situ* conservation of course is that it relies on both political will and an effective security system.

Techniques for producing resistant plants

The techniques that can be used to develop resistant plants are constantly being refined. The most recent refinements are those associated with genetic engineering. Using recombinant DNA technology, breeders can now introduce to crop plants useful genes from completely unrelated species. This technology is described in Chapter 13. Table 10.3 lists some of the best-known plant-breeding techniques. It is important to realise that these are breeding techniques, i.e. they can be used to incorporate any genetic material and not just that which is concerned with host-plant resistance. Indeed, until recently host-plant resistance was not the primary concern of most plant-breeding programmes.

The choice of breeding method used will depend at least partly on the reproductive system of the plant. Plants that are cross-pollinators are

Table 10.3. *Techniques used in plant breeding*[a]

Technique	Procedure
Pure-line selection	Self-pollinating plants (pure lines) with desirable characteristics selected from an initially mixed population
Mass selection	As above, but a population of plants is selected for release
Hybridisation	Relies on initial hybidisation between self-pollinating plants. Followed by pure-line selection or mass selection
Backcross selection	As above, relies on hybridisation. This process is repeated between donor and parent on a recurrent basis
Recurrent selection	Cross-pollinating population repeatedly exposed to pressure. Population of resistant genotypes repeatedly selected
Mutation breeding	Mutagenic agents (X-rays, gamma-rays, chemicals) used to produce resistant plants
Tissue culture	Techniques such as somaclonal variation or mutant selection (discussed in Chapter 13)

Note:
[a] Information collated from a range of sources.

usually highly heterozygous and any forced inbreeding will often lead to a loss of vigour. In contrast, most self-pollinators are largely homozygous. The main plant-breeding technique for self-pollinators used to comprise pedigree breeding, while the main technique for cross-pollinators used to involve recurrent selection. However, in recent years hybrid breeding has become more widely used by breeders to develop both self- and cross-pollinating cultivars. For example, in the mid-1990s it was estimated that hybrid varieties accounted for *c.* 40% of the global seed business. These varieties, of course, are commercially attractive since farmers have to purchase new hybrid seed every year. Typically, the seed of hybrid plants has a much lower yield than that of the parent seed. Brief descriptions of some of the most widely used plant-breeding techniques are given in the following sections.

Pure-line selection

Self-pollinated crops that are phenotypically similar can vary in their innate resistance. In the pure-line selection process, a large number of plants are assayed for resistance (by exposing them to pest pressure) and promising lines are selected for development. This process is repeated as often as necessary. Promising lines usually incorporate resistance as well as agronomic characteristics that ensure they perform as well as, if not better than, the parent line of mixed genotypes. One variation on this technique is to select a number of promising lines and to release these as a bulk. This is called bulk selection. This technique then produces a variety with fewer genotypes than the parent population but with less than those cultivars that are produced by pure-line selection. This approach to breeding produces a marketable product in a shorter period of time (in contrast to pure-line selection) and the genetic variability in the seed that is developed may reduce the selection pressure on pest species to overcome resistance, all other factors being equal.

Hybridisation

This technique is also used for self-pollinating plants but relies on an initial cross between different varieties. Once a cross is achieved, the methods described above are used to develop the line for agronomic release. Rice with resistance to leafhoppers has been developed using this technique and is now grown on millions of hectares worldwide. The most important stage in the process is the initial hybridisation, i.e. getting self-pollinating plants to form hybrids in the first place. This is the most important technique currently used in plant-breeding programmes with self-pollinating crops.

Recurrent selection

This is the main breeding method used with cross-pollinating crops. It is also sometimes called open-field pollination. Plants are selected based upon their ability to resist attack and these plants are allowed to interbreed and to produce the next generation. The cycle of selection and breeding is typically repeated four to five times. It is a technique that is particularly useful in the production of polygenic resistance. For example, maize varieties that incorporate resistance to *Spodoptera frugiperda* have been produced using this technique.

Commercialised plants with resistance to pests

Although the exact resistance mechanism is not always known, there are now an enormous number of varieties with resistance to different insect pests. In these plants, resistance may be monogenic (single gene) or polygenic (many genes). An additional term – oligogenic – has also been used to describe plants with a relatively small number of genes involved in resistance. These terms have also been used to classify resistance as either vertical or horizontal. Vertical resistance is mono- or oligogenic and refers to an established gene-for-gene relationship between the host and the pest. This resistance is the most likely to break down. Horizontal resistance, by contrast, is used in association with polygenic resistance. This resistance is thought to be more long-lasting and less amenable to breakdown. Not surprisingly, breeders find it easier to develop plants with mono- or oligogenic resistance to arthropod pests. That said, it should still be remembered that most host-plant resistance will involve a range of mechanisms and hence, a number of genes, irrespective of what was originally intended. Three examples of extensively planted crops with resistance to arthropod pests are given in the following sections.

The grape phylloxera *Phylloxera vitifoliae*

The grape phylloxera *Phylloxera vitifoliae* is a root-feeding pest of grapes. It was originally known as *Daktulosphaira vitifoliae* and has been given the common name of the grape louse. It is a sap-sucking insect in the order Hemiptera. The control of this pest represents perhaps the best-known example of host-plant resistance in action where the exact resistance mechanism is still not well understood.

The insect first caught the attention of governments when it appeared in England and France *c.* 1863. The pest sucks sap from the roots of grape vines. Infested plants may support colonies of millions, which slowly kill the vine over a number of years. Under ideal conditions there may be up to 10 generations per year, with each female producing 200 eggs by parthenogenesis per generation. Parthenogenesis can occur indefinitely. There is no known control method for this pest, which can feed up to 3 m down in the soil. The only known treatment is to pull up the vine and fumigate the soil, prior to replanting. It has been claimed that soaking (flooding) the soil can control this pest, but this is impractical for most growers.

Following the arrival of the pest in Europe it began to devastate the vine industry in France and in other Mediterranean countries. In France during the 1870s 6 million hectares of agricultural land were involved in growing grapes. There were *c.* 11 billion vines in existence and one-sixth of the French population was involved in the wine industry. Grapes produced one-quarter of the country's agricultural income. A decline in grape production began in the late 1870s, while the worst years developed throughout the 1880s. Clearly the arrival of this pest was devastating for the French economy. The French government offered a reward of FF330000 to anybody who developed a successful control measure.

The grapevine plant is recorded as indigenous to the Americas, Europe and Asia, i.e. it has a mixture of documented origins. It is the very first cultivated plant to be mentioned in the Bible and has a powerful symbolic association with Mediterranean and European civilisations. The grape phylloxera probably entered Europe from the Americas as a result of trade. The pest had existed in North America (in central North America and east of the Rocky mountains) presumably for thousands of years and did not seem to be harmful to American grapevine varieties. Interestingly, Europeans who took their own grapevines for planting in North America during the eighteenth and nineteenth century recorded that their plants always died.

The European grapevines in the 1880s were all derived from a single species, *Vitis vinifera.* By contrast, American grapevines comprised a range of species that included *V. riparia, V. rupestris, V. labrusca* and *V. berlandieri.* All of these American vines appeared to be resistant to the phylloxera. The conclusion was that American vines must be resistant to the phylloxera. During the 1880s trials were undertaken in France with American vines. Although it was possible to grow American vines in French soil, the quality of the grapes was not as good as those produced on phylloxera-free French vines. There was also French reluctance to use American plants!

The first solution to the above was made possible by developments in grafting technology. French vines were grafted on to American phylloxera-resistant rootstock. In order to minimise the American input to the French wine industry, the first rootstocks used were hybrids. The best-known hybrid rootstock was developed from *V. vinifera* and *V. rupestris* and was called AXR#1. This rootstock functioned adequately until 1914 when a new phylloxera biotype[2] developed that was able to overcome the resistance. At that point the French wine growers decided to go for 100% American rootstock. These are the vines

[2] Biotypes are species that are genetically identical with respect to their virulence. The term is most commonly used in relation to vertical resistance where there is a gene-for-gene relationship between a pest and its host plant. Some scientists believe the term to be redundant.

that still support wine production in France today. They are (apparently) completely resistant to the phylloxera, although the exact resistance mechanism is still not understood.

One further addition to the phylloxera story concerns the Californian wine industry. This had begun to develop at the start of the twentieth century and during the period 1910–50 extensive rootstock trials were undertaken. The best rootstock at that time was the hybrid AXR#1 and this was recommended to growers by the government. Initially this rootstock did well and a flourishing wine industry developed in California throughout the 1950s, 1960s and 1970s. As had occurred in Europe, however, the phylloxera biotype with resistance to this rootstock eventually developed and by the mid-1980s the Californian wine industry was in crisis. The solution to this problem – that has been implemented throughout the 1990s – has been to replant every vine with 100% American rootstock. Although successful, it has been estimated that replanting vines in California has cost the wine industry there *c.* $1–2 billion.

The European corn borer *Ostrinia nubilalis*

The European corn borer is a lepidopterous pest that colonised the USA at the beginning of the twentieth century. It is a pest that is native to Europe and Asia. Larvae damage plants by internal feeding (boring). Feeding activity can reduce yield and, in severe cases, can kill the host plant. Depending upon geographical location, this pest can go through one to five generations per year. Since the 1940s, more than 100 insect-resistant lines of corn have been developed. It is estimated that these varieties are grown on *c.* one-third of the US corn area. Research on the resistance mechanisms involved has shown that levels of the secondary plant substance 2,4-dihydroxy-7-methoxy-1, 4-benzoxazin-3-one (DIMBOA) are highly correlated with resistance to first-generation larvae. This chemical decomposes, via intermediates, to release 6-methoxy-2-benzoxazoline (MBOA) and formic acid. MBOA acts as a feeding deterrent and repellent to the corn borer. In addition to studies on DIMBOA, it has also been reported that resistance is linked to levels of silica and lignin in plants. The exact resistance mechanism is therefore likely to result from a blend. Whatever the exact nature of these resistance mechanisms, it has been estimated that the use of resistant corn varieties saves farmers in the USA millions of dollars every year.

The brown planthopper *Nilaparvata lugens*

The brown planthopper (BPH) is the single most important rice pest across much of Asia. Each year, millions of dollars are lost because of damage caused by this pest. Outbreaks of this pest have been reported since 697 AD. Extensive research to develop varieties that are resistant to this pest has been carried out by the IRRI in the Philippines since the 1960s. The first resistant variety to be developed was released in 1963 and since then further varieties have been released as the BPH has evolved to overcome host-plant resistance, a process that has occurred due to the development of new biotypes. Resistant varieties are currently planted over millions of hectares in Asia and have been invaluable in helping to reduce the damage caused by this pest. At present, IRRI is operating a sequential-release strategy of varieties with new resistant genes. It is believed that IRRI has a stockpile of at least six resistant genes. It would seem that these resistant rice varieties incorporate a blend of antixenosis, antibiosis and tolerance. For example, it is known that a high proportion of short-chain hydrocarbons in epicuticular wax will act as a feeding deterrent. It has also been shown that steam distillate extracts of resistant rice varieties are toxic to the pest when applied topically.

Potential problems with host-plant resistance

To date, thousands of cultivars that are resistant to arthropod pests have been developed and released. These cultivars belong to all of the most important food crops worldwide, i.e. they include wheat, rice, sorghum and corn, as well as the world's most important fibre crop, cotton. There is no doubt that the use of these varieties in pest control programmes has led to substantial economic savings.

That said, and phylloxera-resistant rootstock apart, most arthropod-resistant varieties do not represent a complete solution to pest problems. From a practical point of view it is much more likely that a resistant plant variety will be integrated with other control measures such as pesticides or biocontrol agents. In addition, there are some limitations that are associated with the use of host-plant resistance in arthropod pest control. These have been described as problem-trading, yield drag, biotype development and effects on nontarget species. These limitations are discussed in more detail below. Finally, there is also the 10–20-year time period (and the associated cost) that it takes to develop new varieties. It is believed that developments in biotechnology will shorten this time period considerably. However, it is still a problem, and it is one reason why most research on host-plant resistance has focused on major crops that are of global economic significance.

Problem-trading

Most major crops are attacked by a complex of pests, and there is always the potential problem of resistance to one pest being linked to susceptibility to another. For example, pubescence (hairiness) in plants may be attractive to some pests whilst providing resistance to others. In cotton, hairiness provides resistance to jassids *Empoasca* spp. but increases the susceptibility of these plants to the cotton bollworm *Helicoverpa zea*. Similarly, cotton plants that express high levels of gossypol provide protection from *Heliothis* spp. but are attractive to boll weevils. There is no straightforward solution to this dilemma; however, it does highlight the importance of identifying the key pests within a complex.

Yield drag

It has been argued that plant resources which are used for defence will inevitably lead to a reduction in resources available for growth and reproduction (yield). This has been one of the main problems associated with developing soybeans that are resistant to pests. Resistant soybean cultivars tend to produce yields that are unacceptable to growers. Farmers will not use resistant varieties unless their yield is at least as good as susceptible varieties that need chemical protection from pests. Table 10.4 provides some energetic calculations associated with the production of secondary plant substances. It has been argued that the energetic costs involved are small and are consequently unlikely to cause a reduction in yield. However, it is clear that the process does occur in some crop plants, even if it is not universal. From a breeding point of view, the critical point is how much resistance is required before yield drag begins to have a significant effect. It is possible that sufficient resistance may occur long before there is a problem with this process.

Biotype development

Biotypes are a particular problem for host plants that show vertical resistance to pests. Monogenic vertical resistance is highly likely to drive the development

Table 10.4. *Energetic costs associated with the production of some secondary plant substances*

Chemical class	Source	Cost (mg glucose/g tissue)
Diterpene	Needles of *Larix laricina*	3
Glucosinolates	Leaves of *Bretschnedera sinensis*	13
Alkaloid pyridine	Leaves of *Nicotiana sylvestris*	18
Phenolic flavinoid	Leaves of *Isocoma acradenia*	103
Triterpene	Twigs of *Betula resinifera*	307

Source: Modified from Van Emden, H.F. (1997). Host-plant resistance to insect pests. In *Techniques for Reducing Pesticide Use*, ed. D. Pimental, pp. 129–52. New York: John Wiley.

of new biotypes, most notably where the primary resistance mechanism is antibiosis. So far, insect biotypes have developed mostly among hemipterous pests such as the grape phylloxera and the BPH. This is perhaps not surprising since these species reproduce rapidly and can respond to a strong selective force quickly. One of the best-documented cases of biotype development is the BPH. This species is being managed by the selective release of rice varieties by IRRI (see earlier). Overall, problems with biotype development are relatively rare. At present, fewer than 20 species of insect are known to have biotypes.[3]

Effects on nontarget species

Plants that are resistant to arthropod pests may impact on nontarget species in two ways. First, elevated toxin levels may be unpalatable, allergenic or even dangerous for consumers. For example, potatoes with elevated glycoaldehyde levels that were resistant to Colorado potato beetle were withdrawn because of customer complaints. Second, some secondary plant substances may be toxic to beneficial species such as parasitoids. However, pesticides are also a danger to parasitoids, as well as many other species. As above, these concerns

[3] Based upon the constrained definition of biotype (used here) as it applies to vertical resistance development.

point to the importance of producing resistant plants that are fit for the purpose, i.e. with resistance to pest species but with no adverse effects on other species, particularly where the resistance mechanism *per se* involves antibiosis.

Summary

In summary, host-plant resistance proffers an environmentally friendly, simple-to-use technique for controlling arthropod pests. However, because many crops are attacked by a complex of pest species, it is unlikely that control of pests will be achieved by the use of this technique alone. Rather, it is a technique than can be built upon. For example, recent data from China have indicated that when economic comparisons are made between the relative contributions of host-plant resistance and pesticides to pest control, the former may often come out marginally better, i.e. it is economically sensible to blend the two techniques.

Plants that utilise antixenotic or antibiotic mechanisms to deter pest species effectively lower the carrying capacity[4] of the environment for those species. This should make integration of host-plant resistance with other pest control techniques easier. Where this integration has been carried out, it has been a great success.

From a pest control perspective traditional host plant resistance is an example of a cultural method of pest control. Because of the importance of host plant resistance, its discussion merited a full chapter in this textbook. Cultural techniques for controlling pests are discussed in more detail in the following chapter.

Further reading

Ashouri, A., Michaud, D. & Cloutier, C. (2001). Unexpected effects of different potato resistance factors to the Colorado potato beetle (Coleoptera: Chrysomelidae) on the potato aphid (Homoptera: Aphididae). *Environmental Entomology*, **30**, 524–32.

Braman, S.K., Duncan, R.R., Hanna, W.W. & Hudson, W.G. (2000). Evaluation of turfgrasses for resistance to mole crickets (Orthoptera: Gryllotalpidae). *Hortscience*, **35**, 665–8.

Castro, A.M., Ramos, S., Vasicek, A. *et al.* (2001). Identification of wheat chromo-

[4] The carrying capacity of a habitat represents an abstract term that ecologists use to define the maximum number of individuals that the habitat can support. See also glossary for this chapter.

somes involved with different types of resistance against greenbug (*Schizaphis graminum*, Rond.) and the Russian wheat aphid (*Diuraphis noxia*, Mordvilko). *Euphytica*, **118**, 321–30.

Cohen, M.B., Alam, S.N., Medina, E.B. & Bernal, C.C. (1997). Brown planthopper, *Nilaparvata lugens*, resistance in rice cultivar IR64: mechanism and role in successful N-*lugens* management in Central Luzon, Philippines. *Entomologia Experimentalis et Applicata*, **85**, 221–9.

Cole, R.A. (1997). The relative importance of glucosinolates and amino acids to the development of two aphid pests *Brevicoryne brassicae* and *Myzus persicae* on wild and cultivated brassica species. *Entomologia Experimentalis et Applicata*, **85**, 121–33.

Frelichowski, J.E. & Juvik, J.A. (2001). Sesquiterpene carboxylic acids from a wild tomato species affect larval feeding behavior and survival of *Helicoverpa zea* and *Spodoptera exigua* (Lepidoptera: Noctuidae). *Journal of Economic Entomology*, **94**, 1249–59.

Granett, J., Walker, M.A., Kocsis, L. & Omar, A.D. (2001). Biology and management of the grape phylloxera. *Annual Review of Entomology*, **46**, 387–412.

Khan, M.A., Stewart, J.M. & Murphy, J.B. (1999). Evaluation of the Gossypium gene pool for foliar terpenoid aldehydes. *Crop Science*, **39**, 253–8.

Landolt, P.J., Hofstetter, R.W. & Biddick, L.L. (1999). Plant essential oils as arrestants and repellents for neonate larvae of the codling moth (Lepidoptera: Tortricidae). *Environmental Entomology*, **28**, 954–60.

Panda, N. & Khush, G.S. (1995). *Host Plant Resistance to Insects*. Wallingford: CAB International.

Sharma, H.C. (1998). Bionomics, host plant resistance, and management of the legume pod borer, *Maruca vitrata* – a review. *Crop Protection*, **17**, 373–86.

Smith, C.M., Khan, Z.R. & Pathak, M.D. (1994). *Techniques for Evaluating Insect Resistance in Crop Plants*. Boca Raton: Lewis.

Van Emden, H.F. (1997). Host-plant resistance to insect pests. In *Techniques for Reducing Pesticide Use*, ed. D. Pimental, pp. 129–52. New York: John Wiley.

11

Cultural techniques and organic farming

Introduction

Cultural techniques comprise those that involve the manipulation of an agroecosystem in order to decrease the success of pest species within it. As such, many cultural techniques form the basis of what has been called 'augmenting the environmental resistance' or 'preventive pest management'. Such techniques are not new. For example, manipulation of planting dates and burning (as a form of clean cultivation) are both recorded in literature dating to 1000 BC.

From an arthropod pest control perspective, the primary aims of cultural techniques are: (1) to reduce colonisation of a crop by a pest and/or to increase pest dispersal from that crop; and (2) to reduce reproduction and/or survival of a pest in a crop once colonisation has occurred. These main aims can be realised in two ways – either by modifying the host (i.e. the crop) or by modifying the host environment. These two approaches are not mutually exclusive and neither are the techniques that are discussed in this chapter. For example, maximising plant health (host modification) will have an impact on the success of a change in planting density (host environment).

In recent years there has been a resurgence of interest in cultural techniques for arthropod pest control. This revival has occurred both because of an upsurge of interest in organic farming and also because of a more general desire to reduce reliance on chemical methods of arthropod pest control. However, funding for the development of cultural techniques is still largely reliant on government research support. The primary reason for this is that many cultural techniques are, by definition, technique and not product-based, i.e. there is no product to sell. This funding situation may change with the development of good agricultural practice protocols that are currently under discussion between farmers and large food distributors. If supermarkets can develop labelling schemes that communicate good farming practice procedures to consumers, then further funding for the development of cultural techniques for pest control may be made available. Recent (2001) changes to the government set-up in the UK may also help. For example, in 2001 the newly reelected Labour government dissolved the Ministry of Agriculture, Fisheries and Food and introduced a Department for Food, Environment and Rural Affairs (DEFRA). At the time of writing, the indication was that this new department would pay far more interest to countryside stewardship and far less attention to issues associated with mass food production. Since cultural techniques seek to promote sustainable farming and countryside stewardship, this may be good news for the development of new research opportunities. As part of a package of reform measures, the UK government was also discussing how best to implement a proposal for a pesticide 'environmental impact' tax. Such taxes already exist in some northern European countries, e.g. Norway, and there is a belief that a tax would lead to less pesticide use. In North America, many researchers have argued for years that satisfactory pest control can be achieved by focusing on ecological or 'biointensive' methods – cultural techniques. All that is required is a greater degree of government/sociopolitical support for such objectives.

The list of cultural techniques that are discussed in this chapter is shown in Table 11.1. These techniques will inevitably overlap and many farmers will use a mixture of techniques simultaneously. For simplicity, they are discussed separately within this chapter. Host-plant resistance (an example of a cultural technique) was discussed in Chapter 10 and so is not discussed further. Conservation biological control (another example of a cultural technique) is only mentioned briefly, as this subject was discussed in detail in Chapter 6. Following consideration of these techniques, this chapter focuses upon a brief review of organic farming, primarily from a UK perspective. Many cultural techniques are highly appropriate to organic farming systems. Finally, the chapter concludes by making some predictions for future developments, particularly in relation to sustainable farming and to novel cultural techniques that may be developed following advances in biotechnology.

Table 11.1. *Cultural techniques for arthropod pest control*

Host modification	Environmental modification
Manuring/fertilising	Crop rotations
Irrigation	Clean cultivation
Host-plant resistance[a]	Soil cultivation
	Planting dates
	Planting density
	Trap crops
	Abiotic environment
	Intercropping
	Natural enemy conservation

Note:
[a] Discussed in Chapter 10.

Soil cultivation

Tillage or cultivation of the soil can contribute to pest control in a number of ways. For example, exposure of soil pests can lead to their death following desiccation or predation. Death may also occur due to burying or simply as a result of mechanical destruction. Pest species that are the most amenable to control by tillage are therefore usually species that use the soil as a habitat for at least part of their life cycle. For example, in the UK leatherjacket (*Tipula* spp.) larvae can be partly controlled by tillage and in some African countries, grasshoppers (*Oedaleus senegalensis*) can be controlled by exposing egg pods to desiccation. In North America, it has been suggested that wireworms and cockchafer larvae may be controlled by tillage while the European corn borer *Papaipema nebris* has been reported to suffer over 90% larval mortality following ploughing of stubble (see also below) into the soil.

Few economic analyses have been made concerning the value of tillage and this is an area which requires more research. In recent years, the trend in some countries to move toward no- or low-till systems (in an attempt to reduce soil erosion) has led to an increase in problems with some species. For example, the corn pest *P. nebris* increased in numbers in the US midwest following the adoption of conservation tillage measures (Table 11.2). However, it has been claimed that conservation tillage overall tends to decrease pest problems rather than exacerbate them. The reality is that more research is required since the effects (from a pest control perspective) of moving to a minimum tillage system will be highly situation-specific.

Table 11.2. *Pest species that undergo population increases when minimum tillage systems are implemented*

Crop	Pest	Common name
Corn	*Papaipema nebris*	Corn stalk borer
Cotton	*Heliothis* spp.	Bollworms
Oats	*Oulema melanoplus*	Cereal leaf beetle
Soybean	*Anticarsia gemmatalis*	Velvetbean caterpillar
Wheat	*Cephus cinctus*	Wheat stem sawfly

Source: Modified from Speight, M.R., Hunter, M.D. & Watt, A.D. (1999). *Ecology of Insects*. Oxford: Blackwell Science.

Clean cultivation

Destruction of crop residues can control a number of important pests both directly (as above) and indirectly by removal of suitable overwintering sites. This is particularly the case for crops that can be grown as perennials and for crops that can be cultivated continuously (because of an appropriate climate). For example, destruction of cotton plant residues is now a well-established practice for the control of the pink bollworm *Pectinophora gossypiella*. It has also been suggested that clean cultivation should comprise strategies such as the use of high-quality certified seed. Although this is generally the case in most developed countries, it is often not an option in many developing areas. The implication behind the assertion is that certified seed will be free of both pests and diseases at the time of planting/sowing.

Manuring/fertilising

Plants that are well-fertilised (and well-watered) are, in very general terms, better able to tolerate a limited amount of pest damage. This is because healthy plants are usually better able to increase their overall productivity in response to pest attack in contrast to stressed or unhealthy plants (see Chapter 10). In addition, it is known that well-fertilised plants are often able to increase their growth rates and so decrease the time available for pest attack. It is known that some plants become more attractive to pests if they are well-fertilised; however, on balance, healthier plants are better able to withstand attack from crop pests. In addition to external inputs, it is also known that careful management of nursery plants may be critical to their successful survival in the field. For example, acacia trees with deformed roots caused by rough handling during potting have been shown to be susceptible to pest species several years later – trees with reduced vigour (caused, in this case, by handling damage) are more susceptible to pest attack.

Irrigation

Water can be used to assist with pest control both via effects upon plant health and by directly causing the death of pest species. Plants that are water-stressed often have smaller leaf hairs (trichomes), higher nutrient concentrations, warmer[1] leaves and less cuticular wax. All of these changes can make plants more susceptible to crop pests. Water may also help pest control by enhancing the activity of entomopathogenic fungi. For example, in India the use of the fungus *Verticillium lecanii* for the control of planthoppers on rice is closely associated with either natural or artificial rainfall, i.e. irrigation (see Chapter 6).

Direct pest control using water may be achieved by washing pest species from plants or by drowning. This may be particularly the case if detergents are added to the water to reduce surface tension and to assist with degradation of the exoskeleton of target pests. For example, the root-feeding pest of grapevines *Phylloxera vitifoliae* can be controlled by flooding vineyards.[2] In North America flooding is used to control pests of cranberries and to control wireworms and white grubs in sugarcane. Finally, sprinkler irrigation of potato crops can help suppress numbers of potato tuberworms by preventing adult oviposition. Many organic associations recommend water and detergent as an effective method for pest control that is environmentally benign.

Crop rotations

Rotating or changing the crop that is grown in a particular area is an especially useful technique to control pest species with limited dispersal and with limited host ranges. Breaking the cropping cycle prevents a large pest population developing over a number of years. Pests are basically confronted with a nonhost habitat situation. Most rotations involve a cereal

[1] Plants that are water-stressed transpire less and so lose less heat, leading to warmer leaves.

[2] Although it is possible to control *Phylloxera* by flooding, this is not usually an economically viable option (see also Chapter 10).

crop, a legume and a root crop. In general, pests that feed on cereals rarely also feed on legumes or root crops. Historically, the use of rotations represents one of the earliest techniques for pest control. Some North American scientists have referred to this process as 'relay cropping' or intercropping over time.

Most of the pests that can be successfully controlled by using a crop rotation are soil pests. For example, rotation of groundnuts (peanuts) with nonsusceptible crops is one of the most effective methods for controlling soil nematodes in many African and Asian countries. In North America, soybean/corn rotations are used to control the weevils *Graphognathus leucoloma* and *G. peregrinus,* while corn rootworms (*Diabrotica* spp.[3]) and Colorado potato beetle can be controlled by rotating crops. In the latter case, dispersal of the beetle after overwintering is largely by walking and it has been established that separation distances of as little as 200 m between fields can have a major impact on subsequent pest infestations. In Europe, rotations can be used partly to control the soil pests wireworms and leatherjackets.

Trap crops

The primary idea behind the use of trap crops is to concentrate known pests in a specific area. This area may be used to support the build-up of natural enemies or it may simply serve to deter a pest population from colonising a more valuable crop. Finally, the trap crop area may be used to concentrate pests prior to their destruction using chemical or physical/mechanical techniques. For example, lygus bugs (*Lygus hesperus*) can be deterred from colonising high-value cotton fields by trap cropping in strips of alfalfa. Another example is the use of early-planted/early-maturing soybean crops near the primary soybean crop. The early trap crop planted on about 10% of the main crop area has been shown to attract up to 85% of potential pests, which can then be destroyed using an insecticide application. In the USA this technique has been successfully used to control populations of the stinkbug *Nezara viridula.* The use of aggregation pheromones to trap bark beetles on specific trees prior to their (both pests and the host tree) destruction was discussed in detail in Chapter 7.

[3] In areas where the corn rootworm is able to diapause for over a year the use of rotations to control these pests has been shown to break down.

Planting date manipulation

Many arthropod pests have periods during the year when they are at their most dispersive, i.e. when they are most able to disperse and colonise crop plants. The aim therefore with planting date manipulation is to avoid this peak period for crop colonisation. This may be true for both temperate and tropical pest species, although in the case of the former there are more likely to be periods when pests are inactive. One good example is the control of aphid *Rhopalosiphum padi*. This aphid damages cereal crops directly by feeding and by transmitting the cereal-stunting virus barley yellow dwarf virus. The aphid has a peak period of dispersal in the UK during September and October. Delaying planting of winter cereals therefore proffers an opportunity to avoid this pest. Unfortunately, most farmers prefer to plant early and to spray for this pest because of the better weather (for planting) that exists in September.

In cotton, tobacco and rice it has been shown repeatedly that early planting can protect crops against pest attack because plants are able to establish themselves before pests arrive. For example, early-planted rice plants suffer less damage from the water weevil *Lissorhopterus oryzophilus* as a result of improvements in plant health (see also above).

Harvest date manipulation

Many arthropod pests develop resting or overwintering stages before a crop is harvested. These individuals can then infest subsequent crops. One way to prevent this occurring is to harvest a crop before the overwintering or resting stage is reached. This approach has been tried successfully with the cotton pest *Pectinophora gossypiella* and with the potato leafhopper *Empoasca fabae* in alfalfa. Another approach is to harvest a crop in strips. This can provide beneficial predatory species with a continuous habitat for their populations. It is an approach that has worked very well with crops that can be cultivated continuously, such as alfalfa.

Planting density

Crop plant density can have an effect upon pest control in at least two different ways. First, a higher-density crop may be better able to withstand attack by crop pests because the crop as a whole is able to compensate for losses. Clearly there will be a limit to planting density since each additional plant will utilise increasing water and fertiliser resources. Yield per plant may also decline as planting density increases. The second way in which plant density can impact upon pest control is by effects on what has been termed the pest optomotor landing response. This response describes the likelihood of an insect pest flying over a suitable crop landing in that crop. It is thought that one of the triggers for this response is the contrast that exists between the crop and the soil in which it is planted. The higher the crop density, the lower the contrast between crop and soil for a pest flying over the crop. Low-contrast situations are associated with reduced colonisation by crop pests.[4] For example, the sorghum shootfly *Atherigona soccata* deposits fewer eggs on crop plants as the planting density increases from 50 to 200 per square metre.

Abiotic environmental manipulation

Since the development of pest populations is dependent not only upon suitable hosts but also upon suitable environments, the latter can often be manipulated for either full or partial control. Situations where this is possible include greenhouse crops and crops in storage. For example, many refrigeration systems either control or prevent the build-up of pests on crops in storage. Similarly, storage systems in which gas concentrations can be controlled (especially oxygen levels) can be used to control pest species. Clearly, there will be a cost associated with environmental manipulation. However, in many of these systems the primary aim is food preservation, and this process itself will require conditions that are unsuitable for pest population development. Table 11.3 provides examples of storage pests that can be controlled as a result of manipulating the abiotic environment.

Intercropping

Intercropping, strip cropping or mixed farming is an approach that is favoured in many developing countries, particularly among subsistence farmers. It is an approach that is favoured because a diversity of food items can be grown and because farmers are protected against the loss of any one of these if a specific pest outbreak occurs. Clearly, for a highly polyphagous species (e.g. desert locust), a mixed crop will offer little protection. However,

[4] There is an assumption that the crop/soil contrast is a key factor in determining crop colonisation. See glossary for further discussion of the optomotor landing response.

Table 11.3. *Examples of storage pests that can be controlled by manipulating the ambient temperature*

Technique	Pest species controlled
Hot-water immersion	Mediterranean fruit fly *Ceratitis capitata* on bananas, papayas, guavas and mangoes
Hot air	Mediterranean fruit fly, melon fly and oriental fruit fly on numerous fruit crops
Refrigeration	Eggs and larval stages of fruit flies are killed by exposure to temperatures below 3 °C on a range of crops

Source: Data collated from Paull, R.E. & Armstrong, J.W. (1994). *Insect Pests of Fresh Horticultural Produce.* Wallingford: CAB International.

many species have host preferences and so this approach will offer some level of pest control.

For a number of years a great deal of discussion was associated with the utility of mixed cropping systems for pest control. For example, in the 1991 *Annual Review of Entomology*, Andow reported on an analysis of over 200 intercropping experiments. In some experiments pest problems were reduced, in others they were exacerbated and in some, pest problems were unaffected by changes in vegetational complexity.

Where pest problems were reduced (*c.* 50% of experiments) it has been suggested that greater vegetational complexity supports higher numbers of natural enemies and that this is why intercropping can work as a form of pest control, i.e. it is a form of natural enemy conservation. Other researchers however have suggested that an increased vegetational complexity makes host-plant location harder for crop pests. These competing ideas have been referred to as the natural enemy hypothesis and the resource concentration hypothesis, respectively. Overall, it would seem that the latter hypothesis is closer to the truth, since intercropping works to control pests that are mono- or stenophagous, i.e. those with a narrow host range. It is not a suitable technique therefore for the control of polyphagous pest species.

In highly developed agroecosystems there has been a move away from mixed crops to monocrops as a result of the economics of scale that are achievable with specialisation. With the exception of specific situations, i.e. organic mixed farms, this is unlikely to change for the foreseeable future. However, what may change in the future is the genetic diversity of a

Table 11.4. *Examples of pest species controlled by intercropping*

Primary crop	Intercrop	Pest(s) controlled
Beans	Wheat	Potato leafhopper, bean aphid
Cassava	Cowpeas	Whiteflies
Corn	Beans	Armyworms, leafhoppers
Cotton	Alfalfa	Lygus bugs
Melons	Wheat	Aphids, whiteflies
Peaches	Strawberries	Oriental fruit moth

Source: Modified from Horn, D.J. Ecological control of insects. (2000). In *Insect Pest Management*, ed. J.E. Rechcigl & N.A. Rechcigl, pp. 3–24. Boca Raton, Florida: Lewis.

monocrop. For example, for the past 60 years one goal associated with growing crops as monocultures was to reduce genetic diversity to a minimum. This ensured that the seed that was sown would germinate at the same time, grow at the same rate and would flower and set seed at the same time. This then enabled harvest at the same time and a high level of product uniformity. This view of a monoculture has recently been challenged because preliminary data with cereal crops have indicated that yields can be improved by planting varietal mixtures. It is believed that yields (per unit area of soil) are improved in more genetically diverse crops because of varietal variation in terms of rooting demands on the soil. Although blends may only be appropriate to certain types of crop, this is a research area which deserves further investigation. The data collected so far also only relate to yield and not to pest control *per se*. At present, more than 90% of the organic wheat that is grown in Switzerland is sown as a blend of varieties.

Table 11.4 gives a number of examples of situations in which mixed crops have been successfully used to reduce pest populations. One good example of mixed cropping is that of planting blackberries among grapevines in North America. These mixed crops provide effective control of the grape leafhopper because a parasitoid *Anagrus epos* feeds on the pest and on the blackberry leafhopper *Dikrella cruentata*. The blackberries therefore provide an alternative food source that can promote the build-up of the parasitoid population. Clearly, this example could also be considered an example of boosting native natural enemies (see below). However, it is also a good example of the overlap that can exist between cultural pest control techniques.

One final variation on the mixed cropping theme comprises the use of species that are noncrop plants. These intercrops may help promote the build-up of predatory species or they may also target pest species directly. For example, it has been shown that intercropping potatoes with marigolds, *Tagetes* spp., promotes the control of nematodes. It is thought that this may occur because of toxic root exudates produced by the intercrop.

Boosting natural enemy numbers

Conservation biological control was discussed in detail in Chapter 5. As an example of a cultural technique for pest control, the approaches that have been taken include the provision of additional overwintering sites, e.g. beetle banks in the UK, the provision of additional food sources, e.g. nectar sources for parasitoids and understorey management in orchards in order to promote the build-up of natural enemies.

Summary of cultural techniques

There exists a diversity of cultural techniques that can be used to control arthropod pest species. All of these techniques seek to make the crop environment undesirable for pest species, at least to the point where economic damage may be avoided. Many of these techniques overlap and many can be used simultaneously in food production. Figure 11.1 provides a schematic of the cultural techniques that can be used to aid pest control. The figure shows that many of these techniques have at their core the issue of plant health, i.e. freedom from attack by pests and pathogens.

The previous section provided examples of some of these techniques and is by no means exhaustive. For example, the use of protective barriers was not discussed, nor was the use of short-season, rapid-growing varieties (that avoid pest attack). Overall, there has been an upsurge of interest in cultural techniques in recent years, not least because of the growth in the organic food market. This chapter concludes by considering organic (and sustainable) farming in more detail.

Organic farming

Organic farming, as a means of food production, began to be discussed in western literature from the 1920s onwards. Initial ideas were expounded in the UK by Rudolph Steiner, and then later by Lady Balfour, who went on to form

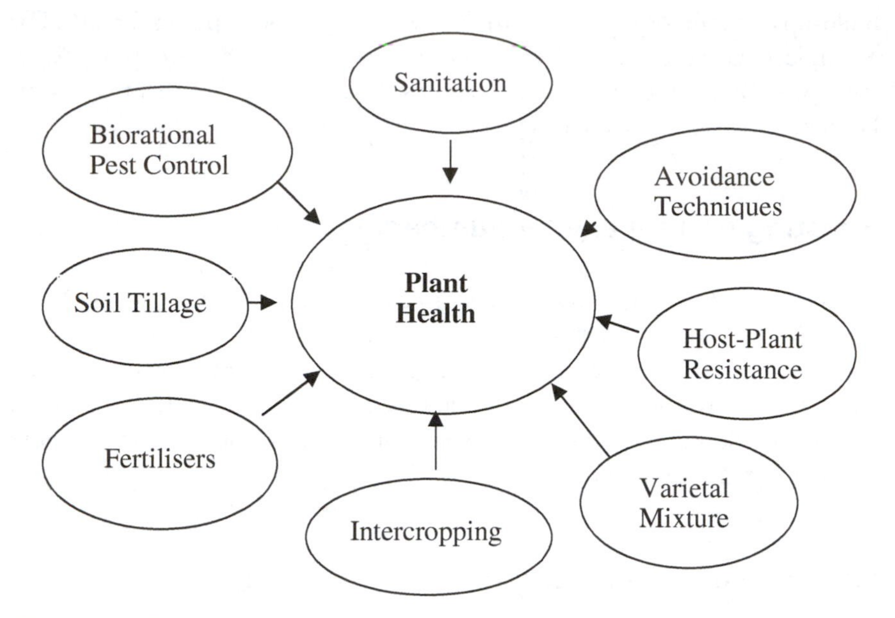

Fig. 11.1. Techniques for improving plant health. All said to be maximised when yields do not suffer because of noxious organisms. Modified from Tamm, L. (2000) The impacts of pests and diseases in organic agriculture. *Proceedings of the BCPC Conference 2000*, pp. 159–166. Farnham, Surrey: BCPC Publications.

the UK Soil Association in the 1940s.[5] Despite this, organic farming did not reach the mainstream during the next 40 years. In recent years, this has begun to change. Throughout the 1990s organic farming as a means of food production developed at a rapid pace, particularly in western Europe. For example, in the time period between 1993 and 1998, the land area devoted to organic food production more than trebled from 0.9 to 2.9 million hectares covering more than 100 000 farms. Although organic farms currently make up only 2% of the total European Union agricultural area, extrapolation of the rate of growth of organic farms suggests that by 2010 approximately 30% of western European agriculture will be organic. These developments have occurred in all European countries, although the rate of change has varied. For example, by 2000 approximately 1.3% of UK agriculture land was organic while comparative figures for Denmark and Austria were 5% and 20%, respectively.[6] In

[5] See glossary for more details.

[6] Comparative data for the USA indicate that between 1992 and 1997 the area of certified organic cropland more than doubled. By 1997 farmers in 49 states had dedicated 52000 hectares to organic food production. This area represents *c.* 0.2% of all US cropland. By 2001 there were more than 7800 certified organic farms in the USA with sales estimated at more than $9 billion.

Figure 9.3 The New World screwworm *Cochliomyia hominivorax* (larva). Photograph by John Kucharski. Reproduced with permission from the USDA/ARS image gallery at http://www.ars.usda.gov/is/graphics/photos/mainmenu.htm

Figure 9.5 Male Mediterranean fruit fly, *Ceratitis capitata,* resting on a leaf. Photograph by Scott Bauer. Reproduced with permission from the USDA/ARS image gallery at http://www.ars.usda.gov/is/graphics/photos/mainmenu.htm

Figure 9.6 The Tsetse fly *Glossina* spp.

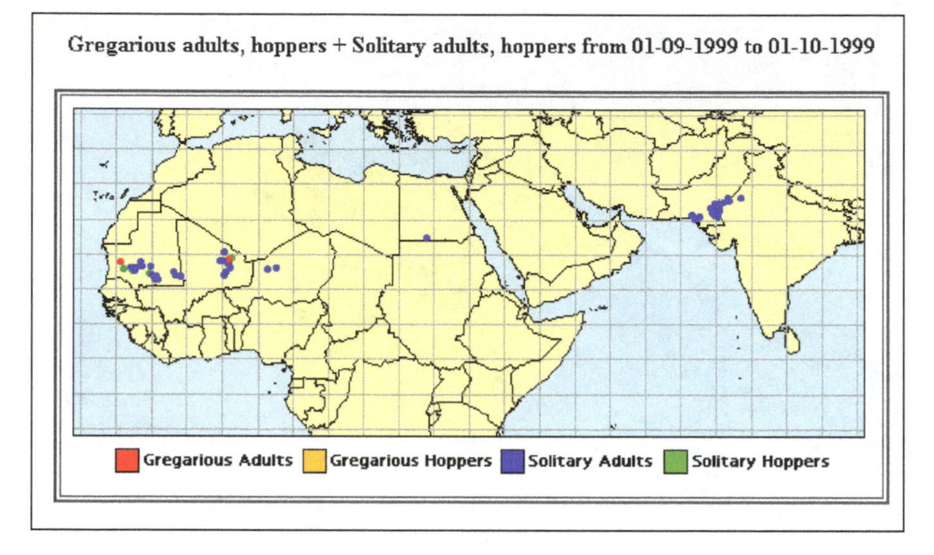

Figure 12.8 Food and Agricultural Organisation locust-monitoring data. Ecosystem is defined as multicontinental. Data abstracted from website at http://www.fao.org.

Table 11.5. *Products for arthropod pest control in organic farming*

Product	Use
Azadirachtin	Insecticide
Gelatine	Insecticide
Hydrolysed proteins	Attractants used in traps in combination with insecticides
Tobacco plant extracts	Insecticide for use against aphids in subtropical and tropical crops
Natural pyrethrins	Insecticide
Quassia	Insecticide
Rotenone	Insecticide
Various plant oils	Insecticide
Microbes	Bacteria, viruses, fungi
Pheromones	'Various'
Fatty acid soaps	Insecticide
Paraffin oil	Insecticide
Mineral oils	Only in tree crops

Source: Data collated from the UK Register of Organic Food Standards (UKROFS). This organisation ensures that EC regulations are implemented and has standards that are less strict than those applied by many certifying authorities, e.g. UK Soil Association, Scottish Organic Producers Association, etc.

2002 DEFRA (UK) announced that it was going to allocate £140 million in funding as part of plans to triple the area of land devoted to organic agriculture by 2006.

Reasons for the upsurge of interest in organic farming are numerous but certainly include changes in customer demands and the implementation of government policy initiatives that are designed to make farming more sustainable. From a pest control perspective, the growth in organic farming has meant a resurgence of interest in cultural techniques for pest control. This is because, within organic production systems, pest control is largely based upon the cultural techniques of crop rotation, host-plant resistance and various agronomic practices that are designed to improve plant health. Although many of these techniques are not product-based, in situations where the cultural technique is product-based, this has also led to new business opportunities. For example, 'biorational' insecticides such as neem, quassia and rotenone (Table 11.5) are all permitted for use in organic production systems. Furthermore, various barriers to pest attack are now marketed on a commercial basis. Such barriers include the use of polyester sheets to protect tomato crops from attack by whitefly during their first 8 weeks of growth. Fine meshes have also been used to protect individual crop plants, e.g. cabbages, from pest attack. All of these products, whose aim is cultural pest control, therefore represent opportunities to business to expand.

Sustainable farming

Throughout the 1980s and 1990s food production in many western countries has developed from a dichotomous to an (almost) inclusive approach. The dichotomy is most clearly seen when high-input, pesticide-intensive production systems are contrasted with low-input, organic production systems. These latter systems have also been referred to as a form of low-input sustainable agriculture (LISA). At the beginning of the twenty-first century, 'sustainability' has increasingly become a factor in mainstream food production units. In short, farmers, governments and policy-makers have begun to appreciate that mainstream farming, based on sustainable farming principles, is an achievable reality. This change has been made possible in part by technological developments. For example, geographic information systems (GIS) technology is now being used in Europe and North America to direct the use of external inputs such as pesticides and fertilisers. These developments have permitted reductions in agrochemical usage on farms. In the UK consumers and retailers have also begun to press for a more responsible approach to food production. At the time of writing (2002) a number of major supermarkets in the UK had begun to work in partnership with farmers to reduce pesticide use on crops. Although there have been criticisms concerning what sustainable farming constitutes, these changes are to be welcomed. For example, while it is relatively easy to define sustainability as 'to keep in existence', the issues of 'for how long?' or 'with what degree of statistical significance?' are rarely addressed.

The difficulties associated with defining sustainability have led some commentators to suggest that, when comparisons with conventional farming systems are made, in fact neither is more nor less sustainable than the other. Farming systems, of whatever type, are essentially artificial (agro) ecosystems that rely on a number of nonrenewable exogenous inputs for their maintenance. In the case of organic farming, some commentators have gone to great lengths to point out some of the dangers associated with organic production systems (Table 11.6). That said, it is clear that farming overall is becoming more environmentally responsible in many European countries.

Table 11.6. *Some issues associated with organic food production*

Factor	Problem
Synthetic fertilisers versus green manure	Organic food appears to have lower nitrate and protein content. Food mycotoxin levels are higher in organic food. Manure degradation can lead to the release of greenhouse gases
Synthetic versus natural pesticides	Approved pesticides used more frequently on organic farms. Approved products can be just as environmentally damaging

Source: Data from Trewavas, A. (2001). Urban myths of organic farming. *Nature* **410**, 409–10.

Summary

It is highly likely that cultural methods for pest control will benefit from developments in biotechnology during the next 10–20 years. One obvious area for development comprises host-plant resistance (see Chapters 10 and 13). However, developments in studies on the genetics and ecology of agroecosystems will also aid with pest and disease control. For example, studies on the genetics of many soil microbes have revealed the importance of coevolution of land plants and associated microorganisms. It has been estimated that there may be somewhere in the region of 1 million different species of bacteria in soil and 1.5 million species of fungi, many of which are critical for soil functioning and for pest and disease management.[7] For example, resistance to fungal pathogens in many crop plants is now known to be mediated by symbiotic associations with arbuscular mycorrhizal fungi. Disease-free plants are also better equipped to protect themselves from arthropod attack. More research is required. However, molecular studies are just beginning to unravel some of the complexities involved in agroecosystem functioning. Where these functions are concerned with pest control, there will undoubtedly be benefits.

At the whole-organism level we now know that in systems where pesticides are restricted (or not used), greater biodiversity is recorded. This biodiversity improvement occurs with microbes, with invertebrates and with a number of vertebrate species, especially birds. Whether improvements in biodiversity

[7] See Brussard, L. (1997). Biodiversity and ecosystem functioning in soil. *Ambio*, **26**, 563–70.

promote more effective pest control is still often thought to be contentious. What is not in doubt is that greater biodiversity may promote an increase in the sustainability of an agroecosystem, all other factors being equal. Sustainability is, in many ways, a buzzword for the twenty-first century. It does suggest a more thoughtful approach to food production and crop protection. In the penultimate chapter of this book the use of cultural techniques within the paradigm of integrated pest management is considered. Integrated pest management is an approach to pest control that is firmly founded on cultural techniques for arthropod pest control.

Further reading

Adeniyi, O.R. (2001). An economic evaluation of intercropping with tomato and okra in a rain forest zone of Nigeria. *Journal of Horticultural Science and Biotechnology*, **76**, 347–9.

Altieri, M.A. (1994). *Biodiversity and Pest Management in Agroecosystems*. New York: Food Products Press.

Andow, D.A. (1991). Vegetational diversity and arthropod population response. *Annual Review of Entomology*, **36**, 561–86.

Brussard, L. (1997). Biodiversity and ecosystem functioning in soil. *Ambio*, **26**, 563–70.

Brust, G.E., Foster, R.E. & Buhler, W. (1997). Effect of rye incorporation, planting date, and soil temperature on damage to muskmelon transplants by seedcorn maggot (Diptera: Anthomyiidae). *Environmental Entomology*, **26**, 1323–6.

Cook, P. & Ramsay, A. (1999). Agenda 2000: farm policies for the future. In *Proceedings Outlook Conference*, pp. 1–29. Edinburgh: Scottish Agricultural College.

Fereres, A. (2000). Barrier crops as a cultural control measure of non-persistently transmitted aphid-borne viruses. *Virus Research*, **71**, 221–31.

Jackson, D.M. & Sisson, V.A. (1998). Potential of *Nicotiana kawakamii* (Solanaceae) as a trap crop for protecting flue-cured tobacco from damage by *Heliothis virescens* (Lepidoptera: Noctuidae) larvae. *Journal of Economic Entomology*, **91**, 759–66.

Karungi, J., Adipala, E., Ogenga-Latigo, M.W., Kyamanywa, S. & Oyobo, N. (2000). Pest management in cowpea. Part 1. Influence of planting time and plant density on cowpea field pests infestation in eastern Uganda. *Crop Protection*, **19**, 231–6.

Karungi, J., Adipala, E., Kyamanywa, S. *et al.* (2000). Pest management in cowpea. Part 2. Integrating planting time, plant density and insecticide application for management of cowpea field insect pests in eastern Uganda. *Crop Protection*, **19**, 237–45.

McPherson, R.M., Wells, M.L. & Bundy, C.S. (2001). Impact of the early soybean production system on arthropod pest populations in Georgia. *Environmental Entomology*, **30**, 76–81.

Mitchell, E.R., Ho, G.G. & Johanowicz, D. (2000). Management of diamondback moth (Lepidoptera: Plutellidae) in cabbage using collard as a trap crop. *Hortscience*, **35**, 875–9.

Oseto, C.Y. (2000). Physical control of insects. In *Insect Pest Management*, ed. J.E. Rechcigl & N.A. Rechcigl, pp. 25–102. Boca Raton, Florida: Lewis Publishers.

Paull, R.E. & Armstrong, J.W. (1994). *Insect Pests of Fresh Horticultural Produce.* Wallingford: CAB International.

Rechcigl, J.E. & Rechcigl, N.A. (2000). *Insect Pest Management.* Boca Raton, Florida: Lewis.

Sequeira, R. (2001). Inter-seasonal population dynamics and cultural management of *Helicoverpa* spp. in a central Queensland cropping system. *Australian Journal of Experimental Agriculture*, **41**, 249–59.

Smith, H.A. & McSorley, R. (2000). Potential of field corn as a barrier crop and eggplant as a trap crop for management of *Bemisia argentifolii* (Homoptera: Aleyrodidae) on common bean in North Florida. *Florida Entomologist*, **83**, 145–158.

Speight, M.R., Hunter, M.D. & Watt, A.D. (1999). *Ecology of Insects.* Oxford: Blackwell Science.

Tamm, L. (2000). The impacts of pests and diseases in organic agriculture. *Proceedings of the British Crop Protection Council (BCPC) Conference – Pest and Diseases*, vol. 1, pp. 159–66. Farnham, Surrey: BCPC Publications.

Trewavas, A. (2001). Urban myths of organic farming. *Nature*, **410**, 409–10.

Vandemeer, J. (1995). The ecological basis of alternative agriculture. *Annual Review Ecology and Systematics*, **26**, 201–24.

12

Integrated pest management (IPM)

Introduction

The term integrated pest control was formally defined for the first time at the end of the 1950s. Since then, a steady stream of similar terms have been coined, including integrated pest management (IPM), integrated crop management (ICM)[1] and, more simply, pest management (PM). Whichever acronym you prefer is not important. The main aim associated with each of these definitions is to get away from the notion of relying on a single pest control method (usually a pesticide application) whose goal is pest eradication.

The concept of an integrated approach that focused on management, rather than control, developed primarily from an awareness of the problems associated with pesticides – resistance, environmental contamination, etc. (see Chapter 4). However, there was also an appreciation during the late 1950s and early 1960s of successful biological control programmes (e.g. cottony cushion scale in California – see Chapter 5) and of the need to conserve beneficial predators in agroecosystems, e.g. by using pesticides selectively. The integration of ecological information (such as predator conservation) in the pest control process can be formally dated to at least 1942, although it has been used for at least 2000 years, for example, the use of ants in China in the second century BC for crop protection (see Chapter 1). The term 'ecological plant protection' was first used in 1957.

[1] Arthropod pest control is just one component of ICM which represents a holistic approach to crop management in which issues associated with crop production and crop protection are considered together. See glossary for more details.

The primary philosophical change that IPM sought to establish was that pest damage is not always economically harmful. Therefore, the goal of pest management should be to reduce pest populations to levels at which economic damage does not happen. It should not be necessary to eradicate most pests. It is believed that the management process, in which protection of beneficial species is often taken to be axiomatic, can best be achieved by using more than one pest control technique, i.e. by integrating techniques.

In the USA IPM was formulated into national policy by President Nixon in 1972. In 1979 President Carter established an interagency IPM coordinating committee and in 1993 President Clinton called for a national commitment to implement IPM on 75% of the US crop acreage by the year 2000.

This chapter begins with some definitions of what IPM is. The steps that are involved in setting up an IPM programme are then described. Specific examples of IPM in action are then considered. These examples include orchard crops, cotton crops, rice crops, vegetable crops and greenhouse crops. The chapter concludes with a consideration of the impacts of biotechnology on IPM programmes and with some predictions for the future. The chapter does not consider mathematical modelling of pest populations within the context of IPM. Although a great deal of time and money has been devoted to this subject, and many useful models have been produced, on a global basis they are not at present widely used.

Defining IPM

The literature associated with pest management is enormous. Figure 12.1 takes data from the Institute of Scientific Information (ISI) and shows that throughout the 1990s approximately 200 research articles on this subject were published each year. This number is undoubtedly an underestimate since it reflects only those journals that ISI lists and it does not include technical publications, popular articles, book chapters, manuals or books. Keeping up with this steady stream of information is difficult.

The first individuals to define integrated pest control did so at the end of the 1950s (Table 12.1). Since then, a number of other definitions have followed. All of the definitions produced so far usually incorporate at least two features. These features are: (1) that pest control involves more than one tactic; and (2) that pests are maintained at levels that are not economically damaging. The third feature that some definitions include is the environment. Words are used that describe minimising environmental damage and/or sustainable agricultural production.

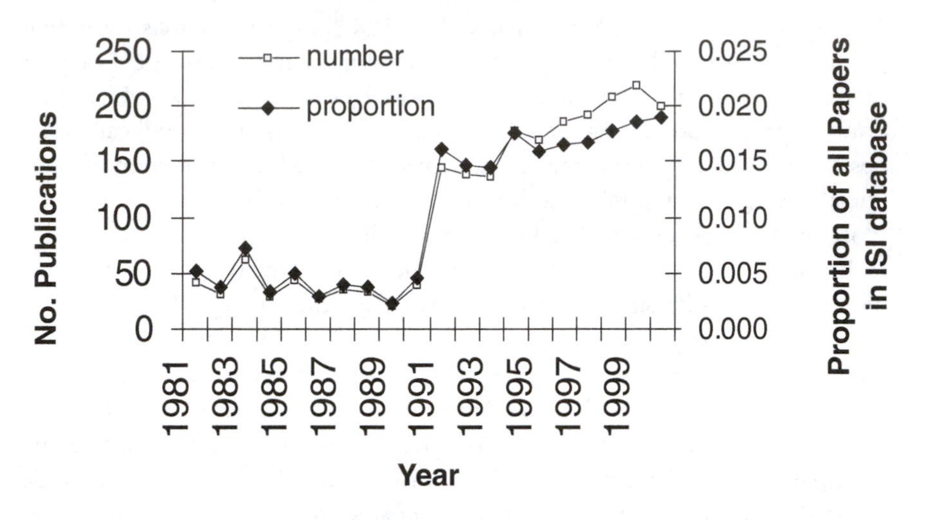

Fig. 12.1. Number of research papers published between 1980 and 2000 which listed 'pest' and 'management' as key words. Data abstracted from Institute of Scientific Information (ISI) Web of Science. See: http://www.isinet.com.

Whichever definition you prefer is not important. What is important is that the philosophy of minimising economic damage is understood and that this should be achieved at minimal environmental cost. The issue of whether this always requires a blend (integration) of techniques is discussed later in the chapter. Although it is difficult to appreciate now, these were not the goals for pest control during the late 1940s and most of the 1950s. When pesticides were first widely used, the goal was to eradicate pests and environmental quality was only a minor interest. The reader should be aware that in many countries today (2002) the aim of pest control is still to eradicate pests using pesticides.

It should be stressed that the concept of IPM is not new. Although it was first formally defined 40 years ago, farmers have been practising various aspects of IPM for millennia, i.e. by using imported natural enemies, by using various cultural techniques and by selecting varieties that are resistant to pest damage. IPM is also not simple and programmes are not developed quickly. To develop an effective IPM programme will take years, as a large volume of ecological data is typically required.

The goals of pest management (minimise economic and environmental damage) can be achieved by doing nothing, by reducing pest numbers (and/or pest fecundity), by reducing crop susceptibility or by using a combination of the latter two. You would do nothing where you know that pest presence is unlikely to affect yield (e.g. one aphid on a pea plant). However, this still

Table 12.1. *Integrated pest management definitions*

Definition	Source
Applied pest control which combines and integrates biological and chemical control. Chemical control is used as necessary and in a manner which is least disruptive to biological control	Stern, V.M., Ray, F.S., Van den Bosch, R. & Hagan, K.S. (1959). The integrated control concept. *Hilgardia* **29**, 81–101.
A pest population management system that utilises all suitable techniques either to reduce pest populations and maintain them at levels below those causing economic injury or to manipulate the populations so that they are prevented from causing such injury	Smith, R.F. & Van den Bosch, R. (1967). Integrated control. In *Pest Control – Biological, Physical and Selected Chemical Methods*, ed. W.W. Kilgore & R.L. Doutt, pp. 295–340. New York: Academic Press
An approach that employs a combination of techniques to control the wide variety of potential pests that may threaten crops	Council on Environmental Quality (CEQ) (1972). *Integrated Pest Management*. Washington, DC: US Government Printing Office
A pest management system that, in the context of the associated environment and the population dynamics of the pest species, utilises all suitable techniques and methods in as compatible a manner as possible and maintains pest populations at levels below those causing economic injury	Glass, E.H. (1975). *Integrated Pest Management Rationale, Potential, Needs and Implementation*. Maryland: Entomological Society of America
An adaptable range of pest control methods which is cost-effective whilst being environmentally benign and sustainable	Perrin, R.M. (1977). Pest management in multiple cropping systems. *Agroecosystems* **3**, 93–118
The use of all available tactics in the design of a programme to manage, not eradicate, pest populations, so that economic damage and harmful side-effects are avoided	Matheny, E.L. & Minnick, D.R. (1981). *Principles of Entomology*. Maryland: Entomological Society of America.

Table 12.1. (*cont.*)

Definition	Source
The best mix of control tactics for a given pest problem in comparison with the yield, profit and safety of alternative mixes	Kenmore, P.E., Heong, K.L. & Putter, C.A.J. (1985). Political, social and perceptual factors in integrated pest management programmes. In *Integrated Pest Management in Malaysia*. ed. B.S. Lee, W.H. Loke & K.L. Heong, pp. 47–66. Kuala Lumpur: Malaysian Plant Protection Society
The optimisation of pest control measures in an economically and ecologically sound manner, accomplished by the coordinated use of multiple tactics to assure stable crop production and to maintain pest damage below the economic injury level whilst minimising hazards to humans, animals, plants and the environment	Office of Technology Assessment, USCongress (1990). *A Plague of Locusts*. Special report OTA-F-450. Washington, DC: US Government Printing Office

requires the existence of a monitoring system for pest presence. A reduction in pest numbers can be achieved by using many of techniques described in this book, i.e. by using predators, pathogens, chemicals, cultural techniques, etc. A reduction in crop susceptibility usually means the development of resistant crop cultivars. Most well-developed IPM programmes now incorporate a combination of these techniques. The range of techniques selected for an IPM programme depends on a number of factors. One of the most important of these is the ecological status of the pest.

Pest status

A useful classification of pest status was provided in the early 1970s when the following categories were described: a subeconomic pest (also called a nonpest), an occasional pest, a perennial pest and a severe pest (Figures

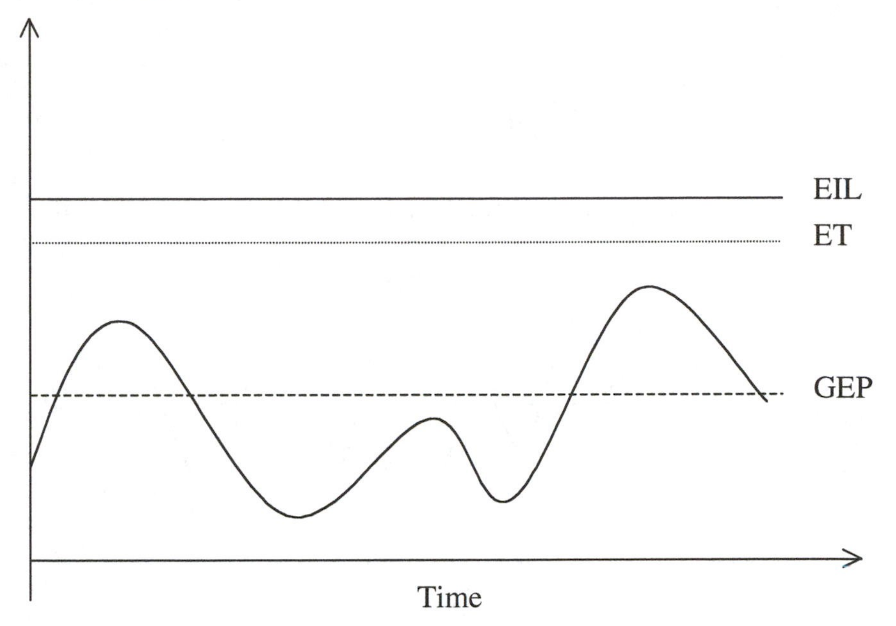

Fig. 12.2. Schematic of the population dynamics of a nonpest. EIL, economic injury level; ET, economic threshold; GEP, general equilibrium position. Adapted from Stern, V.M., Ray, F.S., Van den Bosch, R. & Hagan, K.S. (1959). The integrated control concept. *Hilgardia*, **29**, 81–101.

12.2–12.5). Under this framework a nonpest would fall into the 'do-nothing' category, although monitoring would still be required. There are a large of number of insects that belong to this category. Occasional pests only reach levels that are economically damaging every few years and so monitoring and prediction will be key to their control. No attempt will be made to alter their general equilibrium level.[2] Perennial pests and severe pests, by contrast, need control measures that will serve to reduce their population equilibrium levels. To do this means altering the carrying capacity[3] of the environment, i.e. a range of integrated measures will be required.

The categories identified above are just that. They are useful as a model for pest classification. However, in the real world, the status of pest species changes with the weather, with geographical location, with the market value of produce and with changes in public taste. The categories should therefore be regarded

[2,3] See glossary for Chapters 5 and 10 for more discussion of the terms general equilibrium position and carrying capacity.

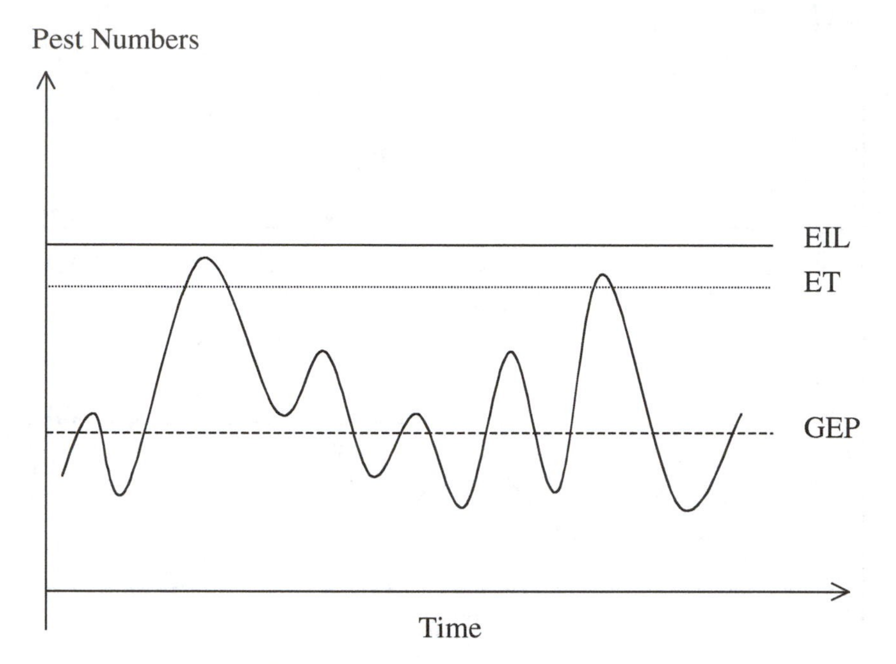

Fig. 12.3. Schematic of the population dynamics of an occasional pest. EIL, economic injury level; ET, economic threshold; GEP, general equilibrium position. Adapted from Stern, V.M., Ray, F.S., Van den Bosch, R. & Hagan, K.S. (1959). The integrated control concept. *Hilgardia*, **29**, 81–101.

as dynamic in the sense that a pest may move between categories depending upon circumstances. Key to these categories of course are the notions of an economic threshold and an economic injury level. These pest densities were introduced in Chapter 1. In the following they are discussed in more detail.

Economic threshold and economic injury level

The economic threshold (ET) and economic injury level (EIL), as applied to pest populations, were first defined in the early 1960s. The definitions given are as follows:

Economic injury level

The lowest pest population density that will cause economic damage.[4]

[4] This is therefore the pest density at which the cost of control is equal to the cost of the damage caused, i.e. the pest density at which economic damage begins. See also Chapter 1.

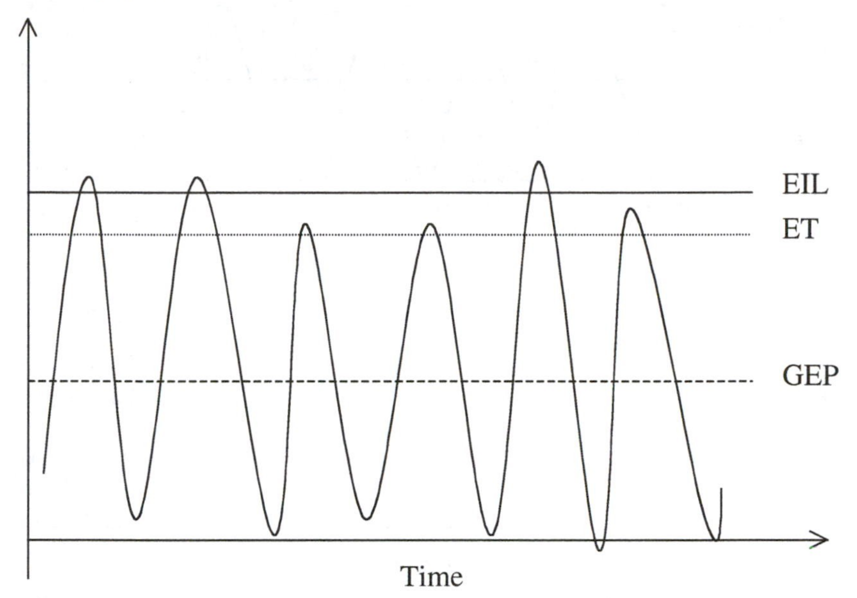

Fig. 12.4. Schematic of the population dynamics of a perennial pest. EIL, economic injury level; ET, economic threshold; GEP, general equilibrium position. Adapted from Stern, V.M., Ray, F.S., Van den Bosch, R. & Hagan, K.S. (1959). The integrated control concept. *Hilgardia*, **29**, 81–101.

Economic threshold

The pest population density at which control measures need to be applied in order to prevent the pest population from reaching the economic injury level.

The ET (sometimes also called the action threshold) and the EIL are therefore pest population densities, and the ET is (theoretically) always lower than the EIL. A schematic of the relationship between these densities was provided in Chapter 1 (Figure 1.5). The difference between the ET and the EIL, which can be measured in time, knowing pest fecundity for different environmental conditions, is dependent upon the time it takes for control measures to work. For example, with a pesticide that is effective almost immediately, the difference between the ET and the EIL may be equated to zero for practical purposes. Where the population dynamics of a pest species are poorly understood the ET may also equate to the EIL since little will be known about population development. Where a biocontrol agent that takes days to reduce pest numbers and pest population dynamics are well known, the difference

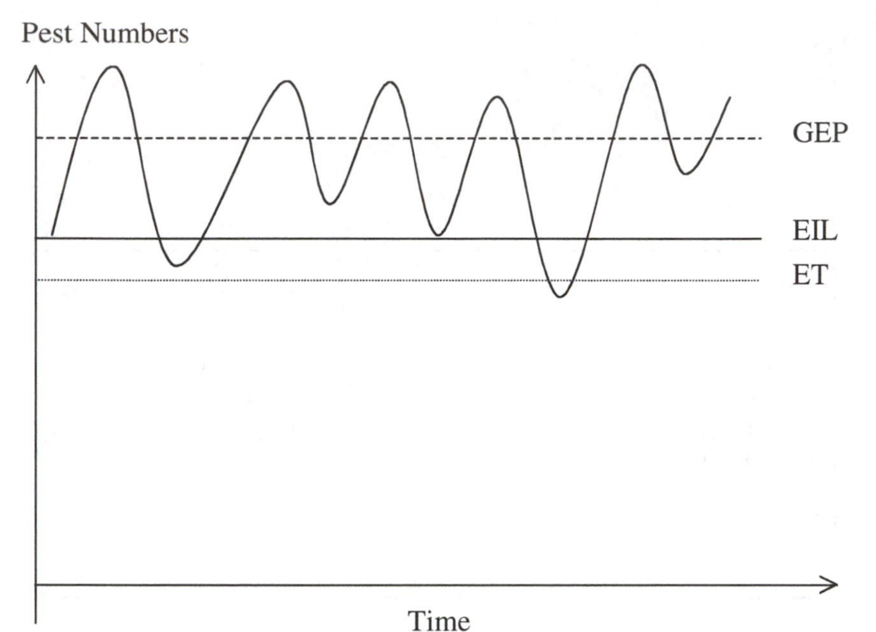

Fig. 12.5. Schematic of the population dynamics of a severe pest. EIL, economic injury level; ET, economic threshold; GEP, general equilibrium position. Adapted from Stern, V.M., Ray, F.S., Van den Bosch, R. & Hagan, K.S. (1959). The integrated control concept. *Hilgardia*, **29**, 81–101.

between the ET and the EIL will be much greater. A selection of economic thresholds for pests on UK crops is provided in Table 12.2.

What should be clear from Table 12.2 is that to use an ET requires that some sort of monitoring or scouting system is in place. This system itself should be economic and simple to use for a grower.

In many situations today the implementation of IPM programmes is still constrained by the cost or the knowledge required to implement a pest-monitoring system. It is simply easier and cheaper to apply a broad-spectrum pesticide. The reader should also be aware that there are specific pest–host situations in which thresholds for control may be inapplicable because the threshold is zero. This will be the case where crop plants are required to be completely pest-free (e.g. plants for export) and where the pest is a vector of disease.

The calculations involved in determining the ET and the EIL for a given pest–host situation range from the simple to the complicated. At the simplest end of the spectrum, the EIL is simply the pest density at which the cost of damage caused is equal to the cost of the control measure applied (see also footnote 4, above). In this simple relationship, cost refers simply to the money

Table 12.2. *Economic thresholds for some UK pests*

Crop	Pest	Economic threshold
Apples	Apple-grass aphid	Aphids on 50% of trusses at budburst
Peas	Pea moth	10 or more moths per pheromone trap in two consecutive 2-day periods
Rape	Blossom beetle	15–20 adults per plant at flowering
Wheat	Rose-grain aphid	30 or more aphids per flag leaf at flowering

Source: Data collated from a range of sources.

spent or lost; it does not include such factors as cost to the environment. A number of equations that can be used to calculate the EIL, are shown in Box 12.1. In more complex models of the EIL, allowance is made for the difference between injury and damage, for environmental costs, for the efficacy of control measures and for different yield–infestation relationships (Figure 12.6).

Despite the above, developing effective ETs and EILs is not straightforward. Even if effective monitoring systems have been developed that are simple and economic to use, there are still a number of factors which often act against the development of effective thresholds. These factors include: market conditions, agronomic practices, geographic location, the costs of control measures and consumer tastes. When the market value of a crop, the damage caused by an insect or the efficacy of a control measure increases, then the ET will decline. Similarly, if the costs of control or environmental damage increase, then the ET will increase. These changes to the ET can happen depending on:

- actual market conditions;
- various agronomic practices (fertilisation, use of tolerant cultivars, etc.) that affect plant quality (and hence value);
- geographical factors that can affect pest species fecundity (and hence propensity to cause more damage);
- the cost of a control measure (which will vary, particularly with government subsidies);
- consumer tastes: in the developed world at least there is now a clear desire to purchase produce that is blemish-free.

As a result of the above, it is therefore often difficult to produce accurate models that can be applied to such a dynamic situation. This is particularly true where food production is a government-subsidised activity and so true

Box 12.1

Simple equations for calculating economic injury levels (EILs). The equations become more complex as additional factors are considered. Modified from Pedigo, (1999). *Entomology and Pest Management*. Upper Saddle River, New Jersey: Prentice Hall.

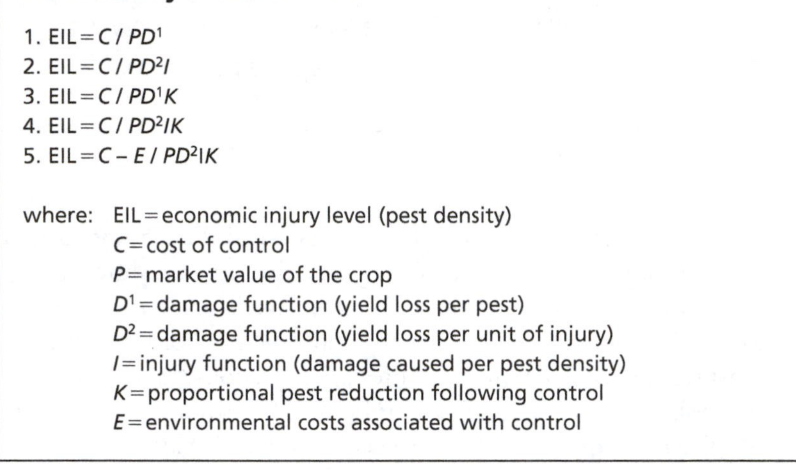

1. $\text{EIL} = C / PD^1$
2. $\text{EIL} = C / PD^2 I$
3. $\text{EIL} = C / PD^1 K$
4. $\text{EIL} = C / PD^2 IK$
5. $\text{EIL} = C - E / PD^2 \text{I} K$

where: EIL = economic injury level (pest density)
C = cost of control
P = market value of the crop
D^1 = damage function (yield loss per pest)
D^2 = damage function (yield loss per unit of injury)
I = injury function (damage caused per pest density)
K = proportional pest reduction following control
E = environmental costs associated with control

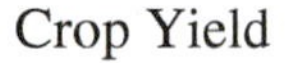

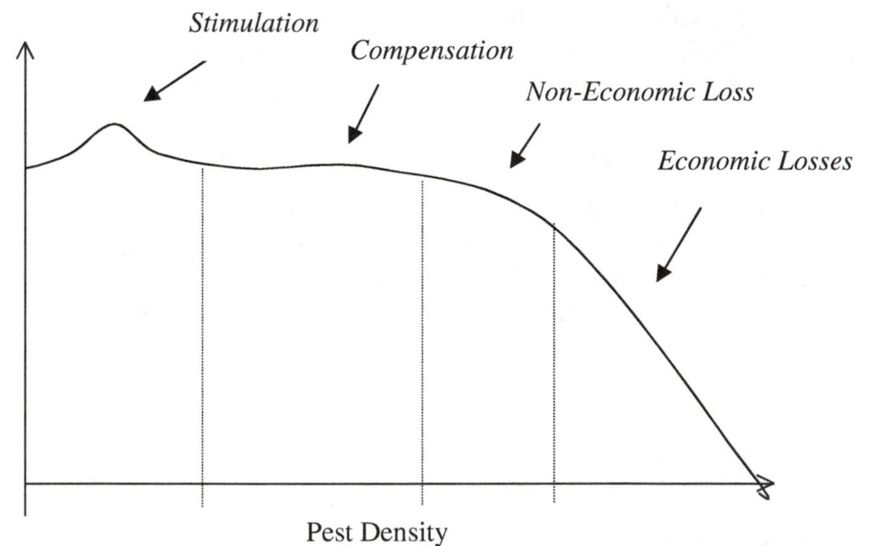

Fig. 12.6. Schematic showing phases in the relationship between pest density and crop yield.

market values are very difficult to quantify. That said, there are a number of examples of successful IPM programmes worldwide. Before these are discussed, a brief description of the general procedures involved in setting up an IPM programme is provided.

Steps involved in establishing an IPM programme

A theoretical framework for establishing an IPM programme is provided in Figure 12.7. The figure shows that the first step is to define the agroecosystem within which an IPM programme is to be applied. Does the agroecosystem comprise a field, a country or even a continent? For example, desert locust monitoring is undertaken by the Food and Agricultural Organisation (FAO) of the United Nations on a multicontinent basis (Figure 12.8, colour plate). The aim of the FAO locust work is provide data to countries at risk from locust outbreaks. This may or may not mean that those countries use an IPM programme to control this pest. However, it is a first step. Once the agroecosystem is defined, the next stage in an IPM programme is to set up monitoring systems for both pests and beneficial species and to establish ETs for pest control. To carry out such work will take time. Establishing thresholds requires data on yield–infestation relationships and on pest population dynamics and is not straightforward. Even when these are established, user-friendly, cost-effective monitoring systems need to be developed. Once thresholds are established and monitoring systems are in place, the next stage of an IPM programme is to make the environment as harsh as possible for pest species. This may involve using resistant cultivars, natural enemies or pathogens and cultural techniques that make pest survival difficult. These stages are akin to reducing the carrying capacity of the environment for a pest or pest complex. Finally, if all else fails, a pesticide application may be needed. However, this may involve reduced rates, spot spraying, use of the least damaging compounds and precision spraying procedures. For example, at the end of the year 2000 technology had been developed that would allow growers to spot-spray pesticides on areas of high pest density.

IPM therefore does not mean that no pesticides are used. Rather, IPM constitutes the sensible use of pesticides, often as a last resort. This distinction has not always been made clearly enough and there are many people who still believe that IPM means 'trying to avoid the use of chemicals'. This is not the case. Indeed, pesticides may be an integral part of an IPM programme. Some authors have even talked about 'integrated pesticide management' as an integral feature of an IPM programme.

IPM programmes have now been put in place for a number of the most

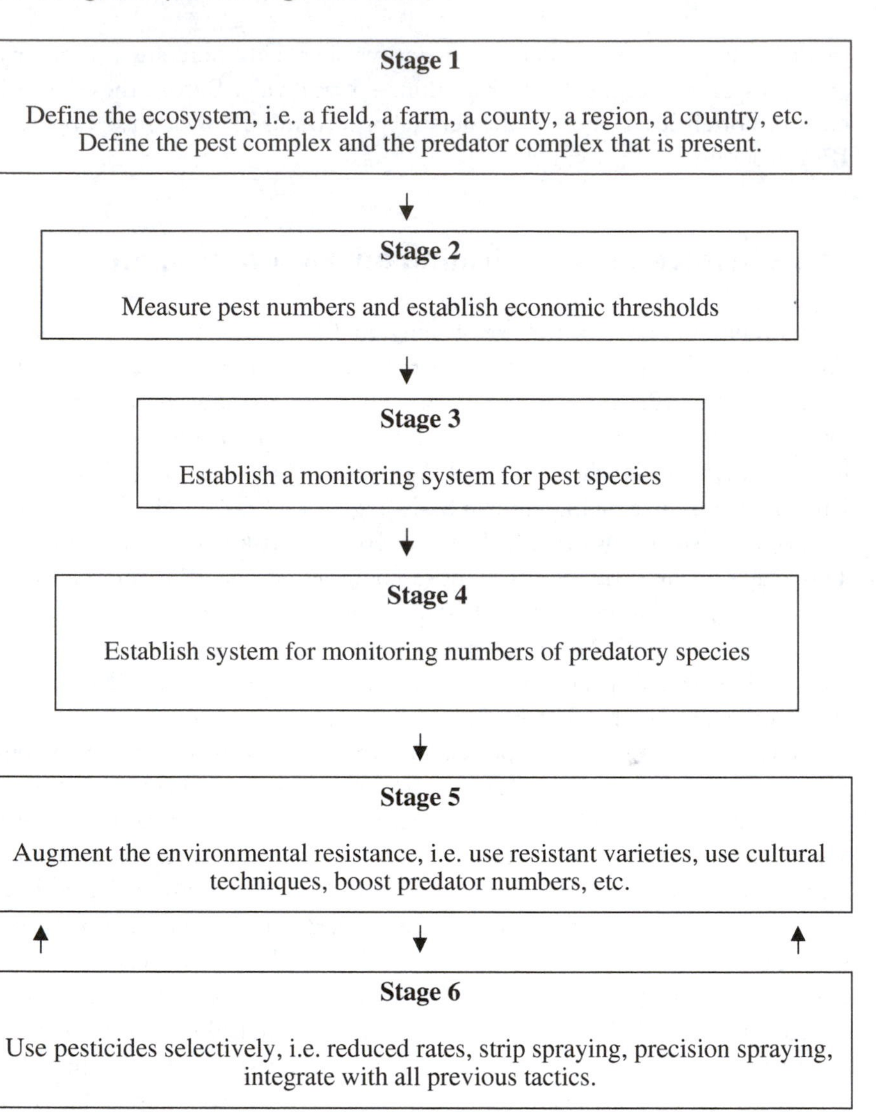

Fig. 12.7. Stepwise procedure involved in setting up an integrated pest management programme. Modified from Van Emden, H.F. (1974). *Pest Control and its Ecology*. London: Edward Arnold.

important crops on this planet. Unfortunately, most of these programmes were established following the failure of conventional pesticide-based approaches. In the future, this is an area from which we can learn, i.e. we do not have to wait for failure to implement an IPM programme on a crop. A summary of some of the best-known IPM programmes globally is provided in the following sections.

Cotton IPM

Cotton is the world's single most important fibre crop and it is also the crop in which IPM programmes were first implemented to any significant extent. Cotton is grown on *c.* 33 000 000 hectares globally and, by value, is sprayed with *c.* 25% of all insecticides. Cotton is not a very 'green' crop. The reasons why cotton is so heavily sprayed are easy to appreciate. It is a high-value crop that is attacked by a complex of pest species. It is a crop that is not consumed, at least not by people. It is grown in tropical and subtropical climates in which pests are able to flourish and where controls over pesticide use are often minimal. Almost all of the disasters that have happened in cotton production systems have occurred following pesticide misuse. Unfortunately, while the development of IPM programmes in cotton has been substantial, it was still the case that, in 2000, more than half the world's cotton crop was treated intensively with insecticides.

The most widely reported cases of ecosystem breakdown that have happened in cotton production systems occurred in the Canete valley in Peru and in the Gezira region of Sudan. The story in both situations is similar. Excessive pesticide use led to the development of resistant and resurgent pest species, and also to the emergence of secondary pest problems. The cotton production systems eventually became economically unviable and remedial measures, largely developed from IPM, were sought. The programmes that were developed were largely sustainable. In the Gezira these programmes have been built upon, particularly more recently with the development of farmer field schools[5] (FFS). However, it is still the case that most of the world's cotton is not under IPM and that many areas which were under IPM have reverted to intensive pesticide use (see following account). In the following sections a detailed description of the Canete valley story is given, followed by a more general description of the current state of cotton IPM.

[5] Developed by the FAO in order to promote the use of IPM in agroecosystems, including cotton (see also later). See glossary for more details.

The Canete valley, Peru

The Canete valley in Peru comprises *c.* 22000 hectares of highly productive cultivated land that was largely planted with cane sugar until the 1920s when a shift to cotton took place. After the Second World War, when synthetic organic pesticides became available, farmers in the valley rapidly adopted this new technology. The main pesticides used were dichlorodiphenyltrichloroethane (DDT), benzenehexachloride (BHC) and toxaphene. During the period 1949–54 yields increased by over 50% and pesticides came to be applied as blanket sprays over the whole valley. However, by 1955 problems began to occur. Aphids developed resistance to BHC, boll weevils began to develop high population levels early in the growing season, and a new moth pest (*Argyrotaenia*) appeared. Bollworm populations developed resistance to DDT and increased in number, and the frequency of treatments went from 8–15 days to every 3 days! The failure of chemical control developed rapidly and by 1956 yields were lower than they had been a decade earlier when these pesticides had not been used. In short, the cotton production system collapsed. In many ways the speed with which problems occurred in the Canete valley are astounding. The story is a classic example of relying on a single pest control tactic (a selection of broad-spectrum pesticides) in an isolated agroecosystem. Because the valley is surrounded by unproductive barren land and because pesticides were applied as blanket sprays, beneficial predatory insects were wiped out rapidly. This led to rapid resistance development in the target pest populations and consequent outbreaks of resurgent pests. Destruction of natural enemies also led to the rapid emergence of secondary pests. A temporary switch to organophosphate pesticides in 1955–56 did nothing to prevent the problems that had occurred.

To rectify the situation, a management plan based on IPM was implemented. This plan included the use of cultural techniques, natural enemies and improved pesticide usage. For example, the use of perennial cotton (which provided overwintering sites for pests) was banned, the use of synthetic insecticides was restricted and farmers were encouraged to use nicotine sulphate. Natural enemies were also reintroduced and insectaries for the production of the bollworm egg parasitoid *Trichogramma* were established. Finally, the government enacted legislation to force grower compliance with a switch to this more balanced IPM programme. The result of this was a decrease in the number of insecticide applications from 16 to 2–3 per crop and an increase in cotton yields which was sustained for the following 20 years. Unfortunately, during the last 10 years, this situation has now changed. In recent years the area planted to cotton has halved following the break-up of large estates, as part of politically motivated measures associated with agrarian reform, and pesticide use has begun to increase again. IPM is still practised on a small area of organic cotton within the valley; however, a large proportion of the cotton crop is now treated intensively with insecticides.

Cotton IPM programmes worldwide

Cotton is grown in both developed and developing countries in North and South America, Africa and Asia. Despite the well-documented cases of the Gezira and the Canete valley, on a global basis most cotton is still produced in non-IPM systems (see also earlier comments). The reasons for this vary but certainly include lack of education, the cost of pesticides, high-pressure sales techniques from pesticide companies and a lack of basic knowledge on natural enemy ecology. For example, it has been estimated that in some areas of India and China cotton receives 14–30 insecticide applications per season. Data from India in fact suggest that cotton uses 50% of all insecticides applied and that this is increasing at *c.* 7% per year. We can contrast this situation with the highly successful cotton boll weevil eradication campaign in North America which has led to a decline in insecticide use by 60–75%. In Australia, cotton production is intensive, and although elements of IPM are used, conventional crops still receive 8–15 sprays a season for pest control. In Burkina Faso and Mali attempts have been made to reduce insecticide applications to cotton using thresholds and reduced rates. Research in Burkina Faso has shown that between 44 and 54% insecticide can be saved while yields remain equal to, or better than, conventional practices. Unfortunately, limited resources in Burkina Faso have meant that uptake of these practices has been limited to only 1000 of the 160000 hectares of cotton that are grown.

Overall, the situation with cotton IPM is varied. In some countries the implementation of cotton IPM is well-developed. In the majority of cotton-growing countries, however, it is not. It has been argued that this situation exists because in countries where cotton production has not yet reached a crisis (through overuse of chemicals) the primary source of advice for growers is the chemical industry (which has a vested interest in promoting its products). Pakistan is a good example. Since 1981 imports of pesticides have escalated substantially (Figure 12.9). These chemicals are used primarily within cotton production. Since average cotton yields in Pakistan are still three times greater than in India, it is difficult to get farmers

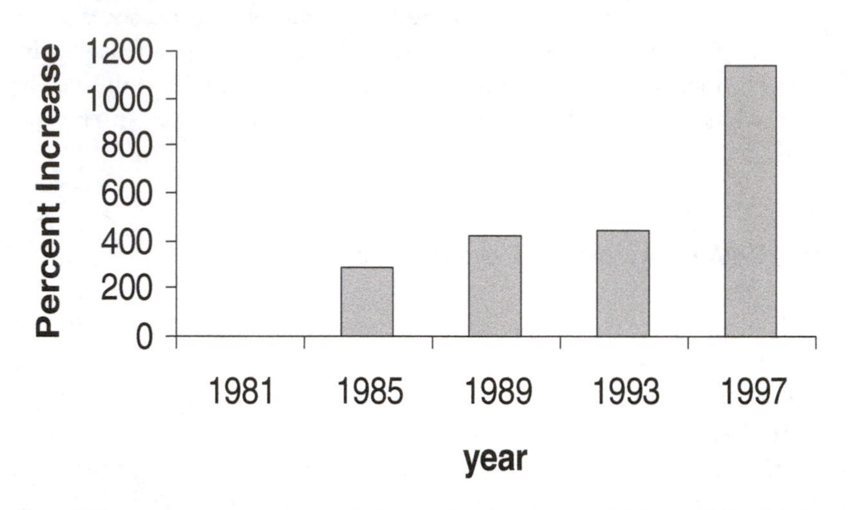

Fig. 12.9. Percentage increase in insecticide imports to Pakistan 1981–97. These figures mirror insecticide applications to cotton crops.

to accept that they have a need to shift to production systems based on IPM. Cotton production based on pesticide use works (at present), even if it has been shown elsewhere to be ultimately unsustainable. The use of FFSs has been put forward as a method to increase the uptake of IPM in cotton. These schools were developed under the aegis of the FAO during the 1990s. Their aim is to empower farmers to make their own decisions based on crop agronomy and knowledge of pests and beneficial species. Whether this happens remains to be seen. However, there is some evidence that the FFS approach can lead to reductions in pesticide usage and to greater implementation of IPM.

Apple IPM

Like cotton, apples are attacked by a number of different pest species. This pest complex varies with geographical location and agronomic practices. For example, Tasmania has five important arthropod pests of apples while the apple crop in the north-eastern USA is attacked by nine major and 20 minor insect pests, in addition to three mite pests (Table 12.3). In Hungary, over 60 phytophagous species are known to be pests of apples. The best-known apple pests include various Lepidoptera, e.g. codling moth *Cydia pomonella,* Homoptera, e.g. rosy apple aphid, *Dysaphis plantaginea*, Coleoptera, e.g. apple

Table 12.3. *Apple pests in Tasmania and north-eastern USA/Canada (common names).*

North-eastern USA/Canada	Tasmania
Codling moth	Codling moth
Oriental fruit moth	Light-brown apple moth
Light-brown apple moth	
Oblique banded leafroller	
Redbanded leafroller	
Apple maggot	
Lesser appleworm	
Tentiform leafminer	
Woolly apple aphid	Woolly apple aphid
San José scale	
Tarnished plant bug	
European apple sawfly	
Plum curculio	
European red mite	European red mite
Twospotted spider mite	Twospotted spider mite

Source: Data from MacHardy, W.E. (2000). Current status of IPM in apple orchards. *Crop Protection*, **19**, 801–6.

blossom weevil, *Anthonomus pomorum*, and Acarina, e.g. European red mite, *Panonychus ulmi*.

After the Second World War, organochloride and then organophosphate pesticides replaced lead arsenate as the chemicals of choice for pest control in orchards in Europe and North America. Widespread use of these chemicals, as in cotton, has led to the development of resistant, resurgent and secondary pests. During the 1970s the predatory mite *Typhlodromus pyri* also developed resistance to organophosphate and carbamate pesticides. This development led to integrated approaches to apple pest management in the 1980s and 1990s in which resistant predatory mites were used in conjunction with pesticides in apple orchards. More recently, apple pest management has developed to include the use of growth regulators, pheromones, *Bacillus thuringiensis* and viruses for Lepidoptera control. For example, in Washington state mating disruption pheromone was used on *c.* 20 000 ha in 1998 to control codling moth and the apple leafroller *Archips podana.*

Overall, pest management in apples, like cotton, is still reliant on the use of pesticides. However, whereas many of these pesticides were once applied prophylactically, it is now common for pesticides to be applied according to thresholds that are based on trap catches. For example, almost 100% (94%) of crop advisers in California now base decisions for Lepidoptera control using pheromone trap catches and phenological models of pest development. The development of IPM in apple crops has also been assisted by favourable ecological characteristics of the agroecosystem. For example, apple orchards are relatively stable and so permit the development of long-term predator populations. In Europe, the implementation of pest management in apple orchards is also being helped by the development of quality assurance schemes. These schemes, which are intended to involve labelling of produce, involve using more expensive, but less pesticide-intense, crop protection methods. It is expected that consumers will pay a premium for the fruit that is produced under these schemes and that this will offset the extra production costs. This is similar to the premium paid for organic produce, although the emphasis is not on pesticide-free produce but rather on 'pesticide-lite' produce that has been produced using best environmental practice. Whether these schemes work remains to be seen. In Europe, at least, restrictions on broad-spectrum pesticide use may force growers to use such production systems anyway.

Rice IPM

Rice is the staple food of half the world's population with a world production (by weight) that is second only to wheat. Approximately 0.5 billion tons of rice were produced in 1998, almost all in tropical and subtropical Asia. Most production is intensive with a new crop established every year by transplantation into standing water. As with cotton and apples, a large complex of pest species attacks rice crops. Taiwanese crops, for example, are attacked by more than 135 insect species. The best-known pests include the brown planthopper, *Nilaparvata lugens*, green leafhoppers, *Nephotettix* spp., the rice stem borer, *Chilo suppressalis* and armyworms, *Spodoptera* spp.

Like cotton and apples, rice is treated extensively with pesticides. Historically, this has created enormous problems with resistance, resurgence and secondary pests. However, since the 1960s a great deal of research has concentrated on the development of resistant rice cultivars that can be used within IPM programmes (see Chapter 10). There are now over 400 resistant rice cultivars planted throughout the world. Most rice IPM combines the use of resistant varieties with monitoring and scouting systems for pests that determine when pesticides are used. For example, pest surveillance centres in

Malaysia monitor 10% of the planted area on a weekly basis. If thresholds are exceeded then farmers are informed and coordinated control programmes can be established. More recent developments with rice IPM utilise FFSs. For example, it is estimated that in Indonesia FAO training of 200 000 farmers in IPM techniques has saved the government *c.* £80 million in pesticide subsidies. In China the use of FFS has proven to be more effective at reducing pesticide applications than conventional approaches based on economic thresholds. In Vietnam during the 1990s the success of participatory experiments where early-season defoliating pests were not treated (allowing the build-up of natural enemies) led to substantial reductions in pesticide use and to increases in profits for farmers.

Vegetable IPM

Vegetables comprise a large and diverse range of crop plants. The most important vegetables worldwide are potatoes, sweet potatoes, cassava, soybeans, peanuts (groundnuts), peas, beans and tomatoes.[6] In Northern Europe the main vegetable crops are carrots and brassicas (rape, cabbage, turnip, etc.). As with the other examples already given, IPM in vegetables has been driven by pesticide misuse and its associated problems, by consumer concern and by legislation. For example, it has been estimated that California's celery growers lost *c.* $20 million during the 1980s following the development of pesticide resistance in the leafminer *Liriomyza trifolii*. Many vegetables, like fruit, are attacked by pest complexes and are heavily treated with pesticides. In developed countries, many of these pesticide applications are driven by consumer demand for perfect produce. In some situations vegetable IPM means little more than controlled use of pesticides based on action thresholds. However, in other situations IPM has developed and includes a range of cultural and biological methods integrated with host-plant resistance and chemicals. For example, in a 4-year study of celery production in the USA a comparison was made between conventional chemical-based crop protection and a novel low-input system based on biological control and rotations of biorational insecticides. The results of this research indicated that pest management costs could be reduced by $250/ha in the low input system, that 25% fewer pesticides could be used (by application) and that crop yield was unaffected. There was therefore a net gain to both the environment and the grower.

[6] Botanically, tomatoes are in fact classified as berries. However, while they are more akin to fruits, it is generally agreed that they are (colloquially) most often considered to be vegetables.

Similar programmes are under development for many other vegetable crops. For example, it is known that dispersal of the Colorado potato beetle *Leptinotarsa decemlineata* can be affected by cultural methods that involve using trap crops prior to crop vacuuming or propane burning. Soybean IPM in South America has developed to include the use of trap crops, host-plant resistance, biological control and the use of biorational insecticides. In Argentina alone insecticide treatments on soybeans have dropped from 2–3 per season to 0.3 (i.e. one season in three), with associated savings of *c.* $1.2 million per annum.

Greenhouse IPM

There are c. 300 000 hectares of greenhouse crops worldwide. Vegetables are produced in two-thirds of these greenhouses and ornamental plants in a third. Although this represents a tiny fraction (0.02%) of the 1.5 billion hectares that are used to grow crops worldwide, greenhouses offer the ability to grow large quantities of high-quality crops on a very small area. For example, in the Netherlands 20% (by value) of agricultural output is produced in greenhouses that occupy only 0.5% of the total agricultural land. The most important greenhouse crops are tomatoes and cucumbers. IPM programmes have now also been developed for sweet pepper, eggplant (aubergine), melon, strawberries and lettuce.

The development of greenhouse IPM has been driven (as above) by pesticide problems, consumer concern and legislation. Despite the fact that chemical costs are usually a low proportion of overall production costs (in contrast to cotton and rice agroecosystems) in greenhouses, the greenhouse environment is ideally suited to IPM. This is because the greenhouse can be treated as an isolated unit with a tailor-made programme. The greenhouse also represents a barrier to dispersal of predators and/or parasitoids (often a problem with field IPM) and so can make their use more cost-effective. The greenhouse barrier can also contain a low-level pest population (often required for biological control). The fact that the crop is protected usually constrains the range of pest species that are able to colonise the crop, further simplifying the IPM process. Finally, because the climate can be controlled in a greenhouse, detailed pest forecasting is possible and concomitant release of predatory species can be a finely tuned process. Climate control may also permit the use of predators and/or parasitoids from warmer climates than would otherwise be possible.

At present, it is estimated that 5% of the world greenhouse area is under

Table 12.4. *Integrated pest management in tomato crops in Europe*

Pests	Control method
Whiteflies	Parasitoids
Bemisia, Trialeurodes	*Encarsia, Eretmocerus*
	Predators
	Macrolophus
	Pathogens
	Verticillium, Paecilomyces, Aschersonia
Spider mites	Predators
Tetranychus urticae	*Phytoseiulus*
Leafminers	Parasitoids
Liriomyza	*Dacnusa, Diglyphus, Opius*
Caterpillars	Parasitoids
Chrysodeixus, Lacanobia, Spodoptera	*Trichogramma*
	Pathogens
	Bacillus
Aphids	Parasitoids
Myzus, Aphis, Macrosiphum	*Aphidius, Aphelinus*
	Predators
	Aphidoletes
Nematodes	Resistant and tolerant cultivars,
Meloidogyne	soilless culture

Source: Data modified from Van Lenteren, J.C. (2000). A greenhouse without pesticides: fact or fantasy? *Crop Protection* **19**, 375–84.

IPM and that there is immediate potential to increase this to 20% by 2010. Almost all of these crops are vegetables. However, *c.* 0.1% of the global ornamental crop is also under IPM. Table 12.4 gives an example of the IPM programme that has been developed for greenhouse tomatoes in Europe. The programme uses a mixture of natural enemies, pathogens, host-plant resistance and cultural control.

The primary tactic in many greenhouse IPM programmes involves the use of biocontrol agents (parasitoids, predators and various microbes). Over 100 biocontrol agents are now commercially available for the control of greenhouse pests and quality control procedures were developed in the 1990s for the 20 most important. At present, the pests that pose the biggest problems are thrips, whiteflies and aphids. There are a few biocontrol agents that have been used to control these pests (e.g. *Encarsia* and *Eretmocerus* for whitefly control); more research is required. More recent developments include the use of 'banker plants' (on which predators can develop) which are introduced to greenhouses to establish early colonies of predators. These plants have been particularly useful for building up aphid parasitoid numbers using wheat plants infested with wheat aphids. The aphids serve as a resource for the parasitoids and are unable to feed on the greenhouse crop.

In addition to biocontrol agents, many greenhouse IPM programmes also use host-plant resistance. Partial resistance may also be developed where plants are bred that improve the activity of natural enemies. For example, the parasitoid *Encarsia* has a higher search efficiency on cucumbers with fewer hairs. It has also been shown that plants producing high levels of predator-attracting volatiles can improve predator performance (see Chapters 5, 10 and 11).

In summary, it is highly likely that greenhouse IPM will continue to expand. It has already been shown that pests in greenhouses can be successfully controlled using a mixture of techniques. These techniques include biological methods, host-plant resistance, pathogens and various cultural methods. It is very likely that in the near future crops will be grown in greenhouses without recourse to conventional pesticides at all.

The impacts of biotechnology on IPM programmes

The impacts of biotechnology on arthropod pest control are discussed in detail in the next chapter and so are only mentioned here in the specific context of IPM. It seems clear that biotechnology in the form of transgenic organisms could be used in many IPM programmes. However, there is a general perception at present that these organisms have been developed as an alternative approach to pest control. Although this perception is undoubtedly incorrect, it does not help.

The primary areas in which transgenic organisms could impact on IPM programmes comprise host-plant resistance and pesticide-resistant natural enemies. The former technology is well-developed and has reached commercial release. The latter technology is still in its infancy. Transgenic crops that express the *Bacillus thuringiensis* endotoxin and are resistant to lepidopterous and coleopterous pests include cotton, corn and potatoes. These crops are now planted across much of North America, and where their use has led to a reduction in pesticide applications there ought to be opportunities for integration with biological control. Indeed, preliminary experiments that have integrated transgenic crops, natural enemies and pesticides have proven successful. In Australia, a model for IPM development that is based on transgenic cotton has

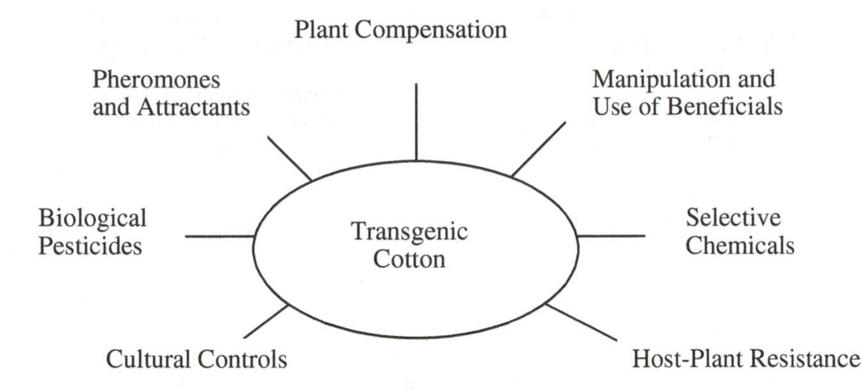

Fig. 12.10. A proposed model integrated pest management system based around the use of transgenic cotton. Reproduced with permission from Elsevier Science from Fitt, G.P. (2000). An Australian approach to IPM in cotton. *Crop Protection*, **19**, 793–800.

already been proposed (Figure 12.10). Field studies in Australia have shown that planting transgenic cotton can lead to reductions in pesticide applications and to concomitant increases in the densities of beneficial species. These increases are thought to hold promise in terms of controlling outbreaks of secondary pests in cotton, especially aphids and mites. In addition, increased natural populations of beneficials should provide more opportunities for their manipulation. At present, much research needs to be completed, but it would appear that transgenic plants ought to integrate well with IPM programmes.

Other areas in which biotechnology may impact upon IPM include pheromones and other semiochemicals, as well as the development of various bioinsecticides. For example, aphid alarm pheromones (Chapter 7) have been assayed for use in integrated control with reduced-dose pesticide applications. Although not economically viable, this may change. Novel bioinsecticides with improved environmental persistence have already been developed, e.g. *Metarhizium anisopliae* for locust control. As technology develops it is likely that further products will come on stream. Some predictions concerning these are made in the final section of this chapter.

Future developments with IPM

A schematic showing the different components of IPM is given in Figure 12.11. The figure shows that IPM, at its simplest, comprises a blend of different techniques. This blend would be what some authors have described as

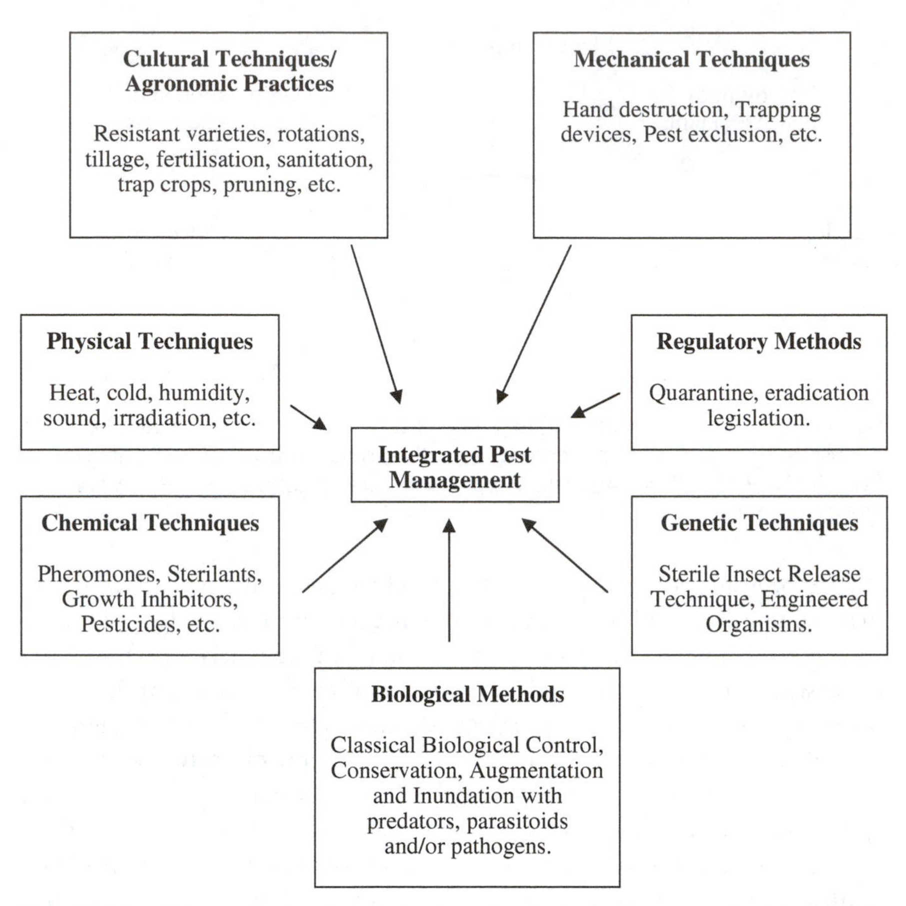

Fig. 12.11. Schematic showing the range of techniques that comprise integrated pest management. Modified from Metcalf, R.L. & Luckmann W.H. (1994). *Introduction to Insect Pest Management*, 3rd edn. New York: John Wiley. Reprinted by permission. Copyright © John Wiley & Sons.

level I IPM and is the level at which most IPM programmes operate today (as described in this chapter). Above this level more detailed systems have been defined (levels II–IV) which incorporate integration of techniques across all pest classes (level II), integration with agronomic practices (level III) and integration with socioeconomic circumstances (level IV). Figure 12.12 represents a schematic modified from the *Annual Review of Entomology* in 1998 that attempts to describe these levels and their associated uptake by farmers. The figure shows that most IPM programmes have not progressed beyond level I. Whether this is a problem is unclear; however it does help with predictions concerning what may develop with IPM in the future.

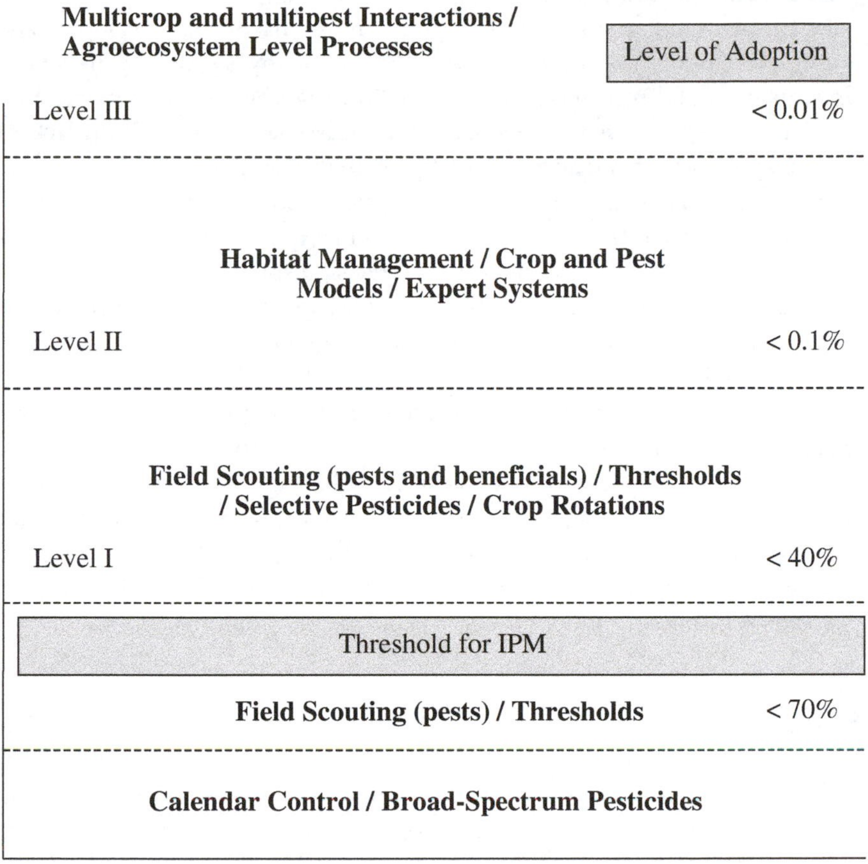

Multicrop and multipest Interactions / Agroecosystem Level Processes

Level of Adoption

Level III — < 0.01%

Habitat Management / Crop and Pest Models / Expert Systems

Level II — < 0.1%

Field Scouting (pests and beneficials) / Thresholds / Selective Pesticides / Crop Rotations

Level I — < 40%

Threshold for IPM

Field Scouting (pests) / Thresholds — < 70%

Calendar Control / Broad-Spectrum Pesticides

Fig. 12.12. Schematic which describes levels of integrated pest management (IPM) and their associated use today (2000). Modified from Kogan, M. (1998). Integrated pest management: historical perspectives and contemporary developments. *Annual Review of Entomology*, **43**, 243–70. Reproduced with permission from Annual Reviews, www.annualreviews.org.

It has recently been argued that, despite legislation, government support, 40 years of funded research and millions of research dollars spent, IPM is really an illusion. Simply blending techniques is not really IPM unless it is based upon a large volume of ecological knowledge and for practical purposes IPM means no more than supervised use of pesticides, a practice that has existed in some areas for at least 80 years anyway. This is what some authors also refer to as integrated pesticide management.

It is certainly true that crop protection worldwide is still dominated by the use of synthetic pesticides. It is also true that the majority of cropping systems are not managed using sophisticated IPM programmes. However, it is also true that first- and second-level IPM programmes have been developed, and are in use within agroecosystems. The IPM acronym has also been used to drive much useful research and, where IPM programmes are in place, concomitant reductions in pesticide use have taken place. To dwell on lack of progress with IPM is not necessarily helpful. In the future it is highly likely that more sophisticated crop management systems will be developed. These systems will involve IPM principles and will incorporate newer biotechnologies such as transgenics, computer-aided precision spraying systems, and geographic information systems (GIS) to monitor the pattern of crop damage and the effectiveness of control measures. The goals of IPM are still valid. If the first highlights in the development of an IPM programme are reductions in pesticide use or increased supervision of pesticide use then these achievements should be lauded. These are stepping stones to the development of more sophisticated programmes and need to be supported as worthy goals. Just because they do not represent a fully fledged IPM programme does not indicate that IPM has failed. It is also not particularly helpful to diminish the progress that a grower has made. It requires a lot more knowledge to use a pesticide sensibly than it does to use a pesticide indiscriminately.

Summary

Historically, the majority of IPM programmes have been driven by crisis in crop production. This is why, for example, there has been limited implementation of IPM in certain crops, e.g. wheat in the UK. This may change in the future. There is now far greater public concern over pesticide usage and with the possible removal of registration for more damaging pesticides this may force farmers to look in detail at alternative options. IPM programmes constitute an alternative. Farmers (and governments) need economic incentives to adopt IPM. In the USA, reregistrations for many commonly used pesticides are currently under way as part of the Quality Protection Act 1996. A similar review is occurring in Europe and it is likely that many widely used pesticides will be withdrawn from use in the near future.

Although initial IPM programmes are likely to concentrate solely on level I solutions, there is evidence that farmers who are exposed to IPM are more likely to be receptive to further improvements. The key first step in the process is to get risk-averse individuals to accept that there are alternatives to calendar crop spraying. It would be preferable if farmers took this first step before a crisis developed in their crop production and protection system.

The argument concerning crisis has also been used from an alternative point of view. This view states that, with the current world population, development is in crisis and food production will need to intensify if mass starvation is to be avoided.[7] Intensive food production requires the use of pesticides. IPM may fit with the sustainable agriculture paradigm but it will not feed the world. This is also an argument that has been used by those seeking to promote the use of transgenic crops within food production (see also Chapter 13).

In many ways this is a spurious argument. IPM is not about not using pesticides, it is about using them sensibly so that sustained agricultural production can occur. One way out of this so-called dilemma would be to recognise that different systems require different solutions, and to recognise that sensible pesticide use is one of the key stages in an effective IPM programme. Highly sophisticated IPM programmes will be appropriate for some crops (e.g. apples in Europe) while pesticides will be critical to ensure food security with other crops (e.g. vegetable crops in many African countries). In many ways, achieving food security is one stage in the development of more sophisticated food production systems. If pesticides are essential to the food production process then they should be used as sensibly as possible. Since IPM programmes are designed to be dynamic, their future for the twenty-first century should be assured.

Further reading

Abate, Y., Van Huis, A. & Ampofo, J.K.O. (2000). Pest management strategies in traditional agriculture: an African perspective. *Annual Review of Entomology*, **45**, 631–59.

Dent, D. (1995). *Integrated Pest Management.* London: Chapman & Hall.

Ehler, L.E. & Bottrell, D.G. (2000). The illusion of integrated pest management. *Issues in Science and Technology Online*. Spring 2000. http://www.nap.edu/issues.

Finch, S. & Collier, R.H. (2000). Integrated pest management in field vegetable crops in northern Europe – with focus on two key pests. *Crop Protection*, **19**, 817–24.

Fitt, G.P. (2000). An Australian approach to IPM in cotton: integrating new technologies to minimise insecticide dependence. *Crop Protection*, **19**, 793–800.

Ivey, P.W. & Johnson, S.J. (1998). Integrating control tactics for managing cabbage looper (Lepidopetra: Noctuidae) and diamondback moth (Lepidoptera: Yponomeutidae) on cabbage. *Tropical Agriculture*, **75**, 369–74.

[7] This is a very old argument. It was first made by Malthus in his *Essay on Population*, written in 1798. See glossary for further details.

Kenmore, P.E., Heong, K.L. & Putter, C.A.J. (1985). In *Integrated Pest Management in Malaysia*, eds B.S. Lee, W.H. Loke & K.L. Heong, pp. 47–66. Kuala Lumpur: The Malaysian Plant Protection Society.

Kogan, M. (1998). Integrated pest management: historical perspectives and contemporary developments. *Annual Review of Entomology*, **43**, 243–70.

MacHardy, W.E. (2000). Current status of IPM in apple orchards. *Crop Protection*, **19**, 801–6.

Matteson, P.C. (2000). Insect pest management in tropical Asian irrigated rice. *Annual Review of Entomology*, **45**, 549–74.

Mengech, A.N. & Saxena, K.N. (eds) (1995). *Integrated Pest Management in the Tropics*. New York: John Wiley.

Metcalf, R.L. & Luckmann, W.H. (1994). *Introduction to Insect Pest Management*, 3rd edn. New York: John Wiley.

Pedigo, L.P. (1999). *Entomology and Pest Management*, 3rd edn. Upper Saddle River, New Jersey: Prentice Hall.

Pena, J.E., Mohyuddin, A.I. & Wysoki, M. (1998). A review of the pest management situation in mango agroecosystems. *Phytoparasitica*, **26**, 129–48.

Perrin, R.M. (1977). Pest management in multiple cropping systems. *Agroecosystems*, **3**, 93–118.

Smith, R.F. & Van den Bosch, R. (1967). Integrated control. In *Pest Control – Biological, Physical and Selected Chemical Methods*, eds W.W. Kilgore & R.L. Doutt, pp. 295–340. New York: Academic Press.

Stern, V.M., Ray, F.S., Van Den Bosch, R. & Hagan, K.S. (1959). The integrated control concept. *Hilgardia*, **29**, 81–101.

Thomas, M.B. (1999). Ecological approaches and the development of 'truly integrated' pest management. *Proceedings of the National Academy of Sciences of the United States of America*, **96**, 5944–51.

Van Emden, H.F. (1974). *Pest Control and its Ecology*. London: Edward Arnold.

Van Lenteren, J.C. (2000). A greenhouse without pesticides: fact or fantasy? *Crop Protection*, **19**, 375–84.

Verkerk, R.H.J., Leather, S.R. & Wright, D.J. (1998). The potential for manipulating crop–pest–natural enemy interactions for improved insect pest management. *Bulletin of Entomological Research*, **88**, 493–501.

Waller, B.E., Hoy, C.W., Henderson, J.L., Sinner, B. & Welty, C. (1998). Matching innovations with potential users, a case study of potato IPM practices. *Agriculture, Ecosystems and the Environment*, **70**, 203–15.

Way, M.J. & Van Emden, H.F. (2000). Integrated pest management in practice – pathways towards successful application. *Crop Protection*, **19**, 81–103.

13

Biotechnology and pest control

Introduction

The word biotechnology is a concatenation (linking) of the words biology and technology. It was first used to describe industrial activities in food processing and agribusiness, that is, industrial or technological activities that involve living organisms. Brewing or cheese-making would be good examples. In the 1970s biotechnology began to be used as a descriptor for a very narrow range of techniques that involved experimentation at the cellular and/or molecular level. Although most of these novel techniques are better described as examples of genetic engineering (an example of a biological technology) many people still regard them as synonymous with biotechnology. In fact, many individuals go even further and regard biotechnology as something that is solely concerned with the creation of genetically modified organisms (GMOs). Depending upon your point of view, the word biotechnology can be used either within a wide or a narrow context. An alternative approach is to make a distinction between 'old biotechnology', e.g. brewing, and 'new biotechnology', e.g. tissue culture, genetic engineering and enzyme technology.

Almost all of the techniques that have been described in this book so far are identifiable as examples of biological technologies. In temporal terms some of these technologies are old while others are new. Pheromones, host-plant resistance and predators are all very good examples of biotechnology in action. In this final chapter the word biotechnology is used in its narrowest sense and the latest scientific developments at the cellular and molecular levels for pest control are described. For the most part this means describing the

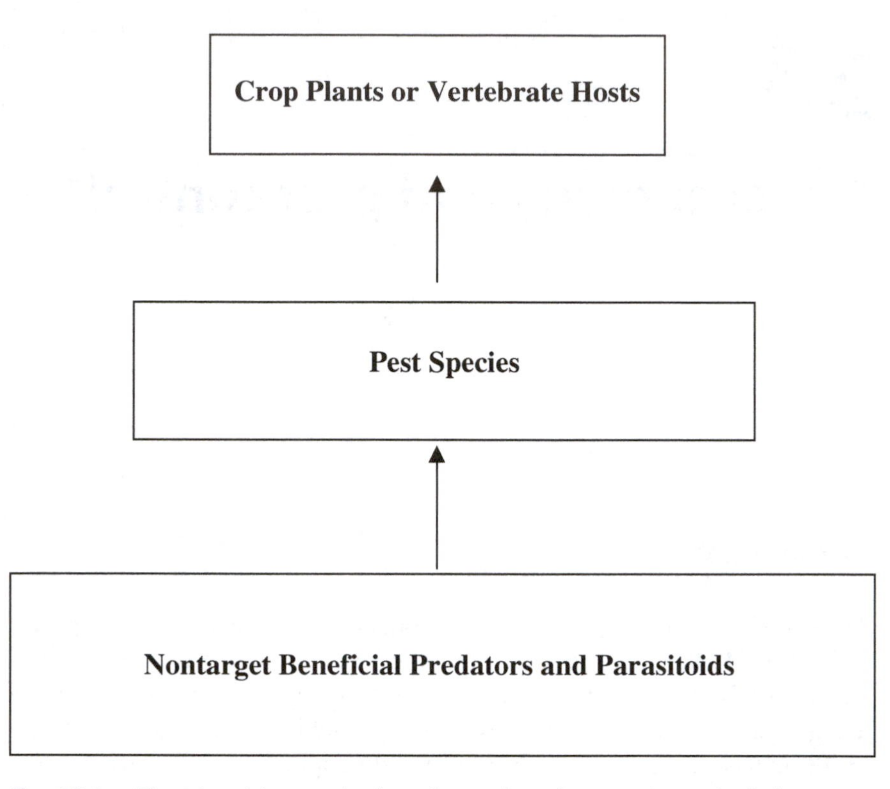

Fig. 13.1. The tritrophic organisation of organisms for genetic manipulation as an aid to arthropod pest control.

creation and use of transgenic organisms and their impacts upon the process of arthropod pest control.

The chapter is organised so that biotechnological techniques are considered first. The applications of these techniques to plants and animals are then described. The advantages and disadvantages of using such organisms for pest control are considered and the chapter concludes by making some predictions about developments that may occur during the next 10–20 years.

Species for manipulation

Within any agroecosystem there are at least three categories of living species that are relevant to pest control. These are: (1) the crop plant or domesticated animal; (2) associated pest species; and (3) associated predatory beneficial species (Figure 13.1). The agroecosystem therefore represents a tritrophic hierarchy. All of these categories of living organism are amenable to biological manipulation using the latest techniques in genetic engineering. Clearly, the best-known group of manipulated species comprises crop plants. However, attempts have also been made to genetically manipulate both pests and predators. Predators can be manipulated either to increase their functional efficiency or to improve their resistance to conventional pesticides. Pests can be manipulated so that their functional efficiency decreases. This is an approach which is particularly useful where vectors of disease are concerned. Specific examples of some of these genetically manipulated species are given later in this chapter.

Biotechnological methods

Biotechnology, as an aid to crop improvement, became scientifically established following developments in our understanding of genetics that took place in the early 1900s. These developments led to the production of novel high-yielding crop varieties that substantially boosted global agricultural output throughout the twentieth century (see also Chapter 10). However, developments in cell and molecular biology that have taken place since the early 1970s have permitted plant breeders to shift their attention from yield to pest control. It is these developments that are discussed in the following sections.

The main biotechnological methods that can be used to produce genetically modified organisms that are of relevance to pest control can be split into two (not mutually exclusive) categories: (1) those that involve tissue culture techniques; and (2) those that involve the use of recombinant DNA. The former techniques have been extensively developed in relation to the production of crop plants while the latter have been tried with plants, pests and predators. In many cases both of these techniques will be used to produce a GMO. For example, tissue culture techniques are often used to regenerate plants that have been produced using recombinant technology. A brief overview of these techniques is given in the following sections.

Tissue culture

Cell or tissue culture techniques have been used most extensively with plant tissue. However, there are invertebrate and vertebrate cell lines that can be cultured and it is likely that *in vitro* regeneration of whole species will become a reality at some point in the near future. One of the principal advantages of using these techniques with plants is that advantageous traits can be selected in the laboratory. The

Table 13.1. *Tissue culture techniques*

Technique	Application	Example
Protoplast fusion	Production of somatic hybrids – novel plants from species that would not cross in the wild	1. Herbicide-resistant potato plants. 2. The pomato – cross between potato and tomato plants
Clonal propagation	Culturing protoplasts to produce uniform plants	Disease-free potatoes and strawberries
Somaclonal variation	Culturing protoplasts with useful traits that are normally hidden	Research with a range of crops currently under way
Mutant selection	Culturing protoplasts and stressing them. Surviving cells are selected	Herbicide-resistant maize.

screening process can therefore be completed relatively quickly in contrast to having to consider whole plants in a field-based situation. One related offshoot of this is that less space is required for experimentation. It is important to realise that novel species that are produced using tissue culture techniques are not transgenic species in the strict sense of the word. Traits that are developed using tissue culture techniques must be selected from those that preexist in the genes of the organisms under manipulation. The overall aim of most tissue culture is to regenerate whole crop plants from a few cells with desirable qualities. A summary of the techniques involved is provided in Table 13.1. At present, the use of these techniques has been primarily restricted to producing plants that are resistant to herbicides or that have improved agronomic qualities. On their own, these techniques have not been used to produce species that are useful to arthropod pest control. However, tissue culture has been used in combination with recombinant techniques to produce crop plants that are resistant to insects.

Recombinant DNA technology

Recombinant DNA technology has now been used to produce transgenic crop plants, transgenic pest species and transgenic predators. Of these, only

Table 13.2. *Recombinant techniques for the production of transgenic plants*

Technique	Application	Example
Agrobacterium-based plant transformation	Ti-plasmid (tumour-inducing plasmid) used to carry novel DNA into plants	Bt – Insect-resistant crop plants (tobacco, corn, cotton)
Particle acceleration	DNA-coated gold particles fired into growing tissue	Used to produce transgenic soybean
Electroporation	Electric current used to alter protoplast membranes permitting DNA uptake	Used to produce transgenic rice
Microinjection	DNA injected into the nucleus or cytoplasm of a protoplast	Used to produce transgenic tomato

Note:
Bt, *Bacillus thuringiensis*.

transgenic crop plants have reached the stage of commercial application. Specific examples of each of these transgenic species will be given later in this chapter. In this section a brief description of the technology itself is provided.

The main techniques that can be used to produce transgenic species comprise *Agrobacterium*-based methods, particle acceleration (sometimes also referred to as biolistics), electroporation and microinjection. All of these techniques can be used with plants but only microinjection has been used successfully with animals. By far the most widely used technique comprises the use of *Agrobacterium* (*c.* 65% of all transformations). This is the bacterium that is responsible for crown gall disease in plants. Typically, the *Agrobacterium* system comprises a binary vector where one vector encodes the genes to be transferred and the other encodes genes for the transfer process itself. The system uses a 'disarmed' Ti plasmid (tumour-inducing plasmid), with the genes for crown gall disease removed. A summary of all the techniques for introducing novel DNA is provided in Table 13.2, whilst Figure 13.2 shows the steps or processes involved in producing a transgenic organism.

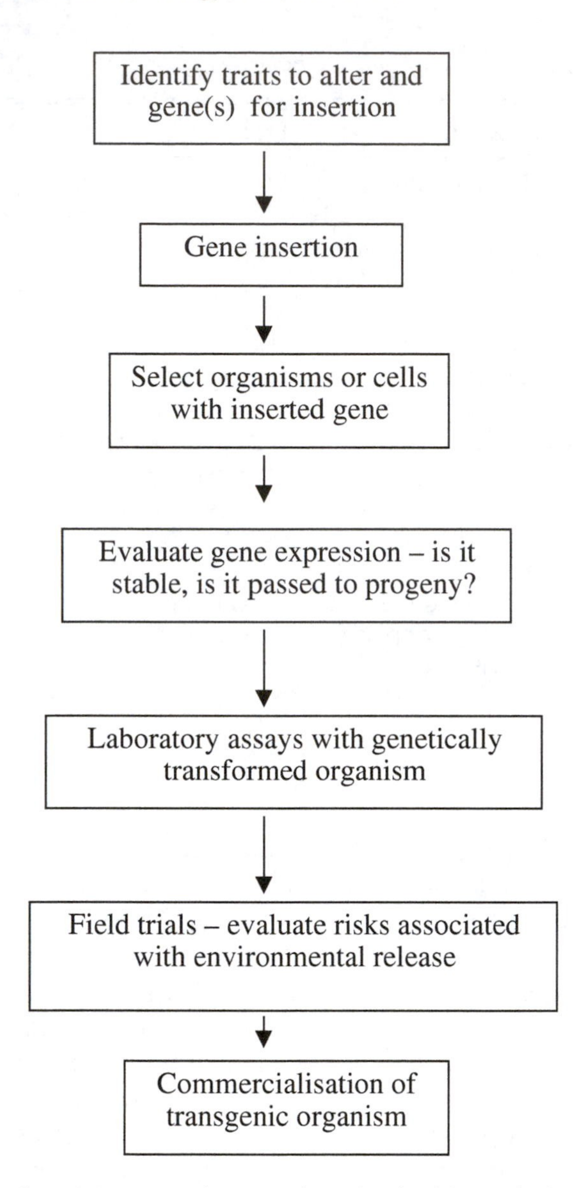

Fig. 13.2. Stepwise procedures involved in producing a transgenic organism.

The first stage is to identify and clone genes that are of interest. The regulatory sequences for the gene of interest must also be identified so that gene expression is controlled. The next stage is then to insert the gene of interest into the organism to be modified. Following insertion, expression of the gene must be assayed, i.e. is expression stable, appropriate and is the expression stably transmitted to progeny? If this occurs then the organism produced is said to be transformed. A typical transgene will be carried by a vector that incorporates promoters, marker genes and regulatory sequences, in addition to the actual gene for the trait desired *per se*. Following transformation, the next stage is to check field behaviour. In most cases this will take a number of years. At the time of writing, concerns about the field behaviour of genetically manipulated crops had delayed their commercialisation in Europe. On a worldwide basis there have been no field releases of transgenic arthropods.

It is important to appreciate that throughout the twentieth century (prior to developments in genetic engineering) plant geneticists and breeders used a range of techniques to develop new gene combinations within plants. These techniques included the artificial manipulation of chromosome number, the development of addition and substitution lines for specific chromosomes, chemical and radiation treatments to induce mutations and chromosome rearrangements and various tissue culture techniques that permitted the recovery of interspecific and intergeneric hybrids. Manipulating the genome of plants is therefore not new. What is novel with recombinant techniques is the speed with which new plants can be developed and the pool from which introduced genes can be selected. Cloned genes can originate from any source of DNA, including plants, animals, microbes or even entirely synthetic sequences. This novel technology has therefore drastically changed what breeders can do and offers enormous potential for crop improvement and pest control. Moreover, the technology also permits far finer control of the whole species improvement process since it is usually single genes (monogenic traits) that are the focus for manipulation.

Transgenic plants

In 1993, the very first transgenic plant to receive approval in the USA for commercialisation was the Flavr Savr tomato. This tomato was engineered using *Agrobacterium*-mediated transfer to contain an antisense gene derived from the bacterium *Escherichia coli*. This gene functions to reduce natural levels of the enzyme polygalacturonase. Reducing levels of this enzyme delays fruit ripening and, it is claimed, improves tomato flavour. Unfortunately, these nascent transgenic tomatoes were not a commercial success because the advantages of delayed ripening were lost during shipping and packing. Subsequent developments with transgenic tomatoes have however proved to be a commercial success.

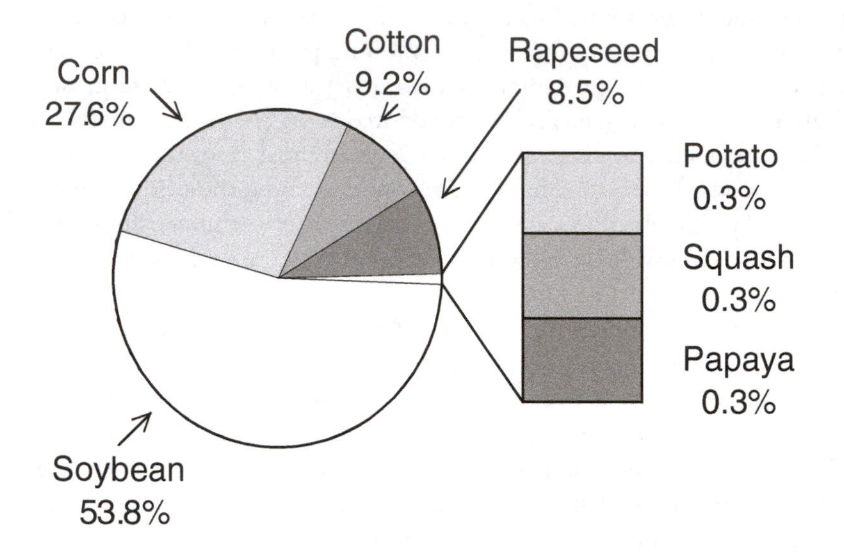

Fig. 13.3. Worldwide production of transgenic crops in 2001 by area. Total area = *c.* 52 000 000 hectares.

Since this first approval more than 50 different transgenic crops have been field-tested in either Europe or the USA and some of these crops have reached the stage of commercialisation. By the year 2000 over 1000 field trials had been carried out in the European Community (EC) and over 3000 had been carried out in the USA. Field trials with transgenic crop plants have also taken place in China, Hungary, India and Russia. Worldwide, approximately 52 million hectares were planted with transgenic crops in 2001. By far the most extensively grown crops were soybean, corn (maize), cotton and rapeseed (canola). In 1999 *c.* 60% of the Canadian rapeseed crop was transgenic, as was 50% of the US soybean crop, 50% of the US cotton crop and 40% of the US corn crop. Figure 13.3 gives a breakdown of the global percentages.

The transgenes that have been introduced to food crops have effects upon product quality, agronomy and resistance to pests and pathogens (Table 13.3). For those transgenic crops that are grown commercially by far the most common trait is for herbicide resistance (Figure 13.4). The next most common trait is for resistance to insects, the focus of this book. In 1999 US farmers alone planted *c.* 8 million hectares with corn plants that were resistant

Table 13.3. *Areas for improvement using genetic engineering with crops*

Research area	Factor
Product quality	Carbohydrate metabolism
	Colour
	Durability
	Fatty acid metabolism
	Fruit ripening
	Processing value
Pest resistance	Insect resistance
	Bacterial resistance
	Fungal resistance
	Virus resistance
	Nematode resistance
Agronomic traits	Drought resistance
	Herbicide tolerance
	Salt tolerance
	Temperature resistance
	Nitrate reduction
	Heavy metal tolerance

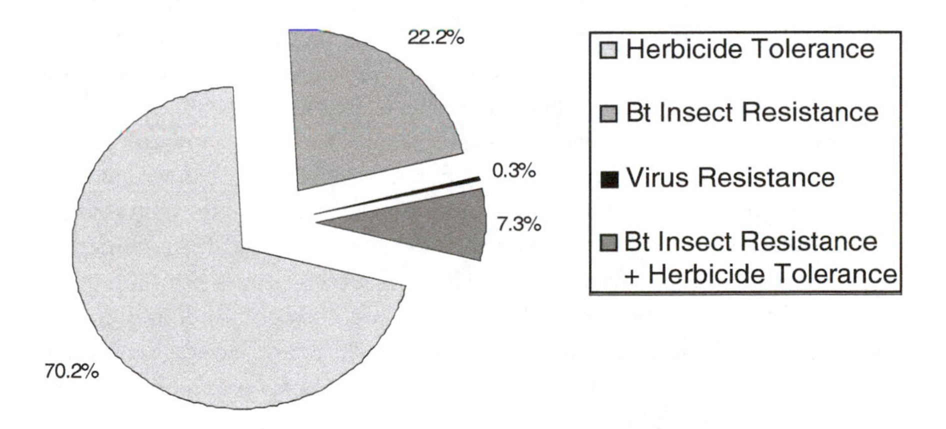

Fig. 13.4. Transgenic traits by category for all crops planted in 1999.

Table 13.4. *Commercially available transgenic crops for arthropod pest control (2000)*

Crop	Pest species
Corn	European corn borer
	Corn earworm
	Southwestern corn borer
Cotton	Tobacco budworm
	Cotton bollworm
Potato	Colorado potato beetle

to arthropod pests. This is about 20% of the total corn hectareage. All of these plants were engineered to express the *Bacillus thuringiensis* (Bt) endotoxin (see also Chapter 6).

Insect-resistant plants

All of the commercially available transgenic plants that are resistant to insects use the deltaendotoxin produced by *B. thuringiensis*. The genes encoding this toxin were first cloned in the early 1980s and the first plants to incorporate these genes were tobacco and tomato. These first examples of plants with engineered resistance to arthropod pests were produced in the mid-1980s. Expression levels of toxin within these first plants were too low to provide adequate protection. However, subsequent research has demonstrated that, by using strong promoter genes, toxin expression levels can be raised by a factor of 100. These are the plants that have now been developed for commercial use. The plants that are currently available on a commercial basis are corn, cotton and potato (Table 13.4). Plants which are currently undergoing field trials include alfalfa, apple, broccoli, chickpea, eggplant (aubergine), larch, peanut, poplar, rice, soybean, sugarcane, tomato and walnut. Most of these plants express the toxin throughout their tissues for the duration of the growing season. This provides season-long protection to both existing and new foliage. Different deltaendotoxins have been incorporated for expression in these plants. For example, Cry1A and Cry1C proteins are toxic to lepidopteran pests while Cry3A proteins are toxic to coleopteran pests (see Chapter 6). Pests that feed on these crops die within a few days of ingesting the toxin. It has been claimed that the use of transgenic cotton plants led directly to a

Table 13.5. *Insecticidal proteins under evaluation for arthropod pest control*

Serine protease inhibitors
Thiol protease inhibitors
Lectins
Alpha-amylase inhibitors
Lipoxygenase
Acyl-hydrolase
Chitinase
Cholesterol oxidase
Ribosome-inactivating proteins

decline in pesticide applications to these crops in the USA. For example, in 1998 *c.* 450000 kg less pesticide was used on transgenic cotton than on conventional cotton.

In addition to the development and commercial release of Bt plants, recent research has looked at the toxicity of lectins. These are sugar-binding proteins that are produced by plants and are believed to provide protection against a variety of species, including important pests. The first demonstration of resistance in a transgenic plant expressing a foreign lectin used the glucose/mannose-binding lectin from a pea plant. Bioassays with tobacco plants that expressed this gene in the early 1990s indicated some toxicity towards lepidopterous pests. However, expression levels were not high enough to justify commercialisation.

More success has been had with lectins that are produced by snowdrops. These lectins are referred to as concanavalin A (Con A) and *Galanthus nivalis* agglutinin (GNA). These lectins are toxic to a number of pests, including homopteran, coleopteran and lepidopteran insects. They inhibit development and decrease fecundity. Recent studies have shown that the lectins bind to midgut proteins and are able to cross the gut lining. Genetically engineered potatoes that express these lectins have now been produced and are currently undergoing laboratory and field trials for efficacy and risk assessment. Similar research is also under way with grapes, rapeseed, rice, sweet potato, sugar cane, sunflower, tobacco, walnuts and tomato. It is highly likely that many of these plants will be released for commercial use during the next decade.

Other genes that have been transferred to plants and that are still in the research phase include those coding for chitinase, cholesterol oxidase and for proteinase and amylase inhibitors. Many of the inhibitory genes have been derived from plants themselves and, although they are able to reduce pest activity, none has been active enough to be deemed worthy of commercial development. A summary of bioassays with transgenic plants expressing insecticidal proteins is provided in Table 13.5.

Transgenic predators

From a pest control perspective, most research on genetically manipulated predators and parasitoids has focused on resistance to pesticides. The reason for this is that it is easier to manipulate traits that are controlled by single genes. Improving the functional efficiency and/or environmental tolerance of predators and parasitoids would benefit arthropod pest control. However, these traits are controlled by gene complexes (polygenic traits) and are therefore difficult to manipulate.

Most of the fundamental research on transgenic arthropods has been carried out with *Drosophila melanogaster*. This research, which began in the early 1980s, showed how transposable elements (mobile units of DNA) could be manipulated to serve as vectors to carry novel DNA into germline cells. Since this work, many genes have been cloned and inserted into *Drosophila*. Most of these genes are not relevant to pest control. A list of pesticide resistance genes that may be of interest in the future is provided in Table 13.6. All of these resistance genes have been cloned and experiments are under way to develop techniques for incorporating them into laboratory-reared insects. For example, the *opd* gene for parathion resistance has been incorporated into laboratory-cultured *Spodoptera frugiperda* (fall armyworm – a pest!). To date, there are no effective examples of genetically engineered predators expressing a resistance gene.

Clearly, the goal of using genetically engineered pesticide-resistant predators would be to develop integrated strategies for pest control that used both pesticides and predators. This is an approach that has already been of use in orchards where pesticide-resistant predatory mites (developed conventionally[1]) can be used with pesticides to control phytophagous mite species. At present there is no effective example of a transgenic predator that can be used in pest control. Because of this, the risks associated with releasing such an organism have not yet been looked at in detail, although it seems likely that

[1] There are now a number of pesticide-resistant predators that are available for use in pest control programmes. These species have been produced by conventional breeding. See Chapter 12 and glossary for Chapter 12 for more details.

Table 13.6. *Some cloned pesticide resistance genes of potential use for the manipulation of beneficial arthropod species*

Gene	Pesticide resistance
Acetylcholinesterase (*Ace*)	Organophosphate/carbamate resistance
β-Tubulin	Benomyl (fungicide) resistance
γ-Aminobutryric acid A ($GABA_A$)	Dieldrin resistance
Cytochrome P450-*B*1	DDT resistance
Esterase *B*1	Organophosphate resistance
Glutathione-*S*-transferase (DmGST 1-1)	DDT resistance
Glutathione-*S*-transferase (MdGST1)	Organophosphate resistance
Parathion hydrolase (*opd*)	Parathion resistance

Note:
DDT, dichlorodiphenyltrichloroethane.

this situation will change. However, it will probably be at least 10 years before the first transgenic arthropods are released into the environment.

Transgenic pests

Although transgenic arthropods have not yet been released into the environment (see above), a great deal of research is currently under way that is concerned with manipulating species that act as vectors (carriers) of plant and animal disease. If single genes can be identified that are critical to a species' vectoring capacity, then these could potentially be manipulated and introduced to wild populations. It is known that the vectoring process itself is not necessarily of benefit to the host, so removal of that trait may not have an impact on fitness. It may even be the case that the fitness of the species concerned improves. The idea of course would be to hinder or prevent successful vectoring of pathogens by pests.

Most studies so far have focused upon resistance to pathogens in vector species and on incompatibility between the vector and the pathogen. For example, it is known that resistance to *Plasmodium* (causal agent of malaria) in mosquitoes is due to an encapsulation reaction. This reaction is apparently bypassed in mosquitoes that are susceptible to the pathogen. Increasing the frequency of the resistance gene(s) in natural populations would therefore be the goal. Another approach that has been suggested is to introduce genes coding for antibodies to *Plasmodium* in the natural populations. If strains of mosquitoes can be developed that are unable to transmit the pathogen then it is proposed that these refractory populations could replace field populations.

At present, the above technologies remain firmly in the laboratory. However, it seems likely that the use of genetically modified pests with reduced vectoring capacity may prove fruitful in the future. Success has already been achieved with replacement experiments using nonvector species produced by conventional breeding. For example, *Aedes* mosquitoes on Polynesian islands that act as vector for dengue fever (a viral disease) have been replaced with nonvector species. It therefore seems certain that similar success will be achieved with engineered species in the future.

Advantages associated with biotechnology for pest control

The advantages that are associated with using transgenic organisms for arthropod pest control cannot be overestimated. If this technology leads to improved control of pests of crops and vectors of disease, then the benefits will be enormous. Higher crop yields and improved disease control are essential for the future survival of many people on our planet (see also Chapter 1). The world population is expected to increase substantially in some of the least agriculturally productive, disease-ridden countries. Food production will therefore have to expand enormously in these countries if mass starvation is to be avoided. Although the planet currently produces enough food to feed the global population,[2] 20 years from now this may not be the case. Simply put, there may not be enough food to go around.

In addition to yield increases, genetically engineered crop plants may also improve the quality of food that people consume. For example, transgenic rice strains have been produced that express the precursor to vitamin A.

[2] Although enough food is produced on a global basis to feed the current world population, there are clearly socioeconomic and political factors that prevent adequate access to this food. In general, many developing countries are food-poor while enormous calorific wastage occurs in most so-called developed countries.

Table 13.7. *The benefits of transgenic organisms for pest control*

Direct exposure of pest species to toxins
Reduced environmental contamination by pesticides
Reduced operator exposure to pesticides
Effective pest control throughout the plant
Effective pest control throughout the growing season
Use of predators and pesticides in integrated control programmes[a]
Reduced vectoring of diseases[b]

Notes:
[a] Transgenic pesticide-resistant predators.
[b] Refractory vector populations replace natural vector populations.

Dietary deficiencies of this vitamin are estimated to cause 5 million cases of an eye disease called xerophthalmia in children every year. At present, about 5% of these children become permanently blind. This transgenic rice has been referred to in the popular press as 'golden rice' and has apparently found favour with some of the pressure groups that have been active in decrying transgenic plant technology! Research is also under way to manipulate the iron content of rice, to manipulate its ability to fix nitrogen and to develop crops that express vaccines. It has been claimed that growing crops that deliver vaccines could do more to eliminate disease worldwide than the Red Cross, missionaries and the United Nations task force combined, and at a fraction of the cost. Transgenic bananas have now been developed that contain the inactivated viruses that cause cholera, hepatitis B and diarrhoea. These crops are currently under evaluation. If the technology also leads to a reduction in environmental contamination by pesticides then the benefits will also be enormous. For example, it is known that during a typical pesticide application a very high proportion of the formulation that is released into the environment is wasted (see Chapter 4). Accurate targeting of pesticides is difficult because most pest species are small. By using transgenic technology, particularly with plants, the toxin is delivered directly to the target organism. The fact that the plant delivers the toxin and not a person also serves as a means to reduce operator exposure to pesticides. Finally, because pesticides are not widely dispersed in the environment their overall effects on beneficial nontarget species are likely to be reduced. Table 13.7 lists some of the benefits that have been associated with using genetically engineered organisms for arthropod pest control.

Although a reduction in pesticide use would be welcomed, recent data have begun to challenge whether this actually occurs. For example, Bt corn, which provides protection from the European corn borer (ECB), was planted on about 10 million hectares worldwide in 1999. Data from the USA however indicate that the ECB is not a pest every year. When an economic analysis is undertaken it indicates that using transgenic varieties is really only sensible when very severe infestations exist, i.e. rarely. In fact, data from the US National Agricultural Statistics Service show that the area treated with pesticides for ECB control in the USA rose from 9.5% to 10.5% between 1995 and 1998, a period during which some 6 million hectares of Bt corn were planted in the USA. It is believed that the use of Bt corn has heightened farmers' perception of this pest and that they are now spraying crops that they would otherwise have left untreated. The use of transgenic crops may therefore have led to an increase in the use of pesticides![3]

Disadvantages associated with biotechnology for pest control

The disadvantages associated with using transgenic arthropods for pest control are at present constrained by the fact that technical issues surrounding their development have yet to be resolved and their efficacy *per se* has yet to be demonstrated. For example, some researchers have questioned whether the use of refractory mosquito populations is likely to be feasible given that the proportion of infective individuals in wild mosquito populations is usually less than 5%, i.e. the selective pressure for refractoriness is likely to be small. There are also risk assessment issues that are associated with the movement of DNA between species. For example, it is known that horizontal transfer of genes can occur in wild populations. The frequency or significance of this remains to be quantified. The upshot of this is that it may be some years before the first field release of a transgenic arthropod happens. However, if transgenic species are developed that are able to make a substantial impact on disease transmission it is likely that they will be rapidly deployed.

In contrast, transgenic plants are already commercially planted and the risks associated with using these organisms are becoming increasingly well-understood. These risks essentially fall into six different categories. These risks

[3] This last paragraph is hardly a description of an advantage associated with using transgenic crops. The paragraph is discursive and is included here because the literature contains no general references that suggest transgenic crops increase pesticide usage (despite the comments made here). In the particular case discussed it should be a relatively straightforward process to educate farmers to reduce their pesticide applications, at least back to the levels they were at before transgenic technology crops were widely grown.

Table 13.8. *Putative disadvantages associated with using transgenic plants for pest control*

The toxicity of transgenic plants to nontarget species
The invasiveness of transgenic plants
The potential for horizontal movement of the transgene
The toxicity of transgenic plants to humans
Concerns surrounding antibiotic marker resistance genes in plants
The development of pest species resistance to transgenic plants

(or putative disadvantages) are listed in Table 13.8 and are discussed in the following sections of this chapter.

Toxicity of transgenic plants to nontarget species

It has been suggested that nontarget (beneficial) species may be adversely affected by exposure to toxins expressed in transgenic plants. Possible routes of exposure include the consumption of poisoned pest species and direct exposure to exudates from plant residues that are ploughed into the soil. This latter route could have harmful effects on both beneficial invertebrates and on soil microbial activity. A great deal of publicity was generated when it was shown that monarch butterflies may be adversely affected by exposure to *B. thuringiensis* toxins in transgenic plants. This result was hardly surprising, since the caterpillars concerned were force-fed in a laboratory and it was well-known that the deltaendotoxin concerned was poisonous to lepidopterous species. Since comparisons with the adverse effects of pesticides were not made in this work, it is not very helpful!

Research in Scotland looked at the effects on ladybirds of consuming aphids that had fed on genetically modified potatoes expressing the GNA lectin. The result was that these ladybirds laid fewer eggs and lived half as long as normal ladybirds. Since ladybirds are important predators in many agroecosystems, this result may be significant. However, as with the studies on the monarch butterfly, no direct comparisons with the adverse effects of pesticides were made. Preliminary studies on the effects of potato GNA lectin on soil microbial activity indicated a transient and unimportant decline in activity. The effects did not persist from one season to the next when controlled field experiments were undertaken. Studies in the USA on Bt corn have shown that toxic root exudates are able to persist by binding to surface-active particles on soils. The significance of this is not yet clear as residues could improve the control of soil pests while also posing a threat to nontarget organisms.

Other studies have shown how *B. thuringiensis*-based transgenic crops can promote biodiversity and help the survival of beneficial predators. For example, studies in transgenic corn have shown that populations of beneficial species tend to be higher than in conventional corn crops where pesticides are used. This is likely to benefit both small mammals and birds, but these measurements remain to be made. In another study it was shown that the parasitic wasp *Cotesia plutella* used chemical cues given off by damaged rape plants to locate its host, the diamondback moth, *Plutella xylostella*. Where this crop was growing and resistant pests were present, the wasp and the crop could effect integrated control. Where susceptible pests were present the wasp was not attracted because too little damage was caused to the crop. At present, it seems that when comparisons with pesticides are made, any adverse effects of transgenic crops on nontarget species will be minimal. There may even be circumstances where transgenic technology can promote the survival of these species.

The invasiveness of transgenic plants

Transgenic plants that are resistant to arthropod pests have a selective advantage that may lead them to become weeds. For example, oilseed rape, the focus of several transgenic projects, survives extremely well in the wild. It therefore seems inevitable that this plant will colonise small-scale horticulture and allotments once transgenic varieties are extensively grown. Once this occurs, the primary issue becomes its ecological significance. Oilseed rape apart, it has been argued that many crop plants have been so manipulated by plant breeders that they have very little chance of survival outside conventional agroecosystems. Furthermore, most studies concerning this issue have concentrated on herbicide-resistant plants rather than arthropod-resistant plants and to date there is no good evidence that invasive plants cannot be controlled using other methods. However, the movement of the transgene itself, through hybridisation with wild relatives, is perhaps of greater concern.

Horizontal movement of the transgene

It is known that horizontal movement of transgenes can occur. If these genes carry a selective advantage (such as resistance to arthropod pests) then new hybrid plants may become invasive. This issue has been of particular concern to organic farmers in the UK who regard transgene movement as a form of genetic pollution. Whether the horizontal movement of transgenes is of any great ecological significance is still unresolved. However, this process may be of sociological and economic significance if consumers, who are unable to avoid transgenic rape oil, simply stop buying this product.

It is easier to measure gene flow than to predict its impact. Studies in the UK with pollen from oilseed rape have shown that it will move at least 4 km from the nearest known source. If regulations concerning the purity of nonmodified crops are stringent then it is likely that opportunities for growing genetically modified crops with high levels of pollen dispersal will be severely constrained. At the time of writing, these issues had still to be resolved. In conventional farming systems gene movement occurs anyway and it is difficult to see how a transgene in an agricultural crop would pose any additional problems, despite its putative selective advantage. Genes for resistance to arthropod pests that impacted on nontarget species feeding on prey species may affect biodiversity. However, this hazard needs to be compared with the harmful effects of pesticides.

The toxicity of transgenic plants to humans

It has been claimed that there are a number of risks to humans associated with eating transgenic food crops. These risks include: (1) novel proteins acting as allergens or toxins; (2) altered host metabolism producing new and unknown allergens or toxins; and (3) reduced nutritional quality leading to dietary deficiencies or health problems, e.g. by reductions in antioxidants in food plants. At present these risks seem alarmist. All of the genetically engineered food plants that have so far been commercialised incorporate proteins that have no structural similarities to those known to cause allergies. These proteins are also heat-, acid- and enzyme-sensitive, i.e. they are easily digestible. Current regulations also require companies to assay food products for their safety. To date, there have been no food safety problems with commercialised genetically modified crops. Where problems are identified, research is usually stopped at an early stage. For example, a project to insert a Brazil nut protein into a soybean plant was halted when tests indicated that the people allergic to the nuts reacted to the modified soy products. Moreover, it has been argued that recombinant techniques are in fact far safer than conventional breeding techniques since specific genes of known character are involved (i.e. monogenic manipulation occurs). Traditional breeding, by contrast, involves the transfer of collections of genes of undefined character (i.e. polygenic manipulation occurs). That said, the fundamental problem is that we simply do not know yet whether engineered food may present more or less risk to consumers.

Antibiotic-resistance marker genes

Marker genes are used by genetic engineers to select plants that have been transformed. Their use precludes the need to wait until a fully formed plant has developed to check whether gene incorporation has occurred. One of the most widely used marker genes has been the *kan-r* gene which encodes an enzyme providing resistance to the antibiotic kanamycin. More specifically, this gene encodes aminoglycoside 3'-phosphotransferase, an enzyme that catalyses the transfer of a phosphate group from adenosine triphosphate to a hydroxyl group of the aminoglycoside antibiotics, including kanamycin. This transfer inactivates the antibiotic. It was this marker gene that was used in the Flavr Savr tomato and it has also been used in the majority of other transgenic crops that have been developed to date.

The use of these marker genes has led to the suggestion that they may be transferred to gut epithelial cells, to gut bacteria and to organisms in the environment. There is a concern that their use may ultimately enhance the development of bacterial resistance to antibiotics. Although there is no evidence that this occurs, scientists have responded and begun to use alternative marker genes such as the green fluorescence protein (GFP) gene which makes plants fluoresce under ultraviolet light. At present, regulatory bodies in Europe and North America have concluded that the use of marker genes is justified because the associated risks (described above) are minimal. Most of these organisations have also indicated that they would be unlikely to approve a food product that contained a marker gene that coded for resistance to an antibiotic that was the only treatment available for a clinical condition, e.g. vancomycin for treatment of certain staphylococcal infections. For the time being broad-spectrum antibiotic marker genes will remain in use.

Pest species resistance to transgenic crops

By far the biggest problem that is likely to occur following the use of transgenic crop plants for pest control is the development of pest species resistance. This occurs of course with conventional insecticides (Chapter 4). However, the critical difference is that a pesticide application represents a selective force that is time-limited, i.e. to the time the application is made.

Current transgenic crops express toxins throughout their tissues continuously and the selective pressure on the target population to adapt will therefore be substantial. Given that most pest species are highly fecund, this will only serve to speed up the process of resistance development. It is already known that pests can develop resistance to the *B. thuringiensis* endotoxin (by reduced affinity of gut receptors to the toxin) so there is no reason not to expect this to happen in the field. Indeed, field populations of diamondback moth that were treated with excessive doses of conventional microbial sprays rapidly developed resistance to the bacterial endotoxin. The problem that will then occur is that a conventional microbial pesticide that has been in use for over 60 years will suddenly become redundant. This will hit organic farmers particularly, as this is one of the few products that they can use for pest control (see also Chapter 11).

It is now generally agreed that resistance development will occur and so the focus of attention has switched to resistance management (not withdrawal of these crops!). At present, the primary strategy comprises the use of refugia where conventional crops are grown. The idea is that continual interbreeding with susceptible pests will slow resistance development. From 2000 the Environmental Protection Agency (EPA) in the USA has mandated that all farmers who grow Bt corn must plant at least 20% of their area with a non-Bt variety. Whether this strategy will work remains to be seen. The refugia strategy is based on compliance and on resistance being recessive. There is now evidence in the USA that compliance is not 100% and, more worryingly, that resistance of ECB to Bt corn is genetically dominant. If this is correct, then resistance will be passed on and the refugia strategy will become ineffective at slowing resistance development. If farmers treat refugia with insecticides to prevent economic losses then the number of susceptible individuals that remain in a population will be reduced anyway.

In the longer term, genetic engineers are proposing that more sophisticated plants are produced. These will be plants that express multiple toxins (the inundative strategy) and/or plants that slow resistance development using tissue-specific or phenologically specific expression systems. These latter plants would not stop resistance development but would behave more like conventional pesticides in that the selective force would be of short duration.

Predictions for future developments

There is no doubt that tremendous progress has been made in applying novel cell and molecular techniques to arthropod pest control. These developments

Table 13.9. *Number of field tests of transgenic crops in the European Community (EC) and the USA as of 1998*

	EC	USA		EC	USA
Maize	192	1019	Barley	0	6
Tomato	45	321	*Arabidopsis*	0	6
Soybean	6	278	*Amelanchier*	0	6
Potato	86	261	Cranberry	0	6
Cotton	1	191	Eggplant (aubergine)	0	6
Squash (marrow)/melon	2	106	Gladiolus	0	6
Tobacco	30	98	Watermelon	0	6
Rapeseed	188	57	Walnut	0	6
Sugarbeet	109	23	Sweetgum	0	6
Alfalfa	2	18	Sweet potato	0	6
Wheat	6	14	Sugarcane	0	6
Rice	0	13	Spruce	0	6
Cucumber	0	12	Pepper	0	6
Sunflower	6	8	Peanut	0	6
Bentgrass	0	7	Pea	0	6
Chicory	32	6	Papaya	0	6
Poplar	6	6	Apple	1	5
Lettuce	4	6	Strawberry	1	5
Grapes	2	6	Plum	0	4
Carrot	1	6	Marigold	8	0
Chrysanthemum	1	6	Cauliflower	5	0
Petunia	1	6	Eucalyptus	3	0
Onion	0	6	Carnation	3	0
Broccoli	0	6	Silver birch	1	0
Belladonna	0	6			

have just reached the stage of commercialisation with the first field releases of insect-resistant plants taking place during the 1990s. Because these novel genetic engineering techniques can make a substantial impact on the time it takes for a plant breeder to develop a new variety, there is no reason to expect that this trend to develop and commercialise transgenic crop plants will change. Table 13.9 lists all of the field trials that had been undertaken in the EC and USA as of 1998. Only a small proportion of these plants are commercially available at present (Table 13.10).

Genetic engineering also gives breeders far more control over the novel

Table 13.10. *Genetically modified crops deregulated by the US Department of agriculture as of January 2000*

Crop	Company[a]	Genetic Modification	Approved
Tomato	Calgene	Fruit ripening altered	1992
Squash (marrow)	UpJohn	Watermelon mosaic virus 2 and zucchini mosaic virus-resistant	1992
Cotton	Calgene	Bromoxynil-tolerant	1993
Soybean	Monsanto	Glyphosate-tolerant	1993
Rapeseed	Calgene	Altered oil profile	1994
Tomato	DNA Plant Technology	Fruit ripening altered	1994
Potato	Monsanto	Coleopteran-resistant	1994
Tomato	Zeneca	Reduced polygalacturonase	1994
Cotton	Monsanto	Lepidopteran-resistant	1994
Corn	Ciba-Geigy	Lepidopteran-resistant	1994
Corn	AgrEvo	Phosphinothricin-tolerant	1994
Cotton	Monsanto	Glyphosate-tolerant	1995
Tomato	Monsanto	Fruit ripening altered	1995
Corn	Monsanto	Lepidopteran-resistant	1995
Corn	DeKalb	Phosphinothricin-tolerant	1995
Corn	Northrup	European corn borer-resistant	1995
Corn	Plant Genetic Systems	Male sterile	1995
Cotton	Du Pont	Sulphonylurea-tolerant	1995
Tomato	Agritrope	Fruit ripening altered	1995
Potato	Monsanto	Colorado potato beetle-resistant	1995
Squash (marrow)	Asgrow	Watermelon mosaic virus 2, cucumber mosaic virus and zucchini mosaic virus-resistant	1995
Corn	Monsanto	European corn borer-resistant	1996
Papaya	Cornell University	Papaya ringspot virus-tolerant	1996

Table 13.10. (*cont.*)

Crop	Company[a]	Genetic Modification	Approved
Soybean	AgrEvo	Phosphinothricin-tolerant	1996
Corn	DeKalb	European corn borer-resistant	1996
Corn	Monsanto	Glyphosate-tolerant and European corn borer-resistant	1996
Soybean	Du Pont	Altered oil profile	1997
Cotton	Calgene	Bromoxynil-tolerant and Lepidopteran-resistant	1997
Corn	Monsanto	Glyphosate-tolerant	1997
Chicory	Bejo	Male sterile	1997
Potato	Monsanto	Colorado potato beetle and potato leafroll virus-resistant	1997
Rapeseed	AgrEvo	Phosphinothricin-tolerant	1997
Corn	AgrEvo	Phosphinothricin-tolerant and Lepidopteran-resistant	1997
Tomato	Monsanto	Lepidopteran-resistant	1997
Beet	AgrEvo	Phosphinothricin-tolerant	1997
Potato	Monsanto	Colorado potato beetle and potato virus y resistant	1997
Corn	Pioneer	Male sterile and phosphinothricin-tolerant	1997
Beet	Novartis	Glyphosate-tolerant	1998
Rapeseed	Monsanto	Glyphosate-tolerant	1998
Soybean	AgrEvo	Phosphinothricin-tolerant	1998
Rapeseed	AgrEvo	Phosphinothricin-tolerant and pollination control	1998
Rice	AgrEvo	Phosphinothricin-tolerant	1998
Flax	University of Saskatchewan	Sulfonylurea soil residue-tolerant	1998

Note:
[a] Company names represent original applicants; many have since changed, following mergers, acquisitions, etc.
Source: Data abstracted from US Department of Agriculture website at http://www.aphis.usda.gov/biotech/not-reg.html. Data do not include extensions to existing applications for deregulation.

Table 13.11. *Current state of developments with transgenic plant technology*

Transgenic plant	Developmental status
Delayed tomato ripening	Commercialised
Herbicide tolerance	Commercialised
Insect pest resistance	Commercialised
Oil-rich Palms	Commercialised
Virus resistance	Commercialised
Drought resistance	Field trials
Salt resistance	Field trials
Biopolymers in plants	Research
Edible vaccines	Research
Lysine-rich cereals	Research
N_2-fixing cereals	Early research

material that they incorporate into plants. Plants of greater sophistication will be produced and commercialisation of these plants will speed up in the next 20 years. In the USA the main arthropod-resistant crops that are under development are corn that is resistant to corn rootworms (*Diabrotica* spp.) and grass that is resistant to various turf pests. Other transgenic crops under development are being manipulated for product nutritional quality and for vaccine delivery. Table 13.11 provides a breakdown of the state of development of transgenic plants and their products.

It seems likely that the USA will lead the way with this technology. From a regulatory point of view, the USA takes a product-oriented approach and requires transgenic plants to undergo food safety tests based on their plant characteristics. Risk assessments of crops in the USA are carried out by the government agencies concerned, i.e. Animal and Plant Health Inspection Service (APHIS), Food and Drug Administration, and EPA. By comparison the EC requires risk assessments for environmental release to be made by applicants. In the EC regulations are based on the EC directive 90/219/EEC. At the time of writing, and despite numerous field trials, no transgenic crops were extensively planted anywhere in Europe. In fact, a landmark judicial ruling in Scotland during 2000 appeared to give genetic protesters the right to destroy genetically modified crops. This ruling was made on the basis that these crops did represent a threat to the environment.

From a pest control point of view, whether this technology will be sustainable remains to be seen. There is a danger that, just as farmers became trapped on 'pesticide treadmills' during the 1950s and 1960s, so farmers in the twenty-first century may become trapped on 'biotechnological treadmills'. On these treadmills, plants of greater and greater sophistication will be required in order to sustain the battle against rapidly evolving arthropod pests. As with pesticides, this is likely to prove unsustainable.

There is no doubt that biotechnology will help food production in terms of yield and nutritional content but whether it will help with arthropod pest control is still not clear. Most crop plants of global significance are attacked by highly fecund pest complexes that have the potential for rapid resistance development. Insect pests will simply evolve to deal with the challenges that transgenic crops present. In western economies at present global food production does not appear to be a substantial issue. The argument that we need to use this technology to boost food production has therefore been difficult to make.

In contrast, substantial progress is likely to be made in the control of vectors of pathogenic organisms. The potential of this technology appears astounding. Although no field releases of transgenic invertebrates have been made, once the potential for disease reduction is demonstrated, these will follow rapidly. In the future, global warming is likely to have a substantial impact on disease transmission. If tropical diseases become more prevalent in western economies then the need for disease control will become more pressing. If the technology exists to control pests by using transgenic arthropods then the risks associated with their release will seem minimal in comparison to the number of lives that may be saved. This is a point of view that is still used to justify using dichlorodiphenyltrichloroethane (DDT) in the control of vectors of malaria. Therefore, there is good reason to expect that, once the technology is developed and if (western) human lives are at risk, it will be rapidly implemented.

Summary

This book has attempted to introduce the reader to the diversity of methods and techniques that are available to control arthropod pests. From this final chapter it should be clear that we are only just beginning to appreciate the benefits that may accrue from developments in genetic engineering. However, substantial progress has already been made in numerous other areas, including the often much maligned chemical industry. From a global perspective, millions, if not billions, of people probably survive because pests that have the potential to feed on crops and to transmit pathogens can be successfully controlled. The approach taken in this book was to introduce the reader to each of the methods separately. The aim was to provide the introductory information that would be needed when operating at the systems level. Clearly, although each chapter deals with a single category of control measure, many of these measures will require integration when it comes to their use in the 'real world'. In the future, the battle with arthropod pests will undoubtedly continue. It is hoped that this text may provide some context for this battle by introducing the basic tools that exist for combat with what is a numerous and formidable enemy.

Further reading

Carozzi, N. & Koziel, M. (1997). *Advances in Insect Control – The Role of Transgenic Plants.* London: Taylor & Francis.

Conner, A.J. & Jacobs, J.M.E. (1999). Genetic engineering of crops as potential source of genetic hazard in the human diet. *Mutation Research*, **443**, 223–34.

Gould, F. (1998). Sustainability of transgenic insecticidal cultivars: integrating pest genetics and ecology. *Annual Review of Entomology*, **43**, 701–26.

Hilder, V.A. & Boulter, D. (1999). Genetic engineering of crop plants for insect resistance – a critical review. *Crop Protection*, **18**, 177–91.

O'Brochta, D.A. & Atkinson, P.W. (1997). Recent development in transgenic insect technology. *Parasitology Today*, **13**, 99–104.

Persley, G.J. (1996). *Biotechnology and Integrated Pest Management.* Wallingford: CAB International.

Schuler, T.H., Poppy, G.M., Kerry, B.R. & Denholm, I. (1999). Potential side effects of insect-resistant transgenic plants on arthropod natural enemies. *TIBTECH*, **17**, 210–16.

Sharma, H.C. & Ortiz, R. (2000). Transgenics, pest management, and the environment. *Current Science*, **79**, 421–37.

Thacker, J.R.M. (1994). Transgenic crop plants and pest control. *Science Progress*, **77**, 207–29.

Van Emden, H.F. (1999). Transgenic host plant resistance to insects – some reservations. *Annals of the Entomological Society of America*, **92**, 788–97.

Glossary and commentary on text

Chapter 1

Abu Mansur

Abu Mansur Muwaffak ibn Ali al-Harawi. Persian phamacologist, who lived in Herat under the Samanid prince Mansur I ibn Nuh, who ruled from 961 to 976. He compiled a treatise on materia medica in Persian and between 968 and 977 wrote the *Book of the Remedies* (*Kitab al-abnyia 'an Haqa'iq al-adwiya*), which is the oldest prose work in modern Persian. It deals with 585 remedies (of which 466 are derived from plants, 75 from minerals, 44 from animals).

Agrochemical companies

As a result of consolidation there were seven agrochemical companies with global sales of over $1 billion in 2000. This contrasts with the 11 companies with sales above this mark in 1995. In 2000 the top seven companies, with sales in billions of dollars, were as follows: Syngenta ($5.9), Monsanto ($3.9), Aventis ($3.7), Du Pont ($2.5), Dow ($2.3), Bayer ($2.3) and BASF ($2.2). In April 2001 Aventis announced that it would sell off its crop science division and sale proposals were sent to most of the companies listed above. In October 2001 Bayer bought Aventis.

Classical scholars

Cato, Marcos Porcius Roman statesman, orator and writer, as well as a large-scale farmer. Wrote *De Agri Cultura* in 160 BC based upon his farming experiences. This is the oldest complete prose work in Latin.

Columella, Lucius Junius Moderatus Roman soldier and farmer who wrote extensively on farming. In the first century AD he wrote *De De Rustica* (on farming and country life) and *De Arboribus* (on trees). In 1745 these were translated into English and published as a textbook entitled *On Husbandry.*

Varro, Marcus Terentius Roman scholar and satirist who was a prolific author. Wrote on more than 70 different subjects in over 600 books. In the first century BC he wrote *Res Rustica*, a treatise that provided practical instruction on farming. This is his only complete work to have survived. This book was later translated into English.

Geological time

Conventionally divided into four major periods, identified from the fossil record as periods when explosive radiations of many new species follow mass extinctions. These major periods are further divided based on distinct, if less dramatic changes in the fossil record. See table below.

ERA	Period	Epoch	Age (millions of years)
Cenozoic	Quarternary	Recent	0.01
		Pleistocene	1.8
	Tertiary	Pliocene	5
		Miocene	23
		Oligocene	34
		Eocene	57
		Paleocene	65
Mesozoic	Cretaceous		144
	Jurassic		208
	Triassic		245
Palaeozoic	Permian		286
	Carboniferous		360
	Devonian		408
	Silurian		438
	Ordovician		505
	Cambrian		544
Precambrian			4600

Neolithic

By convention human development is split into a number of phases. The earliest phase comprises the palaeolithic or old stone age, a period that is approximately coextensive with the Pleistocene geologic era (see above). This time period finished between 40 000 and 10 000 years ago, when it was succeeded by the mesolithic. The mesolithic ends at the point where a decided switch from a hunter–gatherer to a settlement-based farming approach to existence had been made. This period is called the neolithic or new stone age. Mesolithic communities therefore existed in some parts of Europe up until 3000 BC, while neolithic communities had already developed in the Middle East some 7000 years beforehand. Neolithic communities subsequently developed into the urban civilisations that are usually described as bronze, and then iron age, in nature.

Rig Veda

The *Rig Veda*, written in Sanskrit, is the oldest of a set of four books collectively known as the *Vedas*. The other books are the *Yajur Veda*, the *Sam Veda* and the *Atharva Veda*. These books are all considered to be revealed texts that are sacred to the Hindu religion. Opinions vary as to the exact date of compilation or revelation of the four *Vedas*. However, most scholars believe that they are not more than 4000 years old.

r–K continuum

The terms *r* and *K* were first used by the statistician Raymond Pearl in the 1920s in the differential equation that describes S-shaped or 'logistic' population growth. The term 'logistic' had been coined by a Belgian mathematician Pierre-François Verhulst nearly a century earlier, while a Russian ecologist Georgii F. Gause used this equation to model data he had collected for his doctoral thesis during the 1930s. In the differential equation *r* represented the maximum rate of increase of a population, while *K* represented the upper limit to population growth. *K* has subsequently been referred to as the carrying capacity of the environment. These terms were later used by Rober MacArthur and E.O. Wilson in the 1960s to describe what they regarded as different life-history strategies for species. Species could be classified as either more or less *r*-selected or *K*-selected depending upon the durational stability of the habitat they occupied. A theoretical continuum was therefore envisaged from *r*-selecting to *K*-selecting environments. In the 1970s it was realised that this continuum may be useful in selecting different control strategies for pest species. See Southwood (1977; see Further reading) for details.

Trial by ordeal

The seeds of the calabar bean (*Physostigma venenosum*), a plant native to West Africa, contain the poisonous alkaloid physostigmine (also called eserine). This property was exploited by native people in what has been termed trial by ordeal. Accused individuals were made to consume a mixture of pounded seeds infused with water. If the accused vomited then he or she was declared innocent. If the accused died, he or she was declared guilty.

Victor Vermorel

The Vermorel name is most often associated with the cars that they designed at the start of the twentieth century. In addition, in 1880 Victor Vermorel designed one of the first pieces of equipment that could be used for pesticide application. His invention was a sprayer that applied copper sulphate sprays to vineyards in France. It was the design of this and other items of agricultural equipment that allowed the company to expand and develop into the automotive business.

Chapter 2

Ceveratrum alkaloids

To date, over 3000 alkaloids have been identified in over 4000 plant species. These alkaloids can be classified chemically, on the basis of the plant species they are found in, and on the basis of their toxicological effects. The ceveratrum alkaloids (or veratrum alkaloids) are toxic chemicals that have been found in plants in the Liliaceae. These chemicals cause repetitive discharges along nerve axons and appear to mediate their effects by binding to sodium channels.

Deification

Plants, trees and forests have been central to many native cultures as their source of food, shelter and other products. As such, in many cultures they are worshipped. For example, in India, Aranyani and Vana Durga are the goddesses of forests and trees, respectively. Deification is the process by which cultures give or attribute god-like status to other living species, in this case forests or trees.

LD_{50}

Typically, LD_{50} values are used to compare the relative toxicities of different chemicals. The reason that the dose (or concentration – LC_{50}) that causes 50%

mortality in a target population is used is that this can only be a single value. It is exactly halfway between the infinite number of doses that can cause 0% mortality and the infinite number of doses that can cause 100% mortality. Various toxicity classification systems have been devised based on LD_{50} values for pesticides and these are discussed more fully in Chapter 3.

Merck

Primarily a pharmaceutical company. Athough Merck supported the research that led to the discovery of ryanodine, it no longer markets this, or any other, insecticidal products.

Naphthoquinones

In April 2002 *BBSRC Business* (the quarterly magazine of the Biotechnology and Biological Sciences Research Council, UK) reported that manipulation of the gem-dimethyl moiety in naphthoquinones altered their toxicity to whiteflies and mites. This moiety is thought to enhance activity against arthropods and to reduce toxicity to mammalian species. The site of action of these molecules was identified as the mitochondrial complex III. The magazine reported that, because the mode of action and selectivity of these molecules are distinct from those of established insecticides, these discoveries could lead to the development of a new commercial insecticidal class.

NADH

Reduced form of nicotinamide adenine dinucleotide (NAD^+). This is a coenzyme that accepts two protons and one electron from a fuel susbstrate (e.g. glucose) in a dehydrogenase-mediated reaction. Each NADH molecule represents stored energy that can be used to produce adenosine triphosphate (ATP) in the mitochondrial electron transport chain, prior to the release of oxygen. Inhibition of this process reduces oxygen uptake at the cellular level.

Secondary plant substances

Substances that are deemed essential to plant growth and development are often called primary metabolites. Compounds that appear less essential have been referred to as secondary metabolites or secondary plant substances. Many of these compounds are toxic, which is why they have found uses in pest control. As we learn more about plant metabolism, the above division seems less applicable. This is discussed in more detail in Chapter 10.

Voltage-dependent sodium channel

Channels within the nerve membrane whose functioning depends on the axon membrane potential. These channels appear to be disrupted by natural pyrethroids, by veratrum alkaloids (see above), and by various isobutylamides.

Chapter 3

Acetylation, phosphorylation, carbamylation

Names given to the reactions in which the neurotransmitter acetylcholine or an organophosphate or carbamate insecticide combine with the enzyme acetylcholinesterase, respectively.

Atropine

Chemical used in the treatment of poisoning with organophosphate or carbamate insecticides. Atropine is an anticholinergic compound that blocks muscarinic acetylcholine receptors in the parasympathetic nervous system. This prevents acetylcholine from binding to these receptors, i.e. preventing acetylcholine from functioning. The amount of atropine administered requires careful supervision since atropine is a poison in its own right.

Autooxidation

In natural pyrethroids long unsaturated carbon chains can act as sites of autooxidative attack because of a process called resonance stabilisation. In essence, the attacked molecule is stabilised because it can exist in a number of forms. The end result of this process is breakdown of the molecule. Removing sites of autooxidative attack prevents this process occurring.

Cholinergic receptors

Cholinergic receptors are divided and named on the basis of their sensitivity to alkaloids that mimic some of the actions of acetylcholine. These alkaloids are called muscarine and nicotine. Cholinergic receptors are therefore classified as muscarinic or nicotinic.

Diels–Alder reaction

Chemical reaction involving a diene and a dienophile (diene-liker). The reaction is named after the chemists (Kurt Alder and Otto Paul Hermann Diels) who won the Nobel Prize for Chemistry in 1950 for its discovery. From a pest control perspective this reaction is used to produce cyclodiene insecticides. The cyclodienes dieldrin and aldrin were also named after these chemists.

Half-life

The time taken for the concentration of a pesticide in a compartment to decline by one-half. Usually an estimate based on observed dissipation over several half-lives. In the context of pest control many organochlorine molecules have half-lives that are measured in months to years. All other insecticides have half-lives that are typically measured in days to weeks.

Isomerism

Chemicals with the same number and types of atoms as another chemical, but possessing different properties, are referred to as isomers. There are structural isomers, geometric isomers, optical isomers and stereoisomers. From a pest control perspective isomerism is often important because biological activity is often only found in particular isomers.

Knock-down

Rapid cessation of activity following exposure to a pesticide. Typically observed with many natural pyrethroid insecticides. Knock-down does not always lead to death

Leaving group

Organophosphate and carbamate insecticides exert their toxic effects in organisms by binding to the enzyme acetylcholinesterase. Each of these insecticidal groups is composed chemically of a basic structure that either phosphorylates or carbamylates the enzyme (see also above). Attached to this basic structure is the leaving group. The leaving group is so called because it is hydrolysed and splits from the enzyme. The leaving group is analogous to the choline that is produced when the enzyme is acetylated during normal functioning. Manipulation of the chemical structure of the leaving group can produce insecticides with different toxicities.

N-methyl carbamate

Basic chemical structure of all carbamate insecticides. Composed of an *N*-methyl (or dimethyl) structure linked to an ester group which bonds to the leaving group (see also above). This is the basic structure that carbamylates the enzyme acetylcholinesterase.

Oxidative phosphorylation

The mode of ATP (see earlier) synthesis that occurs in mitochondria and which is powered by reactions that transfer electrons from food (e.g. glucose) to oxygen. Accounts for almost 90% of the ATP that is generated during respiration.

Postsynaptic receptors

Receptors for neurotransmitters (e.g. acetylcholine or gamma-aminobutyric acid (GABA)) found on postsynaptic membranes. The synapse is the gap through which a neurotransmitter effects communication between nerve cells and/or between nerve and muscle cells. On one side of this gap is the presynaptic membrane and on the other, the postsynaptic membrane. Receptors on the postsynaptic membrane are targeted by a number of insecticides (e.g. cyclodienes target GABA receptors while neonicotinoids target acetylcholine receptors).

Systemic activity

Insecticides that are translocated within a plant's vascular system are said to have systemic activity. This translocation may occur in xylem, phloem or both. We can contrast compounds with systemic activity with those chemicals that are not translocated (contact insecticides) or with those chemicals that only diffuse a relatively short distance from the point of absorption (quasisystemic). The main advantages with many systemic chemicals are protection from the environment, redistribution of the toxicant from the site of application and protection of new leaf tissues that develop after application.

Chapter 4

Adjuvants

In the UK adjuvants are defined as any substances, other than water, without significant pesticidal properties, which enhance or are intended to enhance the effectiveness of a pesticide when they are added to the pesticide.

Air assistance

A number of sprayers have now been developed that use air assistance in pesticide application. The were originally used most extensively in tree crops but a number of other crop types can now be sprayed with air-assisted sprayers. A review is provided by Hislop (1991; see Further Reading).

Application efficiency

A term that has been widely used in comparisons of the amount of active ingredient applied in relation to the amount that would be needed if pest species could be targeted with 100% precision. The terminology does not pertain to the efficiency of the insecticide or to the design of the equipment that is used. Most authors quote figures below 1%. Describing the application process using these terms is in some ways unfair since there is a great deal of difference between applying a pesticide directly to a pest in the laboratory and applying a pesticide in a field situation. However, such descriptions have generated a great deal of research activity that has led to improvements in the application process.

Biological fitness

Refers to the contribution that an individual makes to the gene pool of the next generation. Species that make a greater contribution (often in the form of offspring) than others are therefore said to be fitter.

Bollworms

Larvae (caterpillars) of various Lepidoptera (butterflies and moths) that feed on cotton bolls. The fruit of cotton plants from which lint is harvested is called a boll. Some bollworms feed on other plant species in addition to cotton.

Broad-spectrum

Chemicals with a broad spectrum of activity will be toxic to a diverse range of pest species. For crops that support a range of pest types these chemicals are favourable to those with a narrow spectrum of activity since only one chemical compound needs to be applied for pest control purposes.

Chemical Manufacturing Association

Association of chemical manufacturers that has a responsible care programme with the vision of ensuring that no accidents, injuries or harm are caused to the environment as a result of chemical usage.

Controlled-droplet application

Sometimes also referred to as CDA. Emphasises the selection of appropriate droplet sizes for different target pests. Since the 1970s a number of CDA sprayers have been developed that can operate at very-low-volume application rates. Most of these sprayers use centrifugal energy to generate their droplets.

Corona discharge

Process by which discharge flows from a pointed surface (e.g. many cereal leaves) and neutralises the charge in a spray cloud. The magnitude (or significance) of this process will depend on a number of factors, including plant type and spacing.

Cross-resistance

Occurs in a pest species when a pesticide selects for resistance to itself plus another, as yet unused pesticide. Likely to occur where pesticides have similar modes of action.

Grey partridge

Game bird whose numbers have declined in the UK from more than 1 000 000 breeding pairs in the 1940s to *c.* 100 000 breeding pairs in the 1990s. Much of this decline has been attributed to changes in farming practices, including pesticide use. In the UK the Game Conservancy has been involved in the development of a species action plan for this bird.

Harvest interval

Time between pesticide application and when it is safe to harvest a crop. This interval is intended to ensure that human exposure to residues in food is minimised. Harvest intervals are usually clearly stated on the product container.

International Labour Organisation

Worldwide organisation of social democratic, socialist and labour parties. It currently brings together 141 political parties and organisations from all continents. Has existed in its current form since 1951.

International Phytosanitary Certificate

The Food and Agriculture Organisation (FAO) of the United Nations (UN) established the International Plant Protection Convention in 1953 to draw up a set of basic rules for a unified global system of plant import and export. These rules are incorporated in International Phytosanitary Certificates which state that exported crops are free from defined pests and diseases and/or that defined treatments have been carried out. The latest set of standards was endorsed in November 1997. For more information see the FAO website at: http://www.fao.org/WAICENT/FAOINFO/AGRICULT/AGP/AGPP/PQ/En/Publ/Ispm/ispm7e.pdf.

Key pests

Pests that cause major damage every season unless controlled. Can be distinguished from occasional pests, potential pests or nonpests. See also Chapter 12 for a discussion of pest categories.

Multiple-compound resistance

Resistance in a pest species to at least two different pesticides brought about by exposure to those two compounds.

Nontarget species

Any living organism other than the intended target. Some nontarget species may be beneficial in terms of pest control, i.e. they may function as biological control agents. However, many nontarget species will not be important from a functional pest control perspective. With the notable exception of chemical warfare, humanity is always a nontarget species.

Organic farming

There is a perception amongst most people that organic farming prohibits the use of chemical pesticides. This is not true. There is also a perception that organic farming is better for the environment. These issues are discussed in more detail in Chapter 11. However, the reader is advised to consult Trewavas (2001) for a discussion of these issues (see Further Reading).

Parasitoid

An organism that is 'parasite-like'. Most parasites do not kill their hosts, at least not quickly. In contrast, parasitoids do kill their hosts by completing their juvenile stages either in or on the host. A number of important parasitoids, from a pest control perspective, are found in the insect orders Hymenoptera and Diptera. These species act as biological control agents for many pests. See also Chapter 5 for more detail.

Pesticide legislation (UK)

Procedures for the national approval of pesticides in the European Union are, at present, being replaced by a system run by the European Commission. Under this new system pesticides are assessed by a committee of member states and, if approved, are placed on the Annex I listing of EC directive 91/414. Once listed, applications can then be made within member states for specified uses for the active ingredient. At present, new pesticides are evaluated under the new system, while many older pesticides are currently under rereview. It is expected that completion of the review of older products will take a number of years to complete. A detailed description of the approval process, including the data required, can be found in the document *A Guide to Pesticide Regulation in the UK*, which can be downloaded from the website run by the Pesticides Safety Directorate (UK) at http://www.pesticides.gov.uk/committees/acp/acpgui1.pdf.

Physicochemical properties

Physical and chemical properties of a pesticide formulation. From a pest control perspective the two physicochemical properties that have frequently been manipulated are surface tension and viscosity. Such manipulation can be carried out by adding adjuvants (see above) to a pesticide formulation.

Rayleigh limit

Maximum electrical charge that a droplet of given volume can sustain. Charging above this limit will lead to spontaneous break-up of the droplet to generate an increase in surface area.

Reduced-rate applications

Pesticide applications in which the manufacturer's recommended rate of application is reduced. Can be very effective in reducing the amount of pesticide used and are typically still effective for pest control. Usually half-rate or quarter-rate applications are made. The problem with these applications is that, because they are not sanctioned by the manufacturer, you have no comeback if your pest control does not work.

Reinvasion

Recolonisation of an area following a species elimination caused by a pesticide application. Many pest species are highly dispersive and are able to reinvade at a faster rate than their predators. This is one reason why pest species resurgence is so common.

Shelf-life

The length of time that a pesticide can be stored for and retain viable activity. Actual shelf-lives will depend not just upon the formulation *per se* but also upon the ambient environmental conditions. Chemical manufacturers typically undertake storage stability tests to evaluate half-lives under worst-case conditions. For most pesticides, 2 years would be a satisfactory minimum for a shelf-life.

Single-compound resistance

Resistance in a pest species to a single pesticide.

Ultra-low-volume (ULV)

Often defined as the minimum volume required per unit area to achieve economic control. Most ULV applications are made at rates from 0.5 to 1.0 l/ha. The development of ULV applications was significantly enhanced by the arrival of CDA (see above).

Vector

From a pest control perspective, an organism that carries a pathogenic organism from one species to another. The pathogen may be harmful to plants or animals. For example, mosquitoes act as vectors for viruses (yellow fever) and protozoa (malaria), while many aphids act as vectors for harmful plant viruses.

World Health Organization (WHO)

International organisation that was first promoted as part of the charter of the United Nations in 1945. Became a formally recognised organisation in 1948. Concerned with the physical, mental and social well-being of all individuals on our planet. Among numerous other activities, produces a *World Health Report* on a periodic basis. See http://www.who.int for more details.

Chapter 5

Augmentation, inoculation, inundation

Within the context of biological control these terms are often used to describe the same process. However, they are not strictly comparable. To augment is 'to enhance' and to inoculate is 'to seed', i.e. these terms apply when natural enemies are present or absent, respectively. The term inundation refers to situations where an area is swamped or 'inundated' with a biocontrol agent. Whether that natural enemy existed in the area beforehand or not is therefore irrelevant. It is probably more correct to use the word inundation in relation to the use of pathogens for biological control since this is certainly what happens.

Cassava

Root crop grown in tropical and subtropical regions. Sometimes bitter and sweet cassavas are referred to as separate species, the former being *Manihot esculenta* and the latter *M. palmata*, but this is incorrect since the toxicity varies according to location. Cassava is the staple food in many tropical countries. It is not traded internationally in its fresh state because tubers deteriorate very rapidly. Eaten in unprocessed form or processed to make flour and tapioca puddings.

Entomophage

Literally 'insect eater'. From the Greek words, *entomon* and *phagein*, meaning insect and eat, respectively.

General equilibrium position (GEP)

Defined as the average density of a population over a period of time (usually lengthy) in the absence of any environmental change. In the context of biological control we are usually trying to reduce this permanently in a classical control programme. By contrast, augmentation/inoculation techniques only seek to reduce the GEP for long enough to avoid any pest-induced economic crop losses.

Host-feeding

Some adult parasitoids are able to puncture other insect species (the host) with their ovipositors. The body fluids that exude from the puncture wound can then be consumed. This is called host-feeding. It has been suggested that adult host-feeding may account for greater host mortality than parasitism in those species where it is found.

Hyperparasites

Literally, parasites of parasites.

Leafminers

General term given to species whose larval and pupal stages are completed inside leaves. As larvae feed between cell layers they leave a clearly visible trail or mine. Found in the insect orders Diptera, Lepidoptera and Hymenoptera. Some are very important pest species.

Mealybug

Insects in the order Hemiptera named because of the waxy or powdery coatings they secrete.

Phytoseid mites

Predatory mites in the family Phytoseiidae.

Scale insects

Insects in the order Hemiptera named because they produce scales under which most females live.

Trophic resource

Any food item.

Chapter 6

Agroecosystem

Literally, an agricultural ecosystem, i.e. a crop.

Anhydrobiotic state

Nematodes that live in Antarctic dry soils are able to survive there by entering an inactive state known as anhydrobiosis. They revive when environmental conditions are improved or when water is added. The exact mechanism of this process is under investigation in order to produce anhydrobiotic nematode formulations (with satisfactory shelf-lives) for use in arthropod pest control.

Baculovirus

Virus from the family Baculoviridae. These viruses have only ever been found in arthropod hosts and so are thought to be completely safe to vertebrates.

Chitin

Structural polysaccharide found in the exoskeleton of arthropods. Also produced by many fungi. Chitin is also the material used in the medical industry as dissolvable threads.

Flagellar serotyping

Technique used to diagnose different bacterial strains. Serum (fluid) from flagella (used for movement) on bacteria is catagorised based on its biochemical properties.

Gram-positive

Bacteria, by convention, can be classified as Gram-positive or Gram-negative. This division is made possible because of differences in the chemical structure in the cell walls of these two groups. Gram-positive bacteria have outer cell walls that are largely composed of peptidoglycan (sugar polymers cross-linked with short polypeptides). Gram-negative bacteria have outer cell walls that are largely composed of lipopolysaccharides. A Gram stain is used to differentiate between these two. Biologically, Gram-negative bacteria are often the most threatening because their lipopolysaccharide walls are often toxic and they also provide protection against host defences.

Haemocoel

Main body cavity of adult arthopods. Contains haemolymph that is propelled by a heart through an open circulatory system.

Inundative releases

When pathogenic organisms are used for arthropod pest control, the pest species targeted is sprayed with the microbe at a very high rate. This is known as an inundative release.

IOBC

The International Organisation for Biological Control (IOBC) was established in 1956 as a global organisation affiliated to the International Council of Scientific Unions (ICSU). IOBC promotes environmentally safe methods of pest and disease control.

Milky disease

A bacterial disease of Japanese beetle larvae and other scarabaeid grubs that eventually turns the grub a milky-white colour. Also called milky spore disease.

Muscardine disease

The very distinctive and noticeable white mummies of caterpillars infected with the fungal pathogen *Beauvaria bassiana* gave rise to the name muscardine, which is derived from the French word for the bonbons which the mummified specimens resembled. Today the term muscardine refers to an insect fungus or disease caused by a fungus.

Pebrine disease

Protozoan disease of silkworms which presents as black spots on the body surface. The word pebrine was first used to describe the plague, which also presents as black spots. Pebrine is derived from the word *pebre* which means pepper.

Plasmid

Extrachromosomal circular DNA that comprises a few genes. In eukaryotic species, plasmids are found in the cytoplasm.

Sotto disease

So called because the disease was first recognised in silkworms in Japan. Sotto means 'sudden collapse'.

Chapter 7

Bark beetles

Beetles in the family Scolytidae that feed as larvae, pupate and mate under the bark of trees. Many are serious pests of forestry.

Ceratocystis minor

A number of bark beetles act as vectors for this wilt-causing fungus. For example, the western pine beetle *Dendroctonus brevicomis* carries spores of this fungus in special pouch-like structures in its head called mycangia. As beetles chew their way through bark, the spores dislodge and begin to germinate. The fungus then invades and blocks the conductive vessels of the inner bark and sapwood. This causes foliage to fade and trees may then die.

Farnesene

The major chemical component of aphid alarm pheromone is a compound called (E)-β-farnesene. At the time of writing this chemical was under commercial development as an adjuvant to improve insecticide applications.

Habituation

Within the context of pheromones this term is usually used to describe any decrease in responsiveness whether or not changes have occurred in the nervous system. Biologically, the term habituation is most commonly used to describe simple learning in which species lose their responsiveness to stimuli that usually provide appropriate feedback.

Isomerism

See definition given in glossary to Chapter 3.

Terpenoids

Large group of secondary plant substances, found in a diversity of plant species. Over 15000 terpenoids have so far been identified. Most are lipid-soluble and they are typically stored or sequestered in specialised glands. Many are aromatic.

Chapter 8

Benzoylphenylureas

In some textbooks these are referred to simply as benzoylureas.

Corpora allata

Glandular bodies, located near the oesophagus, that are connected to insect brains via the corpora cardiaca. Responsible for producing and releasing juvenile hormone.

Insect circulatory systems

In insects and other arthropods their blood bathes internal organs directly. This fluid (blood and interstitial liquid) is referred to as haemolymph. Body movements and a heart typically circulate haemolymph in what is known as an open circulatory system.

Prothoracicotropic hormone (PTTH)

This brain hormone, released in response to growth, causes the prothoracic glands to release moulting hormone.

Thysanura

Primitive wingless order of insects known as bristletails. Colloquially, some species are referred to as silverfish and/or firebrats. Unlike all other insects, adults moult. Up to 60–70 adult moults have been documented for some species.

Chapter 9

Chemosterilants

Chemicals that make treated species incapable of successful reproduction. Many chemosterilants are toxic to vertebrates, including humans, and so are dangerous to work with.

Density-dependent factors

Factors affecting population regulation whose impact is determined by population density. With many species an increase in reproductive effort is a common event following a population decline.

Filariasis

Also known as elephantiasis, this condition is best known from dramatic photos of people with grossly enlarged or swollen arms and legs. The disease is caused by parasitic worms, including *Wuchereria bancrofti, Brugia malayi* and *B. timori,* all transmitted by mosquitoes. At present, the disease affects *c.* 120 million people worldwide (see also Table 1.1, Chapter 1).

Gametogenesis

Gametogenesis is the process of gamete formation from diploid cells of the germline. Spermatogenesis is the process of forming sperm cells, while oogenesis is the process of forming an ovum (egg).

Haldane's rule

J.B.S. Haldane (1892-1964), British geneticist, biometrician, physiologist and populariser of science. In 1922 he appears to have been the first to enunciate that 'when in the F1 [first generation] offspring of two different animal races [lines], one sex is absent, rare or sterile, that sex is the heterozygous [heterogametic] sex'. This has subsequently come to be known as Haldane's rule. It is Haldane who is purported to have made the famous comment that all that biology tells us about the nature of God is that he has 'an inordinate fondness for beetles'.

Natural rate of increase

The per capita rate of increase of a population derived by subtracting the death rate from the birth rate. Also referred to as the intrinsic rate of natural increase. See also glossary for Chapter 1: *r–k* continuum.

Parthenogenesis

Form of reproduction in which offspring develop from unfertilised eggs produced by females. Common in many hemipterous pests, especially aphids.

Rickettsiae

The rickettsiae are a diverse collection of obligate intracellular Gram-negative bacteria found in ticks, lice, fleas, mites and mammals. They include the genera *Rickettsia*, *Ehrlichia*, *Orientia* and *Coxiella*. In mammals they cause Rocky Mountain spotted fever, rickettsial pox, other spotted fevers, epidemic typhus, murine typhus and scrub typhus.

Tobacco budworms

Lepidopterous larvae that cause most damage when feeding on the vegetative buds of crop plants, especially tobacco. In the literature there still appears to be some confusion as to the exact genus of these pests. Some authors refer to *Heliothis* spp. (as here) and others to *Helicoverpa* spp.

Trypanosomiasis

Disease of vertebrates, including humans, in whom it is known colloquially as sleeping sickness. Also called African trypanosomiasis. It is caused by protozoan parasites within the *Trypanosoma brucei* complex. The parasite is spread to humans through the bite of a tsetse fly (*Glossina* spp.). At present, the disease is thought to infect 300 000–500 000 Africans in sub-Saharan regions of the continent. The disease is almost always fatal if not treated. Should not be confused with American trypanosomiasis or Chagas disease. This is caused by the protozoan parasite *Trypanosoma cruzi* which is spread to humans through the faeces of the 'kissing bug' (*Rhodnius* spp.), a hemipterous insect with a needle-like appendage which it uses to obtain human blood. At present, Chagas disease is thought to infect up to 18 million people in South and Central America.

Unguja, Zanzibar

Zanzibar (from the Persian Zendji-Bar, which means 'land of blacks') is located about 35 km off the coast of Tanzania. It comprises the 1464-km^2 main island of Unguja (also known as Zanzibar), the island of Pemba and a number of smaller islands.

Chapter 10

Agronomic characteristics

Agronomy involves the application of science to the process of crop production. The most important agronomic characteristic of a crop would therefore be yield. However, other features of the crop, such as germination, growth rate and fruit set are also examples of agronomic characteristics.

Carrying capacity

The upper limit to population growth as set by the term K in the logistic model of population growth (see glossary for Chapter 1). Ecologists typically use the carrying capacity to describe the maximum stable population that a habitat can support over a relatively long period of time. Largely an abstract concept since the carrying capacity of an environment may fluctuate *per se* while many species, especially crop pests, appear to have population dynamics that are anything but stable.

Cultivar

Literally, abbreviation of cultivated variety.

Germplasm

The part of a cell that is the material basis for heredity and so is transferred from one generation to another. In the context of work carried out and coordinated by the International Plant Genetic Resources Institute (IPGRI) designated germplasm refers to genetic resources designated by IPGRI to be held in trust for beneficiaries within the International Network of *Ex Situ* Germplasm Collections of the Food and Agriculture Organisation (FAO) of the United Nations. See also text for details of organisations involved.

Hydroxamic acids

Hydroxamic acids are naturally occurring compounds that exist in many biological systems. In the context of host-plant resistance the best-studied hydroxamic acid comprises DIMBOA in maize. See text for more information.

Induced resistance

Many plants can be induced to develop resistance when attacked. This resistance has been most studied in relation to attack by pathogenic organisms; it has also been recognised in relation to resistance to arthropod pests. At present, a number of induced resistance mechanisms have been characterised. These include systemic acquired resistance (SAR), induced systemic resistance (ISR) and localised induced resistance (LIR).

Jassids

Sap-sucking homopterous pests in the order Hemiptera. Belong to the hemipteran family Cicadellidae. Also referred to as leafhoppers.

Theophrastus

Theophrastus was a Greek philosopher and the immediate successor of Aristotle in leadership of the Lyceum, which he subsequently presided over for 35 years. He died in 287 BC at an age put between 85 and 107. Some 227 treatises are attributed to Theophrastus. These books deal with religion, politics, ethics, education, rhetoric, mathematics, astronomy, logic, meteorology and natural history. The botanical works of Theophrastus, who has been called the 'father of botany', are the earliest of their kind in world literature. Two surviving books are *Historia de Plantis* (*History of Plants*) and *De Causis Plantarum* (*The Causes of Plants*).

Vavilov

Nikolai Ivanovich Vavilov (1887–1943), Russian botanist and geneticist. Worked as Professor at the Leningrad Agricultural Institute and as Director of the All-Union Institute of Plant Industry. In 1918 he found a variety of wheat that grows at an altitude of nearly 3000 ft (914 m) and is resistant to rust and mildew. His genetic study of wheat variations led to an attempt to trace the locales of origin of various crops by determining the areas in which the greatest number and diversity of their species are to be found.

Chapter 11

Arbuscular mycorrhizal fungi

Mycorrhiza is the term used to describe the symbiosis formed between fungi and the roots of plants. Literally, the word means 'fungus-root' (from the Greek words *mykes* and *rhiza*). One of the most widespread associations involves arbuscular mycorrhizal fungi. These fungi, which form tree-like associations with roots (from Latin *arbuscula*) have been found to interact symbiotically with the roots of about 80% of all plant species. In recent years it has become apparent that many of these fungi may be important in protecting crops, especially from pathogens.

Balfour

Lady Evelyn Barbara Balfour started what has come to be known as the Haughley experiment at her farm in Suffolk in 1939. This experiment was subsequently taken over by the Soil Association in 1947, which for the next 25 years directed and sponsored it. This pioneering experiment was the first ecologically designed agricultural research project on a full farm scale. It was set up to fill a gap in the evidence on which the claims for the benefits of organic husbandry were based. Lady Balfour published a book called *The Living Soil* in 1943 that was to be the catalyst for the formation of the Soil Association in 1946.

Certified seed

In the UK the Seeds (Registration, Licensing and Enforcement) Regulations 1985 require seed processors, packers and merchants, other than those dealing solely in small packages, to be registered by the appropriate UK Agricultural Department. The aim is to ensure that seed of a recognised quality is used in crop production. Similar regulations exist in other countries.

Conservation tillage

System of tillage designed to protect soils that are particularly at risk of erosion (*viz.* high erodibility). Crops are typically grown with minimal cultivation of the soil and stubble or plant residues remain on top of the soil rather than being ploughed or disced. The new crop is planted into this stubble. Different conservation tillage methods have been described as no-till, minimum till, incomplete tillage or reduced tillage. These techniques differ from each other mainly in the degree to which the soil is disturbed prior to planting. However, even in so-called no-till systems the soil is often disturbed prior to planting. To conform to the conventional definition, conservation tillage should leave at least 30% of the soil covered by crop residues.

Diapause

Resting stage in the development of many arthropods. Metabolic processes slow to a minimum. Species usually enter diapause to avoid unfavourable environmental conditions.

Good agricultural practice

In 1997 a number of the major food retailers in Europe formed the Euro-Retailer Produce Working Group (EUREP). One aim of this group is to promote best agricultural practice, particularly in fruit and vegetable production. With that in mind, protocols for good agricultural practice were developed and tested from 1997 onwards. At the time of writing these protocols were still being refined and independent verification systems for growers were under development.

Leatherjackets

Larvae of crane flies in the dipteran family Tipulidae. Adults are colloquially known as daddy-long-legs. Larvae feed in the soil on roots of various plants.

Optomotor landing response

Used to describe the landing response by insects flying over a crop. In a large number of cases, the greater the contrast between the crop and the soil it is planted in (i.e. the lower the planting density), the more likely a pest is to land in that crop. It has been suggested that this could occur because species have evolved to find and exploit plants that exist at low densities.

Pesticide environmental impact tax

Although previously proposed as part of a pesticide use reduction strategy, on 2 February 2000 the UK Prime Minister announced that plans for a pesticide tax were to be put on hold. He announced that the tax would be deferred until proposals for an alternative package of voluntary measures to reduce pesticide use were developed. These plans are being developed by the agrochemical industry, led by the British Agrochemical Association (BAA).

Preventive pest management

Confusing terminology that is sometimes used to describe routine, prophylactic chemical control measures, i.e. measures that prevent pest outbreaks occurring. However, other authors restrict the use of this term to describing environmentally friendly pest control measures (biological, cultural, etc.) that also prevent pest outbreaks from occurring.

Steiner

Rudolph Steiner, another organic pioneer, introduced the tenets of anthroposophy, which combined science with philosophy and spirituality. From his anthroposophic perspective Steiner gave lectures on agriculture that were compiled in his 1924 book, *Agriculture*, just before his death in 1925. His early work formed the basis of what has come to be known as biodynamic farming.

Wireworms

Larvae of beetles in the coleopteran family Elateridae. Larvae feed in the soil on the roots of various plants. Some species are major pests.

Chapter 12

Farmer Field Schools (FFS)

Developed by the Food and Agriculture Organisation (FAO) during the 1990s, the primary aim of FFS is to make farmers experts in their own fields. Farmers are encouraged to define the priority problems and to devise solutions with the help of extension workers. Primary characteristics of FFS are that farmers will learn by doing, that the classroom should be the field, that extension workers are facilitators, not teachers, that farmers meet regularly to discuss issues, and that farmers develop their own learning materials. The FFS approach is now being used by both the FAO and other international organisations (e.g. Cooperative for Assistance and Relief Everywhere (CARE) International) in a diversity of countries throughout Asia.

Institute of Scientific Information (ISI)

Database publisher of scientific, medical and technical information. Product offerings include Current Contents, Science Citation Index and Journal Citation Reports. See website at http://www.isinet.com.

Integrated control

First used by A. E. Michelbacher and O. G. Bacon (1952) to describe methodologies for selection, timing and dosage of insecticide treatments for the control of walnut pests and preservation of beneficial arthropods in California. See Michelbacher, A.E. & Bacon, O.G. (1952). Walnut insect control in northern California. *Journal of Economic Entomology*, **45**, 1020–7.

Integrated crop management (ICM)

This is a term which became more widely used during the 1990s to describe a cropping strategy in which the farmer seeks to conserve and enhance the environment while economically producing safe, wholesome food. The idea behind ICM is to make decisions based on an understanding of the relationships between biological and ecological interactions, in nutrient cycles, pests, weeds and diseases, within the context of farm management practices. In other words, it is a holistic approach towards food production in which arthropod pest control is just one component.

Malthus

The Reverend Robert Thomas Malthus published his *Essay on Population* for the first time in 1798. In this essay he pointed out that, while populations tend to increase geometrically, food supplies tended to increase arithmetically. He thus forecast that famine and war were inherent features of society. Exactly the same arguments have been made at the start of the twenty-first century concerning our need to increase global food production. Clearly, history is repeating itself!

Mathematical models

Not discussed in this chapter. However, an excellent review of modelling as it applies to integrated pest management is given by Way and Van Emden (2000; see Further Reading for details).

Pesticide-resistant predators

A number of pesticide-resistant species have now been selectively bred and are available commercially for use in pest management programmes. The best-known are pesticide-resistant mites, particularly *Phytoseiulus persimilis*. However, pesticide-resistant parasitoids have also been developed for control of leafminers.

Chapter 13

Antisense gene

A short string of nucleotides that can bond to messenger RNA and so block the process of gene expression.

Cholera

Cholera is a severe diarrhoeal disease caused by the bacterium *Vibrio cholerae*. Transmission to humans is by water or food. The natural reservoir of the organism is not known. It was long assumed to be humans, but some evidence suggests that it is the aquatic environment. The bacterium produces cholera toxin whose action on the mucosal epithelium is responsible for the characteristic diarrhoea of the disease. In its extreme manifestation, cholera is one of the most rapidly fatal illnesses known. A healthy person may become hypotensive within an hour of the onset of symptoms and may die within 2-3 h if no treatment is provided. More commonly, the disease progresses from the first liquid stool to shock in 4-12 h, with death following in 18 h to several days. Cholera exists worldwide and there are no completely effective treatments for this disease. Transgenic crops that express inactivated vaccines for this disease are currently under development.

Cloned gene

A gene that is derived from, and is genetically identical to, a single common ancestral gene. Cloning of cells and organisms can also occur. Cloning in the laboratory is now a common procedure in much biomedical research.

Crown gall disease

Disease which is caused by a bacterium called *Agrobacterium tumefaciens*. The disease presents as tumour-like growths on stems of susceptible plants. Research indicated that the tumour-inducing factor was genetic material carried on a smaller mobile DNA unit that was not part of the bacterium's single chromosome, i.e. it was located on a plasmid (see glossary to Chapter 6). Researchers have used the bacterium's ability to integrate its DNA into host plants to transfer genes for pest resistance into plants, so turning the disfiguring disease into a tool for promoting healthy crop plants.

Dengue fever

Disease caused by a flavivirus transmitted to people by mosquitoes of the species *Aedes aegypti*. Dengue virus infections occur in the tropics. The World Health Organization estimates 50 million cases of dengue infection each year. This includes 100–200 cases reported annually in the USA, usually introduced by travellers from endemic regions. Dengue virus causes two diseases – dengue fever and dengue haemorrhagic fever (DHF). Dengue fever can cause severe aches and pains, headaches and high fever. DHF is a more serious illness that includes internal haemorrhaging and dramatic loss of blood pressure. It is often fatal. No virus-specific treatment or vaccine is available.

Escherichia coli

Common bacterium that has been studied intensively by geneticists because of its small genome size, normal lack of pathogenicity and ease of growth in the laboratory.

Functional efficiency

Used in the ecological sense of how well species perform a particular function. For example, how well do predators consume prey items (how many? how quickly?) It is theoretically possible that these characteristics of species could be manipulated by genetic engineering.

Horizontal transgene movement

Movement of transgenes and hybridisation with wild relatives is an important issue for discussion. At the end of 2001 there were reports that transgenic maize in Mexico had hybridised with wild relatives that were at least 100 km from the source crop. The significance of this is still a hotly debated issue.

Kanamycin

A water-soluble aminoglycoside antibiotic that is derived from the bacterium *Streptomyces kanamyceticus.* It is an antiinfective used for treatment of infections when penicillin or other less toxic drugs can't be used. The *kan-r* gene confers resistance to the antibiotic kanamycin; this gene is used as a marker gene as it allows early identification of plant cells that have been successfully modified with a new trait.

Phenological expression

Seasonal expression. The idea with transgenic species is that toxin expression may be turned on and off in response to developmental and/or environmental cues. Such plants have not been developed yet for commercial release.

Promoter gene

A segment of DNA located at the 'front' end of a gene, which provides a site where the enzymes involved in the transcription process can bind on to a DNA molecule and initiate transcription. Promoters are critically involved in the regulation of gene expression.

Recombinant DNA

DNA molecules that have been created by combining DNA from more than one source.

Refractory population

In the context of pest control this is a population that is unable to vector a pathogen successfully.

Tritrophic

Literally, three feeding levels, i.e. autotrophs (plants) photosynthesising, pests feeding on plants and predators feeding on pests. See also Chapter 6.

Vector

In the context of genetic engineering, a process in which foreign gene(s) are moved into an organism and inserted into that organism's genome. Retroviruses such as human immunodeficiency virus (HIV) serve as vectors by inserting genetic information (DNA) into the genome of human cells. Bacteria can serve as vectors in plant populations. See also Chapter 4 for definition at the organismal level.

Xerophthalmia

Disease caused by vitamin A deficiency that is the leading cause of blindness in children in the developing world. The World Health Organization estimates that over 100 million children suffer from vitamin A deficiency across Africa, Asia and South America. Over 350 000 preschool children suffer from partial or total loss of vision as a result of vitamin A deficiency every year.

Arthropod pest index

Beneficial species index

Subject index

CPSIA information can be obtained
at www.ICGtesting.com
Printed in the USA
LVHW080919070121
675887LV00003B/53

9 780521 567879